Land Intelligence and Geospatial Ecosystem

This book provides a comprehensive understanding of how advanced intelligent geospatial technologies are transforming the way we record, register, manage, and use land. It explains the basics of intelligent land administration systems, explores new innovative technologies for capturing, validating, storing, querying, and analysing aboveground and underground land ownership, and discusses how geospatial innovations impact urban planning, sustainability, and legal frameworks. Readers will gain insight into how incorporating new intelligence into land administration and geospatial ecosystems can shape our cities, improve infrastructure management, and address location-based challenges in a rapidly changing world.

Features

- Provides new content on digital, integrated, and intelligent solutions for delivering on-demand knowledge about land and geospatial information and examines the value of this information for sustainability and resilience.
- Includes several case studies, prototype systems, and models that demonstrate the feasibility of intelligent land administration system design and development.
- Explains the foundational role of intelligent land administration systems in improving geospatially enabled decision making for design, planning, construction, and management of complex urban environments.
- Describes the complexity of land and geospatial information in aboveground and underground environments and how digitally intelligent solutions can help address land and geospatial data management and communication challenges.
- Includes new material on digital environments that provides a fully integrated view interlinking the legal ownership of land parcels and physical reality of complex environments, including buildings and urban infrastructure.

This is an excellent book for professionals, academics, researchers, and students in land administration, geospatial information systems, geomatics and surveying, remote sensing and urban planning, sustainability science, disaster management, architecture, infrastructure engineering, and digital construction.

Land Intelligence and Geospatial Ecosystem

Sustainability, Digital and AI Innovations, and Applications

Edited by

Abbas Rajabifard and Behnam Atazadeh

CRC Press
Taylor & Francis Group
Boca Raton London New York

CRC Press is an imprint of the
Taylor & Francis Group, an **informa** business

Designed cover image: Getty Images

First edition published 2026
by CRC Press
2385 NW Executive Center Drive, Suite 320, Boca Raton FL 33431

and by CRC Press
4 Park Square, Milton Park, Abingdon, Oxon, OX14 4RN

CRC Press is an imprint of Taylor & Francis Group, LLC

Library of Congress Cataloging-in-Publication Data
Names: Rajabifard, Abbas editor | Atazadeh, Behnam editor
Title: Land intelligence and geospatial ecosystem : sustainability, digital and AI innovations, and applications / edited by Abbas Rajabifard and Behnam Atazadeh.
Description: First edition. | Boca Raton, FL : CRC Press, 2026. | Includes bibliographical references and index.
Identifiers: LCCN 2026001475 (print) | LCCN 2026001476 (ebook) | ISBN 9781041069898 hbk | ISBN 9781041069904 pbk | ISBN 9781003638308 ebk
Subjects: LCSH: Land use, Urban--Data processing | City planning--Data processing | Geospatial data--Computer processing | Artificial intelligence--Geophysical applications
Classification: LCC HD108.15 .L36 2026 (print) | LCC HD108.15 (ebook) LC record available at https://lccn.loc.gov/2026001475
LC ebook record available at https://lccn.loc.gov/2026001476

ISBN: 978-1-041-06989-8 (hbk)
ISBN: 978-1-041-06990-4 (pbk)
ISBN: 978-1-003-63830-8 (ebk)

DOI: 10.1201/9781003638308

Typeset in Palatino
by KnowledgeWorks Global Ltd.

Contents

Part II Digital Innovations Towards Geospatial Ecosystem and Land Intelligence

Preface

The growing urbanization, climate change, complex infrastructure, and sustainability challenges cannot be addressed by traditional methods for land administration and geospatial information management. These approaches relied on traditional processes, fragmented databases, and static representations. As urban environments expand rapidly and infrastructure development in both underground and aboveground proliferates, there has never been a more urgent need for intelligent, integrated, and digitally enabled land and geospatial solutions in 3D, 4D, and towards nD environments. This book, *Land Intelligence and Geospatial Ecosystem: Sustainability, Digital and AI Innovations, and Applications*, responds to this global need. It provides an in-depth exploration of how land intelligence and geospatial data ecosystems can transform the way we record, manage, and use land and geospatial information in terms of fundamental ideas, cutting-edge technologies, institutional frameworks, and real-world case studies.

The book covers a wide range of topics, from the theoretical foundations of land intelligence and geospatial ecosystems to sophisticated techniques for multi-dimensional and dynamic data modelling, visualization, and GeoAI-driven analytics. In addition, the book addresses the institutional, ethical, and legal facets of land intelligence and geospatial data ecosystem that support accountability and trust. These ideas are tested and used in a variety of jurisdictions and country contexts, as demonstrated by real-world case studies from different jurisdictions, including cases in Australia, China, Finland, and Indonesia.

We hope this book will provide good foundation and latest developments to benefit geospatial and land market and wider practitioners including

- Scholars, researchers, and students looking for a current summary of institutional, legal, and technical developments in the field.
- Professionals and practitioners keen to address the challenges of implementing intelligent land and geospatial ecosystems in real-world practices.
- Policymakers who want to comprehend land intelligence and geospatial ecosystems are transforming inclusive, resilient, and sustainable futures.

We appreciate the contribution of many researchers and passionate practitioners who have enhanced the scope and profundity of this work by sharing their knowledge and perspectives. Their combined contribution exemplifies the collaborative spirit that is necessary to advance theory and practice of land intelligence and geospatial data ecosystems.

We hope this book will provide an exciting and comprehensive foundation to encourage readers to think beyond current paradigms in both modern land administration and geospatial science, to embrace the potential of location intelligence, and to help to create solutions that are socially just, environmentally sustainable, and globally relevant.

Abbas Rajabifard
Behnam Atazadeh
Editors

About the Editors

Abbas Rajabifard is Professor in Geomatics Engineering and Director of the Centre for Spatial Data Infrastructures & Land Administration (CSDILA) at The University of Melbourne. He is a world-renowned expert in geospatial science and land administration whose academic and professional career spans more than three decades. Prof. Rajabifard's scholarly background is in surveying, mapping, land administration, and urban systems, with strong expertise in spatial data infrastructures (SDIs), 3D cadastre, and digital twins. He has authored hundreds of publications, contributing influential research on topics such as modern land administration systems, BIM (building information modelling), land governance, and sustainable spatial planning.

Prof Rajabifard has served as Chair of the Academic Network of United Nations Global Geospatial Information Management—the academic arm of UN initiative—and is currently an Advisory Board member. He is also a member of UN International Advisory Committee of Geospatial Knowledge and Innovation Centre (UN-GGKIC) as part of UN-GGIM that advocates for the strategic use of geospatial data and technology to address major societal challenges: sustainable urban development, resilient infrastructure, disaster management, and sustainable land management, contributing to global efforts towards sustainability and informed policy.

Behnam Atazadeh is Senior Research Fellow and co-leads Land Administration Unit at the Centre for Spatial Data Infrastructures and Land Administration (CSDILA), Department of Infrastructure Engineering, the University of Melbourne. His research focuses on advancing the digital foundations of land administration systems through adoption of emerging technologies. He has extensive experience designing and implementing innovative digital solutions, including modelling, querying, analysing, and validating digital data for land administration systems.

Dr Atazadeh has published widely in leading journals and international conferences in surveying, land administration, and geospatial sciences. His contributions span 3D digital cadastre, geospatial data modelling, and the integration of legal and physical dimensions for urban land administration. He collaborates closely with government and industry to translate research outcomes into practical digital innovations for land and property sectors.

Contributors

Fahimeh Abedi
Centre for Spatial Data Infrastructures and Land Administration
Department of Infrastructure Engineering
University of Melbourne
Melbourne, Victoria, Australia

Salman Makhld M. Aldlbahi
College of Architecture and Planning
King Saud University
Riyadh, Riyadh Province, Saudi Arabia

Mark Allan
Centre for Spatial Data Infrastructures and Land Administration
Department of Infrastructure Engineering, University of Melbourne
Melbourne, Victoria, Australia

Jagannath Aryal
Centre for Spatial Data Infrastructures and Land Administration
Department of Infrastructure Engineering
University of Melbourne
Melbourne, Victoria, Australia

Atie Asadikia
School of Science, Computing and Emerging Technologies
Swinburne University of Technology
Melbourne, Victoria, Australia

Ali Asghari
Australian Urban Research Infrastructure Network
University of Melbourne
Melbourne, Victoria, Australia

Behnam Atazadeh
Centre for Spatial Data Infrastructures and Land Administration
Department of Infrastructure Engineering
University of Melbourne
Melbourne, Victoria, Australia

Glenn Campbell
School of Surveying and Built Environment
University of Southern Queensland
Brisbane, Queensland, Australia

Yiqun Chen
Centre for Spatial Data Infrastructures and Land Administration
Department of Infrastructure Engineering
University of Melbourne
Melbourne, Victoria, Australia

Zahra Gharineiat
School of Surveying and Built Environment
University of Southern Queensland
Brisbane, Queensland, Australia

Dongchen Han
Centre for Spatial Data Infrastructures and Land Administration

Department of Infrastructure
Engineering
University of Melbourne
Melbourne, Victoria, Australia

Serene Ho
Centre for Spatial Data Infrastructures
and Land Administration
Department of Infrastructure
Engineering
University of Melbourne
Melbourne, Victoria, Australia

Hamid Hosseini
Centre for Spatial Data
Infrastructures and Land
Administration
Department of Infrastructure
Engineering
University of Melbourne
Melbourne, Victoria, Australia

Peyman Jafary
Centre for Spatial Data
Infrastructures and Land
Administration
Department of Infrastructure
Engineering
University of Melbourne
Melbourne, Victoria, Australia

Fatemeh Jahani
Centre for Spatial Data
Infrastructures and Land
Administration
Department of Infrastructure
Engineering
University of Melbourne
Melbourne, Victoria,
Australia

Masoud Kamali
Centre for Spatial Data
Infrastructures and Land
Administration
Department of Infrastructure
Engineering
University of Melbourne
Melbourne, Victoria, Australia

Armin Agha Karimi
School of Surveying and Built
Environment
University of Southern Queensland
Brisbane, Queensland, Australia

Alice Kesminas
Centre for Spatial Data
Infrastructures and Land
Administration
Department of Infrastructure
Engineering
University of Melbourne
Melbourne, Victoria, Australia

Katie McDougall
Centre for Spatial Data
Infrastructures and Land
Administration
Department of Infrastructure
Engineering
University of Melbourne
Melbourne, Victoria, Australia

Nikoo Mirhosseini
Centre for Spatial Data Infrastructures
and Land Administration
Department of Infrastructure
Engineering
University of Melbourne
Melbourne, Victoria, Australia

Mahsa Moghadas
Urban Planning and Land
Management Group
Institute of Geodesy and
Geoinformation (IGG)
University of Bonn
Bonn, North Rhine-Westphalia,
Germany

Maryam Nasim
Centre for Spatial Data Infrastructures and Land Administration
Department of Infrastructure Engineering
University of Melbourne
Melbourne, Victoria, Australia

Bipul Neupane
Centre for Spatial Data Infrastructures and Land Administration
Department of Infrastructure Engineering
University of Melbourne
Melbourne, Victoria, Australia

Tuan Ngo
Department of Infrastructure Engineering
The University of Melbourne
Melbourne, Victoria, Australia

Victoria Oberin
Centre for Spatial Data Infrastructures and Land Administration
Department of Infrastructure Engineering
The University of Melbourne
Melbourne, Victoria, Australia

Aravind Poshnath
Centre for Spatial Data Infrastructures and Land Administration
Department of Infrastructure Engineering
University of Melbourne
Melbourne, Victoria, Australia

Ensiyeh Javaherian Pour
Centre for Spatial Data Infrastructures and Land Administration
Department of Infrastructure Engineering
University of Melbourne
Melbourne, Victoria, Australia

Abbas Rajabifard
Centre for Spatial Data Infrastructures and Land Administration
Department of Infrastructure Engineering
University of Melbourne
Melbourne, Victoria, Australia

Kirsikka Riekkinen
Department of Built Environment
Aalto University School of Engineering
Espoo, Uusimaa, Finland

Soheil Sabri
Urban Digital Twin Lab at SMST
University of Central Florida
Orlando, Florida, United States

Bahram Saeidian
Centre for Spatial Data Infrastructures and Land Administration
Department of Infrastructure Engineering
University of Melbourne
Melbourne, Victoria, Australia

Iman Taheri Sarteshnizi
Centre for Spatial Data Infrastructures and Land Administration
Department of Infrastructure Engineering
University of Melbourne
Melbourne, Victoria, Australia

Javad Shahidinejad
Centre for Spatial Data Infrastructures and Land Administration
Department of Infrastructure Engineering
University of Melbourne
Melbourne, Victoria, Australia

Lanxuan Shen
Centre for Spatial Data Infrastructures and Land Administration
Department of Infrastructure Engineering
University of Melbourne
Melbourne, Victoria, Australia

Davood Shojaei
Centre for Spatial Data Infrastructures and Land Administration
Department of Infrastructure Engineering
University of Melbourne
Melbourne, Victoria, Australia

Daniel Steudler
International Federation of Surveyors (FIG)
Zurich, Canton of Zurich, Switzerland

Walter Timo de Vries
School of Engineering and Design
Technical University of Munich
Munich, Bavaria, Germany

Mazdak Ghasemi Tootkaboni
Centre for Spatial Data Infrastructures and Land Administration
Department of Infrastructure Engineering
University of Melbourne
Melbourne, Victoria, Australia

Zhixuan Yang
Dongbei University of Finance and Economics
Dalian, Liaoning, China

Siqing Yu
Centre for Spatial Data Infrastructures and Land Administration
Department of Infrastructure Engineering
University of Melbourne
Melbourne, Victoria, Australia

Acknowledgements

This book presents the latest research and innovation undertaken by the Centre for Spatial Data Infrastructures and Land Administration (CSDILA), Department of Infrastructure Engineering, The University of Melbourne, and affiliated partners. We sincerely acknowledge the invaluable assistance, advice, and support provided by many colleagues and collaborators during the preparation, development, and publication of this book. The collective contributions of scholars, researchers, and professionals have profoundly shaped our thinking and enriched the content presented in this book. We have been privileged to benefit from the extensive international experience and knowledge of colleagues across universities, professional associations, government agencies, and the private sector worldwide. Their insights, shared through scholarly discussions, collaborative research projects, workshops, and professional activities, have significantly informed the ideas and perspectives articulated in this volume. We would like to thank particularly Dr. Katie Mcdougall and Mr. Hamid Hosseini for their invaluable input and support during the preparation of this book.

CSDILA has conducted research in collaboration with partners from various industries over 25 years. We express our sincere appreciation to our colleagues and industry partners whose collaboration, contributions, and commitment have significantly enhanced the practical relevance and impact of this book. Their deep domain knowledge, professional experience, and shared vision for advancing geospatial ecosystems and land administration systems have been instrumental in achieving the objectives of this book. We also extend our gratitude to the University of Melbourne, Faculty of Engineering and Information Technology, and Department of Infrastructure Engineering for continuous support for research and innovation.

We also extend our sincere appreciation to the editorial and production teams at Taylor & Francis Group Press for their exceptional professionalism and commitment during the various stages of manuscript preparation. Their attention to detail, dedication to scholarly excellence, and ongoing support ensured the successful realization of this book. We are particularly thankful to Ms. Irma Britton and Ms. Chelsea Reeves for their guidance and coordination throughout the publishing process.

Finally, we acknowledge the wider academic and professional community engaged in the fields of geospatial science and land administration, whose collective efforts continue to advance the frontiers of research and practice. This book stands as a reflection of their ongoing dedication to innovation, collaboration, and knowledge dissemination for the betterment of geospatial ecosystems and land administration systems at a wider global level.

Abbas Rajabifard
Behnam Atazadeh

The editors and co-authors would like to acknowledge the use of ChatGPT 4.0, OpenAI, and Grammarly to improve grammar and readability for the following chapters: 14, 20, 23, 30, and 31. ChatGPT 4.0 was also used to prepare the abstract for Chapter 5.

1

Land Intelligence and Geospatial Ecosystem for Future Sustainability and Resilience: Setting the Scene

Abbas Rajabifard and Behnam Atazadeh

1.1 Introduction

Land is crucial for promoting economic, environmental, social, and cultural prosperity in any jurisdiction and is fundamental to resilient societies and sustainable development (Williamson et al., 2010). Stated differently, it supports every aspect of sustainability and resilience, ranging from property markets and disaster risk reduction to infrastructure development and natural resource management. Additionally, land has cultural significance, especially for Indigenous communities whose relationships to the land form the foundation of their identities, traditions, and legacy.

Static representations, inefficient manual workflows, and fragmented land records have long been the foundation of land administration systems. This frequently results in cost overruns, major delays, and legal issues in land development projects. Global factors including resource scarcity, climate change, and urbanization have put a lot of strain on land administration systems and made problems with inequality, environmental degradation, and complex governance worse. As a result, land administration systems are becoming less and less equipped to handle complex sustainability and resilience issues.

In response, there has been a paradigm shift in the way we understand and manage land. This shift is driven by technological innovations and digital transformation, creating new opportunities to reform land administration systems via intelligent and integrated approaches for recording, managing, and disseminating land information. These advancements offer potential pathways to transition from traditional systems to dynamic, predictive, and inclusive ones. *Land intelligence* is a critical part of this transition, which is driven by integrated and geospatially enabled digital twins, artificial intelligence (AI), immersive environments, and fit-for-purpose models to manage

DOI: 10.1201/9781003638308-1

all rights, restrictions, and responsibilities (RRRs) and generate actionable knowledge about places, people, and policies.

A critical foundation of land intelligence is the *geospatial ecosystem* (UN-GGIM, 2022), which is a complex, multi-layered infrastructure that facilitates the collection, integration, processing, and application of geospatial data, including land and property data. This ecosystem is shifting from isolated data silos toward interconnected "systems-of-systems" and visually immersive platforms that enable real-time, predictive, and participatory location-based decision-making. It supports a wide range of applications in different sectors, including infrastructure asset management, biodiversity conservation, climate adaptation, smart agriculture, and emergency response. The synergy between land intelligence and the geospatial ecosystem represents a major step forward in land administration systems. It enables smarter and more transparent decision-making that aligns with the principles of sustainability, inclusivity, and resilience. Many jurisdictions around the world are modernizing their land administration systems and responding to the Sustainable Development Goals (SDGs). For these modernization programs, land intelligence embedded within geospatial ecosystems offers a roadmap to address long-standing challenges in land administration. Figure 1.1 provides an integrated paradigm illustrating how foundational frameworks, enabling technologies, and key stakeholders support geospatial ecosystems and intelligent land administration, contributing to improved sustainability and resilience outcomes.

1.2 From Land Administration to Land Intelligence

Land administration systems have long focused on the registration of RRRs and issuing ownership titles. These core functionalities are essential; however, they cannot fully address the needs of modern land administration. In many jurisdictions, land and property data are incomplete, outdated, and fragmented across institutional silos, making it very difficult for policymakers, investors, and communities to gain a real-time, integrated understanding of modern land administration functions in terms of land tenure, land value, land use, and land development.

Recognizing these issues, the United Nations Committee of Experts on Global Geospatial Information Management (UN-GGIM) Expert Group on Land Administration and Management introduced the Framework for Effective Land Administration (FELA) (UN-GGIM, 2020). FELA is developed in line with the broader Integrated Geospatial Information Framework (IGIF) (UN-GGIM, 2023). It outlines nine strategic pathways (governance, policy and legal, financial, data, innovation, standards, partnerships, capacity and education, and advocacy) that serve as a blueprint for nations aiming

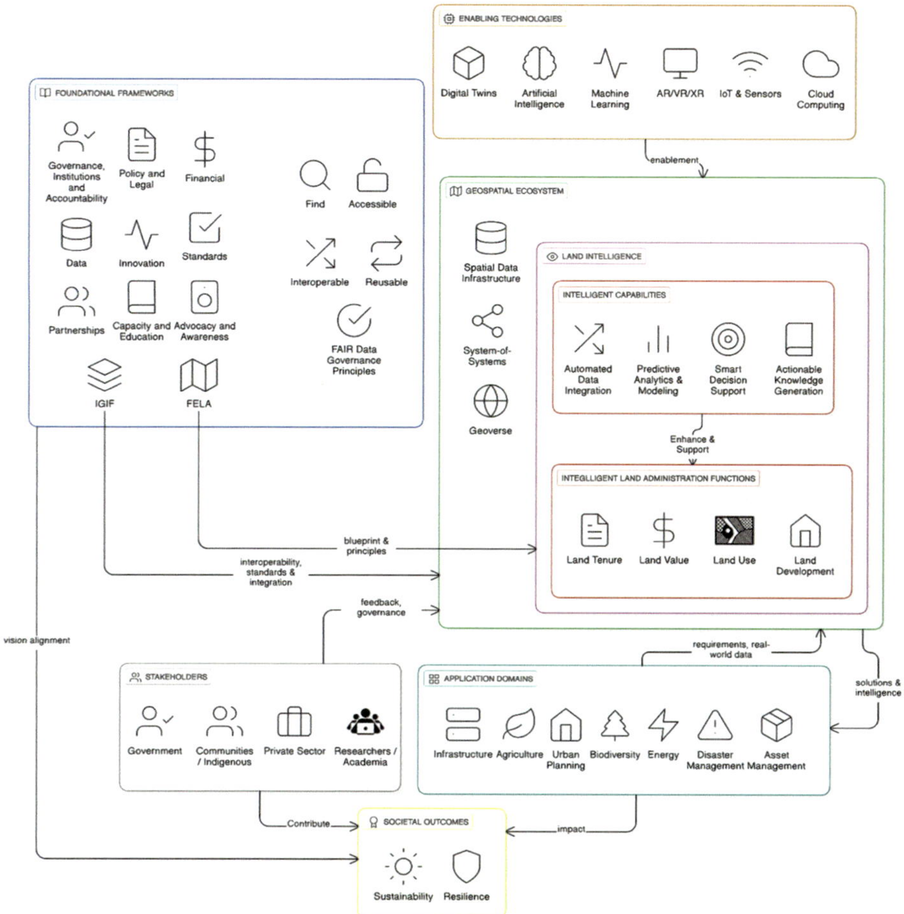

FIGURE 1.1
Land intelligence and geospatial ecosystem for sustainable development and resilience.

to modernize their land administration systems. More than a technical roadmap, FELA positions land administration as both an enabler and indicator of sustainable development, with explicit emphasis on resilience, inclusivity, climate readiness, and the imperative to "leave no one behind."

The critical shift of land administration systems is predicated on the ability to transform disparate, multi-source land and property data into actionable intelligence and meaningful knowledge that can contribute to sustainability and resilience. This is where land intelligence plays a central role in transitioning current land administration systems into an intelligent, dynamic, and integrated digital infrastructure that supports a holistic, evidence-based, and transformative land management paradigm. Land intelligence enables a wide range of stakeholders including governments and communities to make informed decisions by providing a continuous flow of insights that connect land tenure, value, use, and development in real time.

By leveraging advanced technologies such as AI, geospatial digital twins, and immersive visualization platforms such as augmented reality (AR) and virtual reality (VR), land intelligence bridges the gap between raw data and real-world impact. It enables predictive and explanatory analytics, supports scenario modeling, and enhances transparency and inclusivity in land administration processes. Ultimately, land intelligence empowers a shift toward dynamic and transformative land administration systems that are responsive to societal needs and capable of supporting sustainable development and resiliency at both local and global levels.

1.3 Geospatial Ecosystem

Traditional spatial data infrastructures (SDIs) serve as the backbone for national data governance and ensure the storage, discoverability, and distribution of authoritative geospatial datasets. However, due to significant advancements in geospatial technologies and the increasing demand for intelligent decision-making, SDIs are no longer sufficient to meet the requirements for managing complex and real-time geospatial information. The emerging paradigm involves transitioning from static, siloed geospatial systems to a multi-layered geospatial ecosystem that is dynamic, interconnected, and adaptive.

This geospatial ecosystem is characterized by its ability to capture, process, and integrate heterogeneous data sources, including land survey data, satellite imagery, drone captures, sensor data from Internet-of-Things (IoT), crowdsourced geographic information, and administrative records, in near-real time. Such integration facilitates intelligent solutions, greater automation, and enhanced responsiveness to land and geospatial challenges. This shift reflects a broader global trend toward more holistic and intelligent digital ecosystems that underpin smart governance, sustainability planning, and disaster resilience.

UN-GGIM has conceptualized a future perspective on geospatial ecosystems in three interrelated and co-existing elements (UN-GGIM, 2022):

- Spatial Data Infrastructure (SDI): Traditional, server-based platforms that catalogue, manage, and disseminate authoritative spatial datasets. SDIs provide the legal and institutional foundations for trustworthy geospatial data governance.
- System-of-Systems (SoS): A network of interconnected digital services and applications (e.g., smart-city dashboards, climate monitoring platforms, intelligent transport systems) that build upon and enrich SDI datasets. SoSs are characterized by interoperability and cross-domain integration.

- Geoverse: An aspirational, next-generation digital environment, which is often immersive and AI-driven, where geospatial information flows continuously between humans, machines, and decision-support systems. The Geoverse envisions an integrated geospatial user experience with real-time spatial awareness.

Future geospatial ecosystems will evolve toward more open, intelligent, and user-centric environments that go beyond data integration to enable automated, predictive, and adaptive decision-making. These ecosystems will harness digital technologies to enable real-time location-based decisions. Rather than being restricted to the management of static geospatial datasets, future ecosystems will serve as intelligent infrastructure layers underpinning critical applications in urban resilience, land governance, climate action, and infrastructure management. Moreover, future geospatial ecosystems will support semantic integration between spatial and non-spatial data, facilitating more meaningful connections between legal, environmental, social, and economic domains. This evolution aligns with emerging principles for spatially enabled digital twins, which emphasize the integration of real-world data with virtual models to provide dynamic and holistic views of built and natural environments.

Ultimately, the future geospatial ecosystem will act as a foundation for land intelligence, providing the digital and analytical capabilities necessary to address complex land administration challenges via harnessing digitally integrated, intelligent, interconnected, and immersive environments.

1.4 Emerging Trends and Innovations in the Future Geospatial Ecosystem

The geospatial ecosystem is undergoing transformative shifts driven by rapid technological advancements. GeoAI significantly enhances spatial analytics capabilities, allowing rapid processing and analysis of extensive datasets for precise and predictive insights across fields such as urban planning, infrastructure monitoring, agriculture, environmental management, and disaster response. Concurrently, the expansion of open Earth observation data is revolutionizing how spatial data is acquired, processed, and disseminated. This enables real-time monitoring of urban growth, environmental changes, and natural disasters, significantly improving responsiveness and decision-making.

The widespread adoption of open-access policies has facilitated access to high-quality geospatial data, promoting collaboration and innovation among businesses, governments, and communities globally. Additionally, cloud computing and distributed processing infrastructures have facilitated the efficient handling and analysis of large-scale geospatial datasets, overcoming previous computational barriers. Integration with IoT devices further enhances real-time spatial intelligence, allowing for more dynamic and informed decisions. In addition, modern methodologies such as agile prototyping and design sprints are accelerating the innovation lifecycle, streamlining geospatial developments from conceptualization to practical deployment. Adoption of robust governance models guided by FAIR (Findable, Accessible, Interoperable, Reusable) principles ensures data quality, interoperability, and usability across stakeholders. Furthermore, immersive technologies, such as VR and AR, enable intuitive spatial storytelling and enhanced stakeholder engagement. As these technologies advance, addressing ethical considerations, including transparency, bias mitigation, and regulatory compliance, has become crucial for sustainable and equitable geospatial innovation.

1.4.1 Geospatial Digital Twins for Smart City Planning

Digital twins, virtual replicas of physical entities and systems, have emerged as essential tools in smart city planning, providing detailed and dynamic representations of urban environments. These sophisticated models enable stakeholders to simulate and analyze complex urban mobility patterns, optimize utility network operations, enhance infrastructure resilience, and proactively respond to disasters through realistic scenario visualization and predictive analytics. By integrating real-time data from sensors, IoT devices, satellite imagery, and other diverse geospatial datasets, digital twins facilitate informed decision-making and comprehensive urban management strategies. Despite their potential, implementing geospatial digital twins presents several challenges (ANZLIC, 2019). Foremost among these are the complexities associated with integrating heterogeneous datasets from multiple sources, often varying significantly in format, scale, and quality. Additionally, real-time synchronization of diverse data streams demands substantial computational capabilities and robust infrastructure. To address these integration issues, standardized data structures and metadata frameworks have been developed, enabling more efficient and reliable data exchange and interoperability among different systems and stakeholders.

Cloud-based platforms have further streamlined the implementation of digital twins by providing scalable storage, processing capabilities, and analytics infrastructure that can accommodate vast amounts of geospatial information. These platforms offer real-time data synchronization, enabling continuous updates and monitoring, which are crucial for accurate and dynamic urban modeling. Moreover, adopting interoperable and open standards, such as CityGML and BIM (Building Information Modeling), facilitates seamless data integration and enhances cross-sector collaboration (Gröger & Plümer, 2012; ISO, 2013). Future advancements in AI and machine learning (ML) are expected to further enhance the capabilities of digital twins, enabling even more sophisticated predictive modeling, automated anomaly detection, and enhanced decision-support systems to support sustainable and resilient urban development.

1.4.2 Geospatial Applications of Artificial Intelligence

AI-driven predictive analytics have significantly advanced geospatial applications, offering transformative solutions across multiple critical areas (Wang et al., 2024). In land administration, AI models analyze extensive historical and current geospatial datasets, accurately predicting land-use changes and enabling informed urban development planning (Hosseini et al., 2025). Similarly, in flood risk management, ML algorithms utilize terrain, hydrological, and meteorological data to forecast flood events, facilitating timely preventive actions and resource allocation. Infrastructure maintenance also benefits considerably from GeoAI-powered predictive analytics. By continuously monitoring infrastructure conditions through sensors and satellite imagery, AI models identify early signs of deterioration, enabling proactive maintenance scheduling and reducing costly emergency repairs. Deep learning techniques, particularly convolutional neural networks (CNNs), have dramatically improved feature extraction from high-resolution satellite imagery. This advancement enables precise mapping of land-use patterns, identification of critical infrastructure issues, and detection of environmental anomalies such as deforestation or urban sprawl. Moreover, the integration of AI with real-time sensor data and IoT devices has enhanced environmental hazard detection, allowing for rapid response and mitigation strategies in urban and rural contexts alike. AI-driven geospatial applications not only optimize resource management and improve environmental monitoring but also foster proactive decision-making, enhancing the resilience and sustainability of urban environments.

1.4.3 Modern Geospatial Methodologies and Innovation Frameworks

Contemporary geospatial innovation increasingly employs agile methodologies such as rapid prototyping and design sprints, significantly accelerating the development and deployment of effective spatial solutions. This iterative approach facilitates quick testing, refinement, and adaptation based on user feedback, dramatically reducing the time from conceptualization to implementation. Agile practices enhance responsiveness to emerging user needs, enabling rapid adjustments that ensure solutions remain relevant and impactful. The innovation lifecycle for geospatial systems encompasses comprehensive stages, including problem framing, ideation, prototyping, pilot testing, and scaling. Each phase emphasizes collaboration, iterative improvements, and user-centered design principles. By incorporating continual stakeholder engagement and feedback, agile methodologies ensure that geospatial solutions are robust, effective, and directly address real-world requirements. Furthermore, the integration of cross-disciplinary expertise fosters innovation, allowing teams to develop holistic solutions that address complex spatial challenges.

1.4.4 Geospatial Data Governance and FAIR Principles

Effective geospatial data governance is crucial for maintaining data quality, consistency, and interoperability across diverse applications and sectors. Governance frameworks emphasizing clearly defined standards, roles, responsibilities, and accountability mechanisms ensure data integrity, reliability, and accessibility. Effective governance also involves establishing rigorous metadata practices, promoting data standardization, and managing access controls to protect sensitive information. The adoption of FAIR principles and making geospatial data based on these principles further enhances governance effectiveness (GO FAIR, 2023). By ensuring data assets are easily discoverable and accessible while maintaining interoperability across different platforms and use cases, FAIR principles promote wider utilization, facilitate collaboration across disciplines, and maximize the potential impact of geospatial datasets. Additionally, the implementation of automated workflows, standardized APIs, and interoperable formats helps streamline geospatial data exchange, ensuring this data is always current and relevant.

1.4.5 Immersive and Spatial Computing Technologies

Immersive technologies such as VR, AR, and extended reality (XR) are transforming geospatial applications, offering innovative opportunities in land management and spatial storytelling (Bolkas et al., 2021; Shojaei et al., 2019). These interactive technologies enable users to engage intuitively with spatial data, significantly improving comprehension, stakeholder engagement, and decision-making processes. Immersive environments provide stakeholders with realistic and experiential views of spatial scenarios, enhancing collaborative planning and problem-solving capabilities. Integration with advanced location intelligence tools and sophisticated 3D visualization techniques further augments the capabilities of immersive technologies. Combining these technologies allows for more effective communication, clearer insights, and improved participatory planning, supporting diverse applications ranging from urban development to environmental management and heritage preservation. Future developments in immersive technology, such as real-time collaboration platforms and advanced holographic displays, will further amplify their role in geospatial analytics, land administration, and urban planning.

1.4.6 Ethical Considerations in Geospatial Technologies

As geospatial technologies increasingly integrate AI, addressing ethical considerations becomes paramount. Key ethical concerns include transparency in data sources, analytical processes, and decision-making outcomes to foster trust and accountability among stakeholders. Mitigating biases in spatial data and algorithms is essential to ensure fair, equitable outcomes in geospatial analytics, particularly in urban planning, resource allocation, and policy-making contexts. Robust regulatory frameworks and ethical guidelines must be established and continuously refined to guide the responsible and ethical use of geospatial technologies. Protecting privacy, ensuring data security, promoting equity in data-driven decisions, and maintaining transparency throughout analytical processes are critical elements.

1.5 Advancing Sustainability and Resilience Through Land Intelligence and Future Geospatial Ecosystem

Sustainability and resilience have become critically important for land administration systems amid the increasing impacts of rapid urbanization, climate change, biodiversity loss, and socio-economic inequalities. These global challenges demand land administration frameworks and geospatial ecosystems that are not only robust and reliable but also adaptive, predictive, and inclusive. Land intelligence, when embedded within a future geospatial ecosystem, offers transformative potential to address sustainability and resilience challenges by providing actionable insights, fostering more integrated planning, and improving institutional responsiveness.

A core element of advancing sustainability is the capacity to monitor, model, and manage land resources in ways that align with SDGs. Land intelligence enables this by integrating geospatial data on land tenure, use, value, and development with socio-economic, environmental, and infrastructure datasets. Through the dynamic processing of these datasets, land intelligence supports evidence-based policymaking that is responsive to emerging risks and opportunities. For example, predictive modeling can identify areas vulnerable to flooding or drought, informing proactive land-use planning and infrastructure planning decisions that mitigate long-term risks.

Resilience is reinforced through the ecosystem's capacity for real-time monitoring and adaptive management. The future geospatial ecosystem, with its integration of sensor networks, remote sensing, digital twins, and AI, enables continuous assessment of changing land conditions. This capability is vital for disaster risk reduction, climate adaptation, and the management of critical infrastructures and ecosystems. In urban settings, spatially enabled digital twins can simulate the impacts of extreme weather events on infrastructure and populations, guiding more resilient urban design and emergency preparedness strategies.

Moreover, the synergy between land intelligence and the geospatial ecosystem strengthens multi-level governance and community participation. Transparent and interoperable systems allow for inclusive access to land information, empowering communities, indigenous groups, and local authorities to engage more effectively in decision-making processes. This inclusivity supports social resilience by fostering trust, reducing conflict, and promoting equitable resource distribution.

Global initiatives, such as the UN-GGIM's FELA, explicitly link effective land governance to sustainability and resilience outcomes. FELA advocates for systems that can perform under conditions of stress, whether environmental, economic, or political, and adapt through innovation and partnerships. Land intelligence operationalizes this vision by connecting fragmented data, enabling cross-sectoral collaboration, and ensuring that land systems contribute meaningfully to broader resilience agendas. Advancing sustainability and resilience through land intelligence and the future geospatial ecosystem is about embedding anticipatory governance within land administration systems using data, technology, and inclusive practices to foresee risks, plan for long-term needs, and adapt to uncertainty. This integration not only strengthens the capacity of land administration systems to cope with current challenges but positions these systems as a proactive contributor to a sustainable, resilient, and equitable future.

The relationship between FELA and IGIF is especially pivotal in this context. While FELA offers a targeted, sector-specific strategy for modernizing land administration systems, IGIF provides a broader, cross-sectoral blueprint for establishing integrated, national geospatial information management ecosystems. IGIF articulates the technical, institutional, and governance architectures required to achieve interoperability, standardization, and comprehensive data integration across different sectors and jurisdictions. Together, these frameworks create a synergistic foundation for aligning land intelligence initiatives with the evolving demands of the future geospatial ecosystem.

FELA's nine strategic pathways (i.e., governance, policy and legal frameworks, financial sustainability, data management, innovation, technical standards, partnerships, capacity building, and advocacy) set out the institutional and operational reforms needed to modernize land administration. IGIF complements this by specifying the SDIs, interoperability protocols, and governance mechanisms required to support those pathways at scale, ensuring that land information is not siloed but actively contributes to national and global geospatial infrastructures. This integration highlights that land intelligence must not function in isolation but as a core component within an interconnected, multi-sectoral geospatial ecosystem that is capable of leveraging spatial data for sustainable development and resiliency.

1.6 Book Structure

The chapters of this book are structured into three parts as shown in Figure 1.2.

Part I: Foundations of Land Intelligence and Geospatial Ecosystem for Sustainability and Resilience

The book begins with foundational chapters introducing the fundamental principles and concepts behind land intelligence and the geospatial ecosystem.

FIGURE 1.2
Book structure: parts and chapters.

It explores how digital transformation and geospatial innovations have revolutionized land administration practices, paving the way for more efficient and transparent approaches to support sustainable development and enhance community resilience.

Part II: Digital Innovations Toward Land Intelligence and Future Geospatial Ecosystem

This part focuses on the latest technical innovations critical for building intelligent land and geospatial ecosystems. It includes detailed content on advanced data modeling, validation, visualization, and storage. This provides a structured framework for representing complex land and geospatial information in a digital environment. There are also chapters on BIM for high-rise and underground developments to showcase how advanced modeling techniques enhance the management of vertical properties, optimizing space utilization and operational efficiency. Additionally, the use of 3D Geographic Information Systems (GIS) for underground land administration elucidates spatial analysis techniques tailored to subsurface environments, ensuring comprehensive land intelligence and a geospatial ecosystem.

Practical methodologies for data capturing, validation, visualization, and advanced data query and analysis techniques are covered extensively. The integration of AI in land administration processes is also explored in depth, highlighting AI's role in optimizing decision-making and operational workflows.

Part III: Applications and Advanced Services: Empowered by Land Intelligence and Future Geospatial Ecosystem

This part includes chapters on the role of land intelligence and geospatial ecosystem in supporting different application domains such as urban planning, energy management, transport, disaster management, and asset management.

References

ANZLIC. (2019). *Principles for Spatially Enabled Digital Twins of the Built and Natural Environment in Australia*. ANZLIC Barton. https://www.anzlic.gov.au/sites/default/files/files/principles_for_spatially_enabled_digital_twins_of_the_built_and_natural_.pdf

Bolkas, D., Chiampi, J., Fioti, J., & Gaffney, D. (2021). Surveying reality (SurReal): Software to simulate surveying in virtual reality. *ISPRS International Journal of Geo-Information*, 10(5), 296.

GO FAIR. (2023). *Findable, Accessible, Interoperable and Reusable (FAIR) Principles*. https://www.go-fair.org/fair-principles/

Gröger, G., & Plümer, L. (2012). CityGML – Interoperable semantic 3D city models. *ISPRS Journal of Photogrammetry and Remote Sensing*, 71(0), 12–33. https://doi.org/http://dx.doi.org/10.1016/j.isprsjprs.2012.04.004

Hosseini, H., Atazadeh, B., & Rajabifard, A. (2025). Towards intelligent land administration systems: Research challenges, applications and prospects in AI-driven approaches. *Land Use Policy*, 157, 107652. https://doi.org/10.1016/j.landusepol.2025.107652

ISO. (2013). *16739: 2013 Industry Foundation Classes (IFC) for Data Sharing in the Construction and Facility Management Industries*. International Organization for Standardization.

Shojaei, D., Olfat, H., Aien, A., Nimorakiotakis, N., & Rajabifard, A. (2019). Utilising Virtual and Augmented Reality Technologies to Improve the Visualisation of Digital Cadastre – The Victorian Cadastre Experience. *2nd International IAG Workshop on BIM and GIS Integration, 3D GeoInfo Conference 2019*.

UN-GGIM. (2020). *Framework for Effective Land Administration*. http://ggim.un.org/meetings/GGIM-committee/10th-Session/documents/E-C.20-2020-29-Add_2-Framework-for-Effective-Land-Administration.pdf

UN-GGIM. (2022). *Future Geospatial Information Ecosystem: From SDI to SoS and on to the Geoverse*. https://ggim.un.org/meetings/GGIM-committee/12th-Session/documents/Future_Geospatial_Information_Ecosystem_Discussion_Paper_July2022.pdf

UN-GGIM. (2023). *United Nations Integrated Geospatial Information Framework – Part 1: Overarching Strategy*. https://ggim.un.org/UN-IGIF/documents/Part_1_UN-IGIF_Overarching_Strategy_Second_Edition_27Feb2023.pdf

Wang, S., Huang, X., Liu, P., Zhang, M., Biljecki, F., Hu, T., Fu, X., Liu, L., Liu, X., Wang, R., Huang, Y., Yan, J., Jiang, J., Chukwu, M., Reza Naghedi, S., Hemmati, M., Shao, Y., Jia, N., Xiao, Z., … Bao, S. (2024). Mapping the landscape and roadmap of geospatial artificial intelligence (GeoAI) in quantitative human geography: An extensive systematic review. *International Journal of Applied Earth Observation and Geoinformation*, 128, 103734. https://doi.org/10.1016/j.jag.2024.103734

Williamson, I. P., Enemark, S., Wallace, J., & Rajabifard, A. (2010). *Land Administration for Sustainable Development*. ESRI Press Academic Redlands.

Part I

Foundations of Land Intelligence and Geospatial Ecosystem for Sustainability and Resilience

2

Accessing Land for Climate Resilience: Challenges and Solutions

Abbas Rajabifard

2.1 Introduction: The Urgency of Climate Action and the Centrality of Land

The global community faces an unprecedented imperative to address climate change, yet current efforts are significantly off track. The 2023 State of Climate Action Report revealed that 41 out of 42 indicators are not progressing fast enough to meet 2030 targets (Boehm et al. 2022). Achieving these ambitious goals – such as expanding solar and wind power, accelerating rapid transit infrastructure, and reducing deforestation – demands substantial investment and, crucially, significant land transitions. However, this urgency intersects with a pre-existing and worsening global issue: land inequality. Land has historically been, and continues to be, a primary source of power, status, and wealth, often concentrated in the hands of a few. The rapid pace of climate action, if not managed sensitively, risks exacerbating these existing inequalities, directly contravening the principles of 'just transitions' enshrined in the Paris Agreement.

The principle of 'just transitions' in the context of land access for climate actions mandates the recognition of rights for all affected parties, equitable distribution of both positive and negative impacts, and transparent and fair rules and procedures for land access. This chapter delves into the complex challenges inherent in securing land for climate resilience and mitigation efforts, drawing insights from global experience. It then proposes a strategic framework of seven building blocks designed to guide policymakers and practitioners in navigating these complexities, ensuring that climate actions are not only effective but also just and equitable.

2.2 The Critical Link: Climate Actions and Land

Climate actions, whether for mitigation or adaptation, are profoundly land-dependent. Research underscores the immense scale of land needed for climate

DOI: 10.1201/9781003638308-3

initiatives: the Land Gap Report indicates that governments have pledged approximately 1 billion hectares of land for carbon removal, encompassing forest restoration, reforestation, and bioenergy investments. Specifically, the Intergovernmental Panel on Climate Change (IPCC) projects that 0.2 billion hectares will be required for bioenergy and afforestation investments in the East Asia and Pacific Region alone by 2050, an area approximately two-thirds the size of Africa. A World Bank analysis of Nationally Determined Contributions (NDCs) further revealed that nearly one in two climate actions in the East Asia Pacific and South Asia regions will involve land, with half of these potentially necessitating land-use change or interfering with existing land and resource access rights. Seventy-five percent of these land-related actions are concentrated in the forestry, energy, transport, environment, and agriculture sectors.

The implications of these land requirements are significant:

- **Energy Sector**: As the most common type of mitigation action, energy projects (37% of all global mitigation actions) often require extensive land for new renewable energy infrastructure like solar and wind farms and their associated transmission networks. For instance, Bangladesh's goal to source 30% of its energy from renewables by 2030 faces challenges due to legislation preventing agricultural land conversion (Climate Change Laws of the World 2024).
- **LULUCF (Land Use, Land-Use Change, and Forestry)**: Actions in this sector, such as reforestation and forest protection, typically span vast territorial scales and often occur in ecologically fragile settings or areas with existing land use and local livelihood dependencies. This necessitates careful consideration of existing land rights and resource dependencies.
- **Agriculture**: While many agricultural adaptation actions focus on climate-smart techniques, they often require understanding land users. A significant challenge arises from the prevalence of smallholder farms (84% globally, accounting for 35% of the world's food supply), many of which lack formal documentation of land ownership.
- **Transport**: Low-carbon transport transitions, such as new rail lines and electric vehicle charging infrastructure, demand contiguous land corridors. This can be particularly complex in urban areas with high informality, where information on rightsholders and informal settlers may be missing, potentially exacerbating socio-spatial inequalities.

Beyond these traditional land uses, climate actions are also broadening the very concept of 'land.' Attention is increasingly turning to new 'land' frontiers, such as seabeds for critical minerals essential for clean energy technologies like electric vehicle batteries. These areas are often poorly regulated, leading to a rush for access by multinational companies despite limited governance frameworks and potential long-term implications for resource management, as seen in the Pacific nations with their rich deep-sea resources.

2.3 Challenges in Accessing Land for Climate Resilience

Accessing land for climate action is fraught with multi-faceted challenges, stemming from entrenched socio-economic and political factors:

1. *Land Inequality and Unclear Rights*: Despite significant development interventions, global land inequality is worsening. The top 10% of rural populations now control 60% of agricultural land value, while the bottom 50% possess only 3%. The number of land deals has surged globally, with over 40% in African countries, often targeting agricultural land. This concentration of land is compounded by insecure tenure, particularly for vulnerable groups:
   - **Informal Tenure**: A significant proportion of land users, especially smallholder farmers and informal settlers in cities, lack formal documentation of their rights, making land access negotiations complex and often inequitable.
   - **Women's Land Rights**: Studies from Asia and Africa reveal substantial gender gaps in land ownership and documented tenure, leaving women particularly disadvantaged in land-related decisions and benefits.
   - **Indigenous Peoples and Local Communities**: Their traditional land rights are often not formally recognized, making them highly vulnerable to displacement or dispossession when land is sought for climate projects.
2. *Environmental and Cultural Sensitivity*: Climate actions frequently target areas with high ecological value or cultural significance, posing a dilemma. For example, commercial brushwood harvesting for bioenergy in Sweden could benefit biodiversity but also threaten other species. Similarly, high restoration potential areas in Madagascar often lie on untitled land, where unclear or contested tenure hinders conservation efforts. Navigating these sensitive areas requires careful assessment of impacts and the development of alternative strategies that respect both environmental integrity and cultural heritage.
3. *Policy and Data Gaps*: Many countries grapple with inadequate policy and data landscapes that impede effective land access for climate action:

- **Fragmented Planning Systems**: Development plans are often fragmented, outdated, or lack integration with climate change objectives. This makes it difficult to align land-use planning with national climate goals and ensures coordinated action across different governance levels (national, sub-national, local).
- **Lack of Quality Data**: A pervasive challenge in developing countries is the absence of comprehensive, accurate, and accessible land-related data. This hinders informed decision-making regarding site selection, impact assessment, and the identification of rightsholders. Many land administration systems (LASs) are in need of modernization and digitization to provide precise details on land rights and interests.
- **Institutional Coherence**: Rules governing land often lack coherence across different jurisdictions or between land-related and climate policies. This can lead to 'grey areas' where exploitation occurs, creating uncertainty and conflict.

4. *New 'Land' Frontiers and Regulation Gaps*: The emergence of new land frontiers, such as deep-sea mining for critical minerals, presents novel challenges. These areas are often poorly or entirely unregulated, leading to a rush to access resources without robust governance frameworks. This rapid expansion risks long-term implications for environmental good governance and sustainable resource management, potentially repeating past mistakes of resource extraction that have devastated landscapes.

2.4 Solutions: A Framework for Just Land Transitions

Addressing these challenges requires a systematic and integrated approach. This chapter proposes a framework of seven interconnected building blocks designed to support practitioners and policymakers in enabling just and equitable access to land for climate actions. These blocks are categorized into four 'core' blocks, which directly influence land identification, control, and use, and three 'amplifying' blocks, which provide wraparound support for technical and social constraints.

Core Building Blocks: These four areas are critical for strategic development and directly shape how land is accessed and managed for climate actions.

- **Building Block 1 – Enabling Policy and Regulatory Frameworks**: The aim of this building block is to provide the strategic rules and regulations that facilitate just and equitable land access, establishing priorities for land use and development and governing land conversion or acquisition by the state. It should also enable effective dispute resolution. Key actions of this building block include: assessing existing spatial plans and policies to identify available land access options, coupled with on-ground consultations to understand their local impact; developing clear, consistent, and coherent policies and regulations that equitably govern land access and manage conflicts, ensuring no one is left behind; and considering the increasing recognition of the 'rights of nature' in climate policies, which can significantly impact land access depending on their legal structure. For example, the U.S. Bureau of Land Management's Renewable Energy Rule updated regulations to reduce rents and fees, promoting wind and solar development on public lands.

- **Building Block 2 – Robust LASs**: The aim of this building block is to provide the fundamental land data infrastructure necessary for informed land access strategies and decisions. LAS must deliver precise details on land rights and interests for land-based climate actions. Key actions of this building block include: utilizing accessible and comprehensive LAS data to assess existing interests and stakeholders where climate action affects land, including rights, restrictions, and responsibilities; where LAS is not accessible or comprehensive, manually integrating available data to determine land interests and identify rightsholders. Critically, understanding who is not included in existing LAS and working to identify all affected stakeholders; seeking finance to develop technically and institutionally integrated and accessible systems; and improving data coverage through collection processes. This includes digitizing, validating, and updating land records, linking them to cadastral maps, and developing unified land information systems. In Germany, geospatial analytics based on comprehensive digital land records are used to identify suitable areas for wind development, overcoming regulatory constraints.
- **Building Block 3 – Strong Land Tenure Systems and Institutions**: The aim of this building block is to recognize and leverage legitimate land tenure systems (formal, informal, customary, and Indigenous) to negotiate land access, regardless of their formal recognition status. Key actions of this building block include: establishing the level of community and statutory recognition of existing land tenure systems, their relevance, and how they enable or constrain land access; identifying the needs, values, and capacities of local land tenure systems and institutions for negotiating land access, and working with and capacitating these local institutions; paying specific attention to strengthening women's land rights, the rights of Indigenous Peoples and local communities, and informal settlers in cities. In Myanmar's Tree Planting and Agroforestry Initiative, addressing tenure insecurity is crucial to enhance effectiveness, especially given the risk of

deforestation if land rights are not secured. The success of national restoration plans in Malawi, for instance, is predicated on investing in policy frameworks and local institutions that protect legitimate forest tenure rights.

- **Building Block 4 – Sustainable Land Management (SLM)**: The aim of this building block is to ensure land resources are utilized in ways that mitigate climate change and adapt to its impacts, enhancing carbon sequestration, protecting ecosystems, and reducing vulnerability to climate hazards. Key actions of this building block include: recognizing and accounting for trade-offs for local land-dependent communities when existing land use is changed; building incentives and capacity to transition agricultural systems to more SLM practices, such as agroforestry. In Kenya, clarifying land rights and ownership is crucial for empowering communities to invest in long-term land management practices like agroforestry, supporting carbon sequestration and biodiversity; establishing appropriate governance structures for monitoring and evaluation of SLM practices; and assessing and identifying global and local data to inform SLM, and leveraging resources from organizations like FAO.

Amplifying Building Blocks: These three areas provide crucial support and enable the effective implementation of strategies defined under the core blocks.

- **Building Block 5 – Governance**: The aim of this building block is to provide oversight and accountability for land access strategies, including coordination, reporting, conflict resolution, and ongoing monitoring, evaluation, and learning activities. Key actions of this building block include: structuring governance systems that recognize and leverage legitimate land-related authorities to implement and monitor land access strategies; identifying appropriate governance indicators (e.g., from tools like the Global Land Governance Index: https://www.landexglobal.org/en/ or LandMark: http://www.landmarkmap.org/); developing local capacity to ensure governance systems operate with public confidence, a vital aspect of just land transitions. An example from Ecuador shows how a participatory process fostered territorial governance, resulting in a municipal ordinance for conservation and sustainable use, highlighting effective governance for resource protection.
- **Building Block 6 – Community Consultation**: The aim of this building block is to establish an iterative, two-way communication process between affected communities and stakeholders involved in climate action, ensuring agency and participation in land access negotiations and decisions. Key actions of this building block include: developing

transparent and trustworthy consultation processes that enable meaningful participation, adhering to relevant standards around Free, Prior, and Informed Consent (FPIC); engaging in proactive community consultation, as demonstrated in Bangladesh's Renewable Energy projects, which can significantly reduce conflicts during land acquisition by enhancing acceptance and participation. USAID also advises employing local organizations trusted by the community to represent their interests in negotiations.

- **Building Block 7 – Capacity Development**: The aim of this building block is to enhance the competency and capabilities of people, processes, systems, technologies, and available resources to deliver common objectives related to climate action and land access. Key actions of this building block include: recognizing that many NDCs and long-term strategies seek increased human, institutional, and technological support for climate actions, often conditional on external support; investing in upskilling workforces and updating data systems for planning, assembly, allocation, and permitting of land for climate actions. In Afghanistan, strengthening land tenure systems and institutions is highlighted as a critical step to ensure women's participation in forest regeneration, which requires significant capacity development interventions. Similarly, in Kenya, promoting and educating communities on the benefits of agroforestry is key to encouraging participation.

2.5 Putting Solutions into Practice: Case Studies

The practical application of these building blocks can be seen in diverse global contexts:

- **Myanmar's Tree Planting and Agroforestry Initiative (LULUCF Sector)**: Facing land tenure insecurity and the risk of deforestation, Myanmar's efforts aim to increase tree canopy cover. Success hinges on capacity building (promoting tree planting) and community consultation (engaging in land-use changes). Crucially, a Robust LAS is needed to mitigate conflicts, and Strong Land Tenure Systems and Institutions must be strengthened to incentivize long-term investment.
- **Afghanistan's Regeneration of Degraded Forests (LULUCF Sector)**: To regenerate degraded forests, the case highlights the need for Strengthening Land Tenure Systems and Institutions, particularly to ensure women's land rights are recognized and that benefits reach women equally, boosting overall effectiveness and inclusivity.

- **Bangladesh's Renewable Energy Projects (Energy Sector)**: For grid-connected solar installations, challenges include securing land tenure and managing land assembly. The solution emphasizes Robust LASs for site identification, Enabling Policy and Regulatory Frameworks for legal acquisition and compensation, and critically, community consultation to reduce conflicts and enhance participation.
- **Kenya's Agroforestry Program (Agriculture Sector)**: Addressing a lack of formal land rights for pastoralists and small-scale farmers, Kenya's initiative to plant agroforestry trees requires Robust LASs for site identification. Capacity building and community consultation are vital for educating communities about benefits and mitigating conflicts The Strong Land Tenure Systems and Institutions block is essential to provide formal land rights, enabling long-term investment in tree planting.

These examples collectively underscore that successful and just land access for climate actions requires a multi-faceted approach, integrating legal reforms, robust data systems, inclusive governance, and active community engagement.

2.6 Conclusion

Accessing land for climate resilience and mitigation presents one of the most pressing and intricate challenges of our time. The imperative to meet climate goals by 2030 necessitates unprecedented land transitions, but this must be achieved without exacerbating existing land inequalities or imposing undue burdens on vulnerable communities. This chapter has highlighted that around one in two climate actions will directly impact land, and the current landscape of land relations, often characterized by insecure tenure, data deficiencies, and fragmented policies, poses significant barriers.

The proposed framework of seven building blocks – comprising four core blocks (Enabling Policy and Regulatory Frameworks; Robust LASs; Strong Land Tenure Systems and Institutions; SLM) and three amplifying blocks (Governance; Community Consultation; Capacity Development) – offers a comprehensive approach. This framework recognizes that coherent land access strategies must reflect broad policy aspirations, acknowledge local tenure systems, be supported by robust data, and be underpinned by strong governance, genuine community consultation, and continuous capacity development.

By adopting this integrated approach, policymakers and practitioners can ensure that climate actions are implemented in a manner that is not only effective and sustainable for the planet but also just and equitable for all people involved. This will foster informed decision-making, genuine community engagement, and the equitable distribution of benefits, ultimately aligning climate actions with the fundamental objectives of the Paris Agreement. The time for responsible and inclusive action is now, acting in ways that do not create harms to people and planet.

References

Boehm, Sophie, Louise Jeffery, Judit Hecke, et al. 2022. *State of Climate Action 2023.*

Climate Change Laws of the World. 2024. *Mujib Climate Prosperity Plan 2022–2041.* https://climate-laws.org/document/mujib-climate-prosperity-plan-up-to-2030_88cc

Dooley, Kate, Heather Keith, Anne Larson, et al. 2022. *The Land Gap Report.* https://www.landgap.org/

3

Digital-Twin-Enabled Land Administration and Its Connections to Social Sustainability

Kirsikka Riekkinen and Abbas Rajabifard

3.1 Introduction

The increasing complexity of urban areas along with advancements in digital technologies, multi-dimensional data models, such as Land Administration Domain Model (LADM), CityGML, and Building Information Model (BIM), and spatial data infrastructures as a fundamental enabler have created an opportunity for the next generation of urban land administration and 3D spatial planning. Constantly developing technologies in location intelligence and capabilities such as digital twins (DTs) enable the seeking of new solutions for administrative and visualization purposes, in various forms of applications enabled by digitalization and modernization of land administration systems. The aim of this chapter is to explore the connections between land administration and social sustainability, by advancing the concepts of social sustainability, social inclusion, and social exclusion, and especially their indicators.

The holistic relevance of land administration in achieving or supporting the achievement of sustainable development goals (SDGs) is and has been recognized by organizations themselves, even before the current SDGs (e.g., UN Millennium Goals). Also, this is recognized by scholars, and the current need to integrate land administration data with other sources is noted by Unger et al. (2021). LADM has been studied and further developed to support several aspects of sustainability, such as women's access to land (Unger et al., 2023), green urban areas or infrastructures, or sustainability of disaster risk areas (Unger et al., 2019). The applications of LADM have especially supported social sustainability by providing a basis for the development of the Social Tenure Domain Model (STDM), which is promoted by organizations such as the Global Land Tool Network. Still, little attention has been paid to the possibilities provided by digital applications for land administration, beyond LADM and its extensions, to administer and support sustainability.

DOI: 10.1201/9781003638308-4

Therefore, the literature used for this study uses concepts beyond LADM and STDM studies to provide a new aspect for land administration studies.

The motivation of this study lies in the need for people-centered tools in land administration and discussions raised recently by Park et al. (2023) on the importance of social issues related to digital technologies. Where one of the reasons for socially unsustainable structures lies in the embeddedness of the economy into all aspects of society (Lawson, 2010), we propose to search for tools and measures to enhance social sustainability and inclusion. Our focus is on the administrative perspective, following Labonté's (2004) suggestion on tackling exclusion within communities.

The aim of this chapter is to explore the connection between digital land administration systems and social sustainability. The objective is to first study the concepts and connections between social sustainability, social inclusion, and social exclusion, and DT as a platform for modern land administration systems with existing literature. After that, the indicators for measuring social inclusion and exclusion are examined more closely, and a thematic analysis using the lens of land administration is done. The goal is to find those indicators that can be influenced with the land administration tools (tenure, value, use, development, see Enemark, 2004). Finally, the importance and possibilities of land administration to support social sustainability using a DT platform are discussed.

Analysis and results will demonstrate the topical approach of connecting land administration and social sustainability, and discussing the potential of technological development in advancing this connection by examining the sustainability from social inclusion and exclusion perspectives. The results contribute to the discussion of sustainable development in society, future land administration, and administrative and policy-level decision-making.

3.2 Digital Twin for Modern Land Administration

DT refers to the construction of digital information of a physical system, which is embedded into the system and creates a "twin" of that physical entity through its entire lifecycle. As such, the concept of DT in planning and urban land administration suggests that two virtual and real systems are connected through four phases: (a) creation; (b) production; (c) operation; and (d) disposal (Grieves, 2011).

Recent advancements in the internet of things (IoT) technology and other real-time data streaming and data collection methods as well as cloud computing opportunities have created an ecosystem ready for the implementation of DT in different disciplines. In recent days, DT is being defined and developed in the context of data infrastructure and data life-cycle, spatial data infrastructures, urban land administration, building and facility management (Lu et al., 2020), and several other application domains. Several jurisdictions have already adopted or implemented the concept of DT by developing proof of concept, and they have done defined a series of principles persuading their future digital transformations (ANZLIC, 2019).

According to the DTs consortium, a DT is a virtual representation of real-world entities and processes, synchronized at a specified frequency and fidelity (Oclott & Mullen, 2020). DTs use real-time and historical data to represent the past and present and simulate predicted futures, and DTs are motivated by outcomes, tailored to use cases, powered by integration, built on data, guided by domain knowledge. The Centre for SDIs and Land Administration at the University of Melbourne (CSDILA, 2022) has developed a DT platform showcasing a series of applications for urban land administration, urban planning, and analytics. The DT platform provided an opportunity to integrate the BIM with spatial data as a proof of concept to address the challenges of 2D land administration and moving toward 3D digital cadaster.

As part of the CSDILA-DT system, they have facilitated the cities' productivity, sustainability, and livability, and also have demonstrated a series of capabilities to support and monitor SDGs and their selected indicators in particular in the context of SDG 11. The CSDILA-DT development is based on the relationships between SDGs, urban system for smart cities, and the 3D land and property in DT as three components of a framework. This framework supports the delivery of resilience, inclusiveness, and connectivity, and shows their interconnections with each other.

In addition, the system architecture considered FAIR (Findable, Accessible, Interoperable, Reusable) Data Principles promoted by the Australian National Data Service (ANDS). By adopting a standards-based approach, the CSDILA-DT platform enables the spatial information to be searched, discovered, aggregated, and visualized seamlessly, irrespective of database management system, data model, data content (raster, vector, metadata, etc.), data server, or preferred (client) application (as indicated in specification requirements). As such, in general, in the CSDILA-DT platform, a group of connectors is considered to tackle each type of accepted data format. The platform allows data registration with the capability of metadata enrichment (ISO 19115-3) and access control (Figure 3.1). It should be mentioned that access to data is defined based on organizations/users. The data can then be utilized with built-in tools such as "Integrating 2D/3D Cadaster" and "Quantitative Shadow Analysis"; third-party tools such as "Building Energy Analysis"; and cross-platform applications.

While there is still little research around DTs in land administration, digital technologies can potentially improve the accuracy and efficiency of land administration processes, which can lead to better outcomes for communities. For example, digital data capturing, validation, and visualization can be used to create more accurate and up-to-date land information, which can help to prevent disputes over land ownership. These technologies can also

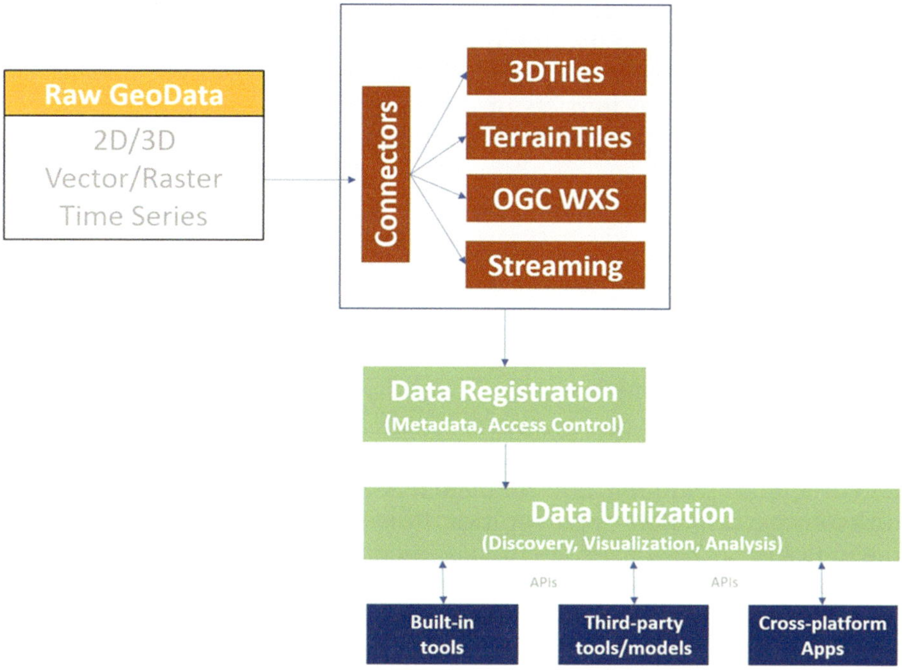

FIGURE 3.1
General workflow of CSDILA – digital twin – platform (CSDILA 2022).

promote greater transparency and accountability in land administration processes. For example, digital systems can be used to track land transactions and ownership, which can help to reduce corruption and ensure that land is allocated fairly. In addition, using digital technologies can lead to greater access to land administration information for communities and stakeholders. For example, DT platforms can be used to share information about land ownership, land use, and zoning regulations, which can help to promote greater community participation in land development processes. Another important point is that there will be improved collaboration among government agencies, communities, and other stakeholders in land administration processes if digital land administration systems are fully implemented. For example, digital-twin-enabled land administration platforms can be used to facilitate stakeholder engagement and consultation, which can help to ensure that land-related decisions reflect the needs and priorities of communities.

DTs provide several potential applications for land administration. Because they are realistic depictions of the physical environment, they enable communities to participate in collaborative decision-making processes linked to land development processes. This can ensure that development projects are socially inclusive and meet the requirements of the community. Being a real-time representation of infrastructure, services, and amenities, they are making information more accessible. Finally, DTs are supporting disaster management and resilience, enabling the simulation of the potential impacts of disasters on communities, providing a way for decision-makers to mitigate these risks and protect vulnerable communities.

3.3 Social Sustainability and Land Administration

Williamson et al. (2010) have identified the tools for modern LAS. These include the capability of a system to handle separately traded land-related interests, an accurate and transparent system, an organized controlling system to support planning, environmental issues, and risk management, and different participation opportunities. Based on our analysis, we propose the indicators related to economic, socio-cultural, and institutional dimensions of social inclusion/exclusion to be used to promote social sustainability with the tools of land administration. There are indicators that have a direct connection with the land administration, and indicators that support decision making (indirect). Most of the indirect indicators can be mapped, thus used for spatial analyses and spatially enabled decision-making (Scott & Rajabifard, 2019), whereas those putting direct requirements for the land administration system remain at the policy and document level.

For indicators of the economic dimension, we propose using different local affordable housing indices. Using spatial tools, we can also map these indices to reveal areas where people are struggling with affordability and differences between areas. This information can be used for decision-making in spatial planning and investment decisions. One very specific indicator of social exclusion is the distance between home and work (Silver, 2015). This can be used either to support public transportation planning on local level, or to map risk for individual exclusion and possibilities to locate new workplaces Our third proposed indicator for supporting social sustainability by economic land administration means is the number to what extent local or national land administration system supports trading of other rights of land than ownership rights (Williamson et al., 2010). This indicator reflects the transformation of society from ownership-led toward a more flexible structure, where other interests for land are also tradable, and rather than the land market, we discuss the commodities market.

Indeed, societies all over the world are transforming into more flexible and, due to globalization, more mobile. People are moving at a local and global level. Due to globalization and urbanization, there are risks of gentrification and segregation, providing fewer opportunities for those who are in need. To detect and visualize gentrification, we propose to map people's housing patterns. This enables us to recognize areas of gentrification for decision-making to allocate adequate measures to the area. It should be noted that gentrification is not always seen as negative (Chaskin & Joseph, 2013). Mapping of people's housing patterns may reveal some other important aspects for decision-making. In the same vein, we propose to examine the heterogenetic structure of forms of housing and income to reveal the possible risk of segregation. Finally, segregation, social exclusion, and safety are often bundled together (Silver, 2015). This is the motivation for why we propose to map urban safety. This can be done by combining crime statistics, but also by using so-called soft-GIS tools to involve citizens and let them point out perceived unsafe areas (Kahila & Kyttä, 2009).

Institutional, sometimes referred to as societal or legislative, dimension relates to social sustainability by supporting social cohesion, which is related to trust in society. Transparent administrative and governing processes are also identified as part of social sustainability by Boström (2012). For this reason, we propose to use as an indicator of social cohesion the degree of transparency and openness of governance, in this case, land administration. Transparency is one of the positive impacts of opening administrative data (Janssen et al., 2012), thus an indicator for that would be the degree of accessibility and open data provided within the land administration. Providing open data includes several benefits to society, such as new services for citizens and businesses and promotion of democracy (Williamson et al., 2006; Magis & Shinn, 2009).

Examining the connections between social inclusion/exclusion and land administration, we notice that the relationship is twofold: there are some direct requirements for the land administration system welling from the social inclusion indicators, and some indirect welling from the basic functions of land administration. The direct requirements for the land administration system refer to actual capabilities of the system, such as enabling a complex commodity market rather than just facilitating the land market (Williamson et al., 2010), or providing transparency, access to data, and good governance, including equality. Enabling the presentation of spatial data and visualizing the data to produce indirect impacts on social inclusion is the other dimension. The visualization is needed to support decision-making and policymaking to support social inclusion, for example, in forms of tackling gentrification and segregation.

Therefore, we propose to use the concept of spatially enabling society, and moreover, digitally and spatially enabling systems. This means that the use of spatial information and digital tools is a presupposition rather than a requirement set by the support of social inclusion. Digitalization adds another indicator for inclusion, though. The ability of citizens to participate and access information is described by digital inclusion/exclusion (Scott & Rajabifard, 2019). This measures access to electricity, computers, or mobile devices with internet access and thus serves as an indicator of participation possibilities.

3.4 Discussion

Fostering sustainable development in every action of society is critically important for the world's future. Sustainable development often comprises three dimensions of economic, environmental, and social sustainability. Connections between economic and social development with registering and formalizing land titling have been an object of previous studies (e.g., Feder & Nishio, 1998; de Soto, 2000); the development of multi-purpose land administration systems (Ting & Williamson, 1999; Ho & Rajabifard, 2012) has given new possibilities for incorporating other aspects of sustainability in the research. The land administration system reaches into many functions of society. Not only is it needed for securing rights to land and registering the land, but it also provides tools for supporting and creating land valuation, land use in forms of monitoring and planning, and development of urban and rural areas. This gives a fruitful starting point for studying the connections between land administration and social sustainability.

The aim of this chapter was to demonstrate the connection between land administration and sustainable development, and shed light on how DTs are enablers in this equation. First, we provided an overview of the dimensions and concepts of social sustainability and social inclusion, and their indicators drawn from the existing literature. After that, we elaborated the indicators further from a digital land administration point of view. Then, we were able to draft indicators that measure social inclusion and social sustainability to which land administration tools can contribute.

Our findings show that land administration tools, when used correctly, can promote (social) sustainability in several areas. The multi-purpose functions of land administration systems connect the economic, sociocultural, and institutional dimensions of social sustainability. The interconnection mechanism between social inclusion/exclusion and land administration is twofold: the indicators for social inclusion set direct requirements for land administration systems, such as the requirement for enabling complex commodity trading and transparent governance. On the other hand, land administration functions provide tools to support social inclusion, either by enabling spatial analysis to support decision-making and policymaking, or by responding to the direct requirements of social inclusion indicators.

The findings show that the indicators require a holistic understanding of social issues in society, as well as multiple data sources from various data sets. This calls for a holistic view of land administration, moving away from its "traditional" core function of securing ownership rights. We can conclude that land administration needs to renew itself by providing integrated services and including private partners in the ecosystem of land administration, to achieve digital advancement but at the same time support social sustainability. This finding is supported by other recent studies. For instance, Unger et al. (2021) call for an integrated data modeling approach for land administration, where land administration has a direct relation with the SDGs.

It seems that digital-twin-enabled land administration systems can increase transparency and accountability in communicating land information by providing real-time access to this information. This can potentially reduce corruption and increase trust in responsible authorities. Furthermore, DTs can aid in the monitoring of changes in land tenure, land use, and land value, providing valuable information for land management decision-making. Moreover, digital-twin-enabled land administration can increase citizen participation in land management processes. Citizens can become more informed and engaged in decision-making processes. More specifically, DTs can be used to collect feedback from citizens and allow them to report land-related issues, such as encroachments and illegal activities. This can help to improve the efficiency and effectiveness of land management processes. Furthermore, DTs can be used to simulate the impact of different land development scenarios, allowing citizens to provide input during the land development projects.

3.5 Conclusion

The megatrend of digitalization has recently been a major driver for change and transformation in land administration (Riekkinen & Zetterqvist, 2020; Riekkinen et al., 2022). It provides several opportunities from various aspects of land administration, but additionally raises challenges (Krighsolm et al., 2020), especially regarding security and data privacy. The scientific discussion of the future of land administration can be divided into two veins: the one concentrated on technical solutions of accelerated digitalization and its applicability to land administration and cadastral data (e.g., Jaljolie et al., 2021), the second focusing on legal and cadastral data interoperability challenges (Aien et al., 2013; Riekkinen et al., 2022) from an administrative perspective.

This chapter provides a basis for a new vein of research, where expanding the concept of land administration with accelerated digitalization to support new social applications is emerging, and even reaching a new, virtual dimension (Park et al., 2023). The evolution from fiscal tool to multi-purpose cadaster has been the trend of land administration alongside the societies to adapt to new megatrends. Nowadays, the concept is expanding to multi-purpose and beyond. The importance of up-to-date cadastral and land registration data becomes even more important, as digitalization enables smart contracts, artificial intelligence, and automated data collection. These drivers in technological development create new playgrounds for scientists and professionals, where 3D cadastral data integrated into DTs can be used to create social benefits, in the forms of connecting social sustainability, land administration, land use planning, and other interests in one single platform.

References

Aien, A., Kalantari, M., Rajabifard, A., Williamson, I., Wallace, J. 2013. Towards Integration of 3D Legal and Physical Objects in Cadastral Data Models. Land Use Policy, Vol. 35, No. 2013, pp. 140–154. https://doi.org/10.1016/j.landusepol.2013.05.014

ANZLIC. 2019. Principles for Spatially Enabled Digital Twins of the Built and Natural Environment in Australia. The Australia and New Zealand Land Information Council (ANZLIC).

Boström, M. 2012, 2012. A Missing Pillar? Challenges in Theorizing and Practicing Social Sustainability. Sustainability: Science, Practice, and Policy. https://doi.org/10.1080/15487733.2012.11908080

Chaskin, R. J., Joseph, M. L. 2013. 'Positive' Gentrification, Social Control and 'Right to the City' in Mixed-Income Communities: Uses and Expectations on Space and Place. International Journal of Urban and Regional Research. Vol. 37, No. 2, pp. 480–502. https://doi.org/10.1111/j.1468-2427.2012.01158.x

CSDILA. 2022. Digital Twin Platform and Capabilities, Centre for SDIs and Land Administration-CSDILA website. https://eng.unimelb.edu.au/csdila/projects/digital-twin (accessed April 9, 2022).

de Soto, H. 2000. The Mystery of the Capital. Why Capitalism Triumphs in the West and Fails Everywhere Else. Black Swan.

Enemark, S. 2004. Building Land Information Policies, Proceedings of Special Forum on Building Land Information Policies in the Americas, 26–27 October 2004, Aguascalientes, Mexico. https://www.fig.net/resources/proceedings/2004/mexico/papers_eng/ts2_enemark_eng.pdf

Feder, G., Nishio, A. 1998. The Benefits of Land Registration and Titling: Economic and Social Perspectives. Land Use Policy, Vol. 15, No. 1, pp. 25–43.

Grieves, M. 2011. Virtually Perfect: Driving Innovative and Lean Products Through Product Lifecycle Management. Space Coast Press.

Ho, S., Rajabifard, A. 2012. 3D Land and Property Information System: A Multilevel Infrastructure for Sustainable Urbanization and a Spatially Enabled Society. In Rajabifard, A., Coleman, D. (Eds.), Spatially Enabling Government, Industry and Citizens: Research and Development Perspectives. GSDI Association Press. Pp. 103–124.

Jaljolie, R., Riekkinen, K., Dalyot, S. 2021. A Topological-Based Approach for Determining Spatial Relationships of Complex Volumetric Parcels in Land Administration Systems. Land Use Policy, Vol. 103, p. 105291. https://doi.org/10.1016/j.landusepol.2021.105637

Janssen, M., Charalabidis, Y., Zuiderwijk, A. 2012. Benefits, Adoption Barriers and Myths of Open Data and Open Government. Information Systems Management, Vol. 29, No. 4, pp. 258–268. https://doi.org/10.1080/10580530.2012.716740

Kahila, M., Kyttä, M. 2009. SoftGIS as a Bridge-Builder in Collaborative Urban Planning. In Geertman, S., Stillwell, J. C. H. (Eds.), Planning Support Systems Best Practice and New Methods. © Springer Science+Business Media B.V. https://doi.org/10.1007/978-1-4020-8951-0_19

Krighsolm, P., Riekkinen, K., Ståhle, P. 2020. Pathways for a Future Cadastral System. Land Use Policy, Vol. 94, p. 104504. https://doi.org/10.1016/j.landusepol.2020.104504

Labonté, R. 2004. Social Inclusion/Exclusion: Dancing the Dialectic. Health Promotion International, Vol. 19, No. 1. https://doi.org/10.1093/heapro/dah112

Lawson, V. 2010. Reshaping Economic Geography? Producing Spaces of Inclusive Development. Economic Geography, Vol. 86, No. 4, pp. 351–360. https://doi.org/10.1111/j.1944-8287.2010.001092.x

Lu, Q., Xie, X., Parlikad, A. K., Schooling, J. M. 2020, October. Digital Twin-Enabled Anomaly Detection for Built Asset Monitoring in Operation and Maintenance. Automation in Construction, Vol. 118, p. 103277. https://doi.org/10.1016/j.autcon.2020.103277

Magis, K., Shinn, C. 2009. Emergent Principles of Social Sustainability. In Dillard, J., Dujon, V., King, E. (Eds.), Understanding the Social Dimension of Sustainability. New York: Routledge. Pp. 15–44.

Oclott, C., Mullen, C. 2020. Digital Twin Consortium Defines Digital Twin – Digital Reflections, Digital Twin Consortium, Digital Twin Consortium, Dec. 2020. https://blog.digitaltwinconsortium.org/2020/12/digital-twin-consortium-defines-digital-twin.html (accessed Apr. 19, 2021).

Park, J., Choi, W., Jeong, T., Seo, J. 2023. Digital Twins and Land Management in South Korea. Land Use Policy, Vol. 124, p. 106442. https://doi.org/10.1016/j.landusepol.2022.106442

Riekkinen, K., Penkkilä, O., Krigsholm, P., Ekroos, A., Häkkänen, M. 2022. Security of Supply Perspective on Real Property and Control Thereof (English Abstract). Publications of the Government's Analysis, Assessment, and Research Activities, Vol. 2022, No. 8, p. 83. https://urn.fi/URN:ISBN:978-952-383-134-6

Riekkinen, K., Zetterqvist, F. 2020. Responsive Land Administration – Future Scenarios. The Annual World Bank Conference on Land and Poverty, 2020, proceedings. 14 p.

Scott, G., Rajabifard, A. 2019. SDGs Roadmap. In Rajabifard, A. (Ed.) Sustainable Development Goals Connectivity Dilemma. Land and Geospatial Information for Urban and Rural Resilience, CRC Press, pp. 13–44.

Silver, H. 2015. The Contexts of Social Inclusion. DESA Working Paper No. 144, Department of Economic and Social Affairs, United Nations, https://www.un.org/esa/desa/papers/2015/wp144_2015.pdf

Ting, L., Williamson, I. 1999. Cadastral Trends: A Synthesis. The Australian Surveyor, Vol. 4, No. 1, pp. 46–54.

Unger, E.-M., Bennett, R., Lemmen, C., Zevenbergen, J., Dijkstra, P., de Zeeuw, K. 2019. LADM Based Models for Sustainable Development LA-DRM for Disaster Prone Areas and Communities. 8th International FIG Workshop on the Land Administration Domain Model, 1-3 October 2019, Kuala Lumpur, Malaysia. https://doi.org/10.4233/uuid:d1023d5c-83a8-419b-b221-e95789b8defd

Unger, E.-M., Bennett, R. M., Lemmen, C., Zeverbergen, J. 2021. LADM for Sustainable Development: An Exploratory Study on the Application of Domain-Specific Data Models to Support the SDGs. Land Use Policy, Vol. 108, No. 2021, p. 105499. https://doi.org/10.1016/j.landusepol.2021.105499

Unger, E.-M., Lemmen, C., Bennett, R. 2023. Women's Access to Land and the Land Administration Domain Model (LADM): Requirements, Modelling and Assessment. Land Use Policy, Vol. 126, No. 2023, p. 106538. https://doi.org/10.1016/j.landusepol.2023.106538.

Williamson, I., Enemark, S., Wallace, J. 2006. Incorporating Sustainable Development Objectives into Land Administration. Proceedings FIG XXIII Congress, Shaping the Change, Munich, Germany.

Williamson, I., Enemark, S., Wallace, J., Rajabifard, A. 2010. Land Administration for Sustainable Development. ESRI Press, 487 p.

4

Knowledge Integration in Geospatial Intelligence for Sustainability Science from the Lens of Citizen Science

Siqing Yu

4.1 Introduction

Sustainability science addresses complex and interconnected socio-environmental challenges such as climate change, biodiversity loss, and resource depletion (Kates et al. 2001). These challenges are marked by high levels of uncertainty, nonlinearity, and contested values, demanding approaches that integrate multiple forms of knowledge, also sometimes referred to as wicked problems (Head 2022). Unlike "tame" problems with clear boundaries and solutions, wicked problems defy straightforward technical fixes due to their dynamic and multifaceted nature. They involve multiple interacting systems—ecological, social, economic, and political—that influence one another in nonlinear ways, making it difficult to predict outcomes or isolate causes.

Addressing these challenges requires more than disciplinary expertise; it demands integrative approaches that synthesize diverse knowledge systems, including scientific data, local and indigenous knowledge, and experiential insights. Such integrative efforts enable a more holistic understanding of the problem context, fostering adaptive and participatory strategies that can evolve over time. Moreover, wicked problems often involve trade-offs and value-laden decisions, underscoring the need for inclusive governance structures where multiple voices contribute to defining both the problem and the solutions. Sustainability science thus emphasizes collaboration across disciplines and sectors to co-produce knowledge that is actionable, equitable, and sensitive to the needs and aspirations of affected communities. In response, researchers have increasingly turned toward collaborative, interdisciplinary, and transdisciplinary practices. One emergent area of interest is the convergence of geospatial intelligence (GEOINT) and citizen science as tools for advancing sustainability science through knowledge integration.

DOI: 10.1201/9781003638308-5

Traditionally, GEOINT originated within military and strategic contexts, serving as a critical tool for national security through the collection, analysis, and interpretation of spatial data to support defense operations and tactical decision-making (Clark 2020). Over time, however, the scope of GEOINT has broadened significantly, expanding into a wide array of civilian and environmental applications. These include urban planning, disaster management, environmental monitoring, natural resource management, and public health, where spatial analysis informs policy, risk assessment, and sustainable development efforts. This evolution reflects the growing recognition that geospatial data and technologies are essential not only for security but also for addressing complex societal challenges that transcend traditional boundaries.

Concurrently, citizen science has emerged as a powerful and increasingly legitimate mode of public participation in scientific research. By engaging non-professionals in activities such as environmental monitoring, biodiversity surveys, and dissemination of findings, citizen science initiatives democratize knowledge production and expand the scale and scope of data available to researchers (Bonney et al. 2016). These initiatives not only enrich scientific datasets with local and experiential insights but also foster community awareness, empowerment, and stewardship of environmental resources.

This chapter investigates the dynamic interplay between these two evolving domains, emphasizing how integrating citizen-generated data with GEOINT frameworks can enhance both the quantity and quality of spatial information. By combining the technical precision and analytical rigor of GEOINT with the localized, diverse contributions from citizen scientists, this integration promises to improve knowledge production, promote more inclusive and transparent decision-making processes, and ultimately support more effective and equitable sustainability outcomes.

4.2 The Role of Geospatial Intelligence in Sustainability Science

GEOINT encompasses the collection, analysis, and interpretation of spatial data using imagery, mapping technologies, and advanced spatial analysis techniques to understand and respond to geographic phenomena. Traditionally focused on the acquisition and processing of satellite imagery and aerial photographs, GEOINT has evolved beyond purely technical and military applications to embrace more comprehensive approaches that integrate social, political, and environmental dimensions (Clark 2020). This shift acknowledges that geographic phenomena are not only physical but also shaped by human activities, governance structures, and socio-economic factors, requiring interdisciplinary perspectives for meaningful interpretation.

Within sustainability science, GEOINT plays a crucial role in supporting a range of critical functions, including monitoring changes in land use and land cover, assessing risks and vulnerabilities related to natural disasters, and informing strategies for climate change mitigation and adaptation. By providing spatially explicit data and analytical tools, GEOINT enables researchers, policymakers, and communities to visualize complex environmental processes, track ecosystem health, and evaluate the impacts of human interventions over time.

Despite these advances, many traditional GEOINT approaches have been characterized by top-down data acquisition methods and expert-driven analyses. This technocratic model often privileges centralized institutions and specialist knowledge, which can limit the inclusion of diverse knowledge holders such as local communities, indigenous peoples, and non-expert stakeholders. The resulting exclusion can undermine the responsiveness of GEOINT to localized contexts, social dynamics, and community needs, potentially reproducing epistemic injustice that is detrimental to public trust. Epistemic injustice refers to the unfair treatment of individuals or groups in their capacity as knowers (Fricker 2007), often through the dismissal, devaluation, or marginalization of their knowledge and experiences. In GEOINT contexts, epistemic injustice can manifest when local or indigenous knowledge is overlooked in favor of official or expert data, leading to incomplete or biased spatial analyses. This exclusion not only undermines the legitimacy and accuracy of geospatial products but also perpetuates power imbalances, reducing trust and hindering effective, equitable decision-making. Addressing epistemic injustice is therefore crucial to ensuring that GEOINT integrates diverse perspectives and produces knowledge that is both socially just and practically relevant.

To overcome these limitations, there has been a growing movement toward participatory and open geospatial systems that better align with the principles of sustainability science (Goodchild 2007). These participatory approaches emphasize the democratization of spatial data collection, analysis, and interpretation by actively involving local actors and citizens as contributors and collaborators. By valuing local perspectives, experiential knowledge, and grassroots expertise, participatory GEOINT fosters more inclusive, context-sensitive, and socially just outcomes. Moreover, open geospatial platforms and volunteered geographic information (VGI) initiatives harness new technologies, such as mobile devices and crowdsourcing, to enable broader public engagement and the co-creation of geospatial knowledge. Such approaches not only enhance data richness and granularity but also build trust, transparency, and shared ownership among diverse stakeholders, thereby strengthening the societal impact of GEOINT in addressing sustainability challenges.

4.3 Knowledge Integration: Concepts and Challenges

Knowledge integration involves the deliberate synthesis of insights from diverse disciplines, sectors, and societal actors to develop comprehensive understandings of complex, multifaceted issues. In sustainability science, this often means bridging scientific knowledge with local, indigenous, and experiential ways of knowing, recognizing that each offers unique perspectives and forms of evidence essential for addressing "wicked problems" that cross disciplinary and institutional boundaries (Cash et al. 2003). However, this process is not simply additive; it requires careful attention to epistemic justice—the fair recognition and valuation of different knowledge systems and knowers. Epistemic injustice occurs when certain groups, especially marginalized communities, are systematically discredited or excluded as legitimate knowledge holders, often due to prevailing power asymmetries or entrenched disciplinary biases.

In GEOINT and sustainability contexts, such injustices can marginalize valuable local or indigenous insights, leading to incomplete or skewed understandings that undermine both equity and effectiveness. These dynamics highlight the need for approaches that do more than just combine knowledge—they must foster genuine co-production, where stakeholders collaboratively generate knowledge through inclusive, reflexive, and iterative processes. Knowledge co-production emphasizes mutual respect, shared authority, and ongoing dialogue, helping to bridge epistemic divides and disrupt hierarchies that privilege scientific expertise over other ways of knowing.

Yet, realizing meaningful integration remains challenging. Epistemic differences between natural sciences, social sciences, and community knowledge can create friction, compounded by institutional structures that reinforce disciplinary silos and limit sustained collaboration (Turnhout et al. 2012). Overcoming these barriers requires intentional frameworks and supportive institutions—such as boundary organizations, transdisciplinary research platforms, and policy-oriented knowledge networks—that facilitate trust-building, reciprocal learning, and knowledge translation among diverse actors. These arrangements create spaces where multiple epistemologies can coexist, enabling more holistic, just, and actionable knowledge production crucial for tackling sustainability challenges.

4.4 Citizen Science as a Catalyst for Integration

Citizen science—defined as the involvement of non-professionals in scientific research—has gained recognition as a valuable source of data and insight, particularly in environmental science (Haklay 2012). Its applications in biodiversity monitoring, air and water quality assessment, and disaster response have demonstrated its potential to enrich scientific understanding and foster community engagement.

Citizen science contributes to knowledge integration in several ways. First, it brings local and experiential knowledge into scientific processes, creating more grounded and context-sensitive insights. Second, it encourages interdisciplinary collaboration by linking environmental data collection with social science perspectives on participation, trust, and governance (Irwin 1995). Third, it enhances the legitimacy and usability of knowledge by involving stakeholders in defining problems and interpreting results.

Participatory data collection initiatives such as the Global Learning and Observations to Benefit the Environment (GLOBE) Program or iNaturalist illustrate how citizen-generated data can meaningfully contribute to large-scale research. Additionally, citizen science can serve as a platform for informal science education, increasing scientific literacy and public awareness of sustainability challenges.

A recent study by Yu et al. (2025) further illustrates how citizen science can act as a site for knowledge co-production and integration. Drawing on comparative case studies from environmental and urban citizen science projects, the research highlights how participatory design, mutual learning, and epistemic trust facilitate the integration of scientific and lay knowledge. Rather than treating volunteers merely as data collectors, the projects enabled volunteers to shape research questions, contribute methodological insights, and interpret findings in ways that were locally relevant. This co-productive dynamic not only enhanced the quality and contextual relevance of the knowledge generated but also addressed underlying issues of epistemic injustice and power asymmetry. Yu's study thus exemplifies how citizen science, when thoughtfully designed, can serve as a catalyst for equitable and integrative knowledge production (Figure 4.1).

Nevertheless, citizen science faces challenges regarding data quality, inclusivity, and sustainability. Efforts such as standardized protocols, training, and digital platforms have helped address these issues. The emergence of mobile applications and online repositories has improved data reliability and accessibility, enabling more robust integration with institutional science.

4.5 Linking Geospatial Intelligence and Citizen Science

The integration of GEOINT with citizen science represents a promising avenue for enhancing sustainability science. This linkage is particularly evident in the rise of VGI, where individuals contribute georeferenced data that feed

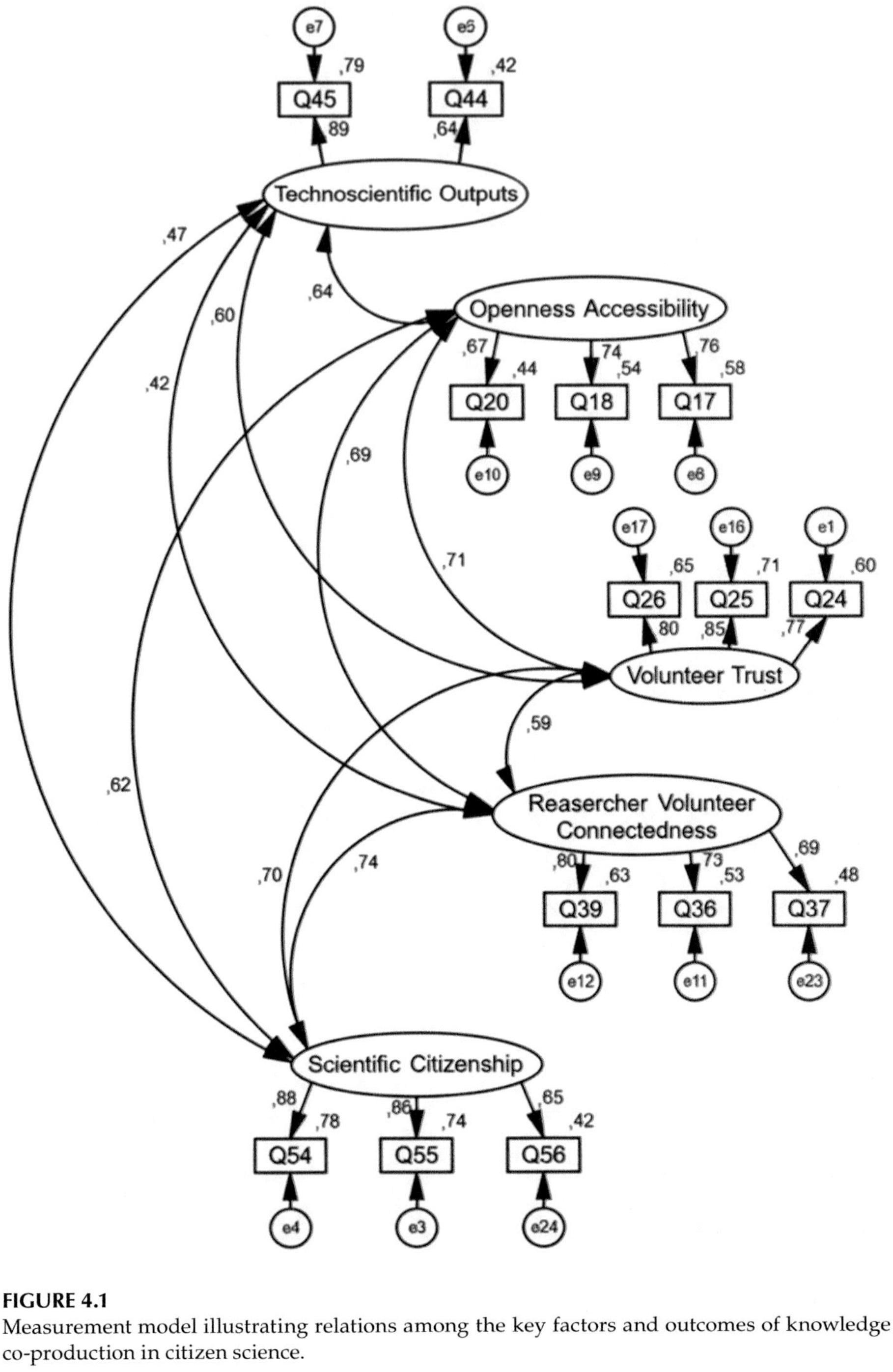

FIGURE 4.1
Measurement model illustrating relations among the key factors and outcomes of knowledge co-production in citizen science.

into spatial analyses (Goodchild 2007). Examples include real-time mapping of floods, tracking of urban green spaces, and community-led water quality monitoring.

Technological advances in mobile devices, sensors, and open-source platforms have made it easier for citizens to generate and share geospatial data. These contributions can complement remote sensing and official data sources, filling spatial and temporal gaps, and offering ground-truth perspectives (Goodchild and Glennon 2010; Haklay and Weber 2008).

In addition to data contributions, citizen science can play a role in validating and refining geospatial models. Ground-truthing satellite imagery through local observation improves the accuracy of land-use classification, for example. Community mapping initiatives have also been used to visualize informal settlements and environmental hazards that are often overlooked in official datasets.

Beyond data collection, integrating GEOINT and citizen science requires mechanisms for joint problem framing, data interpretation, and decision-making. Co-designing monitoring programs and visualization tools can help democratize the geospatial domain and enhance its responsiveness to community needs. Participatory Geographic Information Systems (PGIS) are one such tool, allowing citizens to engage in spatial decision-making processes directly.

4.6 Toward a Co-Productive Framework

To fully realize the transformative potential of GEOINT and citizen science in advancing sustainability goals, we advocate for a co-productive framework that effectively bridges the natural and social sciences. This framework emphasizes the collaborative and iterative generation of knowledge, wherein both researchers and societal actors actively participate as equal partners. Such co-production fosters mutual learning, shared ownership of knowledge, and enhanced relevance of scientific outputs to real-world challenges (Jasanoff 2004). Key elements of this co-productive framework are described in the following sections.

4.6.1 Inclusivity

Meaningful engagement requires involving a wide spectrum of participants, especially those whose perspectives have historically been marginalized or excluded—such as indigenous communities, minority groups, and low-income populations. These groups often hold critical local knowledge and experience, yet their voices are frequently absent from data collection and policy dialogues. Ensuring their inclusion not only enriches the data but also enhances the legitimacy and fairness of sustainability initiatives.

4.6.2 Reflexivity

Reflexivity entails a conscious and ongoing examination of the power dynamics, assumptions, and biases embedded within knowledge production processes. Researchers and participants must critically reflect on how their own worldviews, institutional positions, and social contexts influence the framing of problems, data interpretation, and decision-making. This awareness helps prevent the reproduction of existing inequities and opens space for alternative ways of knowing, fostering epistemic justice.

4.6.3 Iterativity

Sustainability challenges and environmental conditions are inherently dynamic and complex. A co-productive framework must therefore be flexible, allowing for continuous cycles of feedback, learning, and adjustment. This iterative process ensures that knowledge remains relevant and responsive as social priorities and ecological realities evolve. It encourages adaptive management and fosters resilience both in communities and ecosystems.

4.6.4 Usability

The ultimate goal of co-produced knowledge is to be actionable—informing policies, planning strategies, and community practices that drive tangible sustainability outcomes. Usability emphasizes translating scientific insights into accessible formats and decision-support tools that meet the needs of diverse stakeholders, including policymakers, practitioners, and local residents. This focus enhances the practical impact and real-world relevance of GEOINT and citizen science initiatives.

4.7 Conclusion and Future Directions

Integrating GEOINT with citizen science offers powerful opportunities for advancing sustainability science. By combining technical precision with local knowledge and participatory processes, this integration enhances the relevance, legitimacy, and effectiveness of knowledge production. The convergence of these fields allows for richer spatial analyses, more inclusive data ecosystems, and stronger linkages between science and action.

Future research should explore methods for evaluating the impact of integrated approaches, including metrics for knowledge co-production, data quality, and policy uptake. There is also a need to develop inclusive technologies that account for digital divides and to institutionalize co-productive practices within research and governance structures. Questions around data ethics, ownership, and access remain central to this agenda.

As sustainability challenges grow in urgency and complexity, embracing diverse ways of knowing and innovative collaborations will be essential. The integration of GEOINT and citizen science, guided by a co-productive framework, stands as a promising example of how to build more resilient, adaptive, and just knowledge systems for the future.

References

Bonney, Rick, Tina B. Phillips, Heidi L. Ballard, and Jody W. Enck. 2016. "Can Citizen Science Enhance Public Understanding of Science?" *Public Understanding of Science* 25 (1): 2–16. doi:10.1177/0963662515607406.

Cash, David W., William C. Clark, Frank Alcock, Nancy M. Dickson, Noelle Eckley, David H. Guston, Jill Jäger, and Ronald B. Mitchell. 2003. "Knowledge Systems for Sustainable Development." *Proceedings of the National Academy of Sciences* 100 (14): 8086–91. doi:10.1073/pnas.1231332100.

Clark, Robert M. 2020. *Geospatial Intelligence: Origins and Evolution*. Georgetown University Press, Washington, DC, USA.

Fricker, Miranda. 2007. *Epistemic Injustice: Power and the Ethics of Knowing*. doi:10.1093/acprof:oso/9780198237907.001.0001.

Goodchild, Michael F. 2007. "Citizens as Sensors: The World of Volunteered Geography." *GeoJournal* 69 (4): 211–21. doi:10.1007/s10708-007-9111-y.

Goodchild, Michael F., and J. Alan Glennon. 2010. "Crowdsourcing Geographic Information for Disaster Response: A Research Frontier." *International Journal of Digital Earth* 3 (3): 231–41. doi:10.1080/17538941003759255.

Haklay, Muki. 2012. "Citizen Science and Volunteered Geographic Information: Overview and Typology of Participation." In *Crowdsourcing Geographic Knowledge: Volunteered Geographic Information (VGI) in Theory and Practice*, edited by Daniel Sui, Sarah Elwood, and Michael Goodchild, 105–22. doi:10.1007/978-94-007-4587-2_7.

Haklay, Mordechai, and Patrick Weber. 2008. "OpenStreetMap: User-Generated Street Maps." *IEEE Pervasive Computing* 7 (4): 12–18. doi:10.1109/mprv.2008.80.

Head, Brian W. 2022. "The Rise of 'Wicked Problems'—Uncertainty, Complexity and Divergence." In *Wicked Problems in Public Policy*, 21–36. Cham: Springer International Publishing. https://link.springer.com/10.1007/978-3-030-94580-0_2.

Irwin, Alan. 1995. *A Study of People, Expertise and Sustainable Development*. 1st ed. London: Routledge.

Jasanoff, Sheila. 2004. *States of Knowledge: The Co-Production of Science and Social Order*. London ; New York: Routledge.

Kates, Robert W, William C Clark, Robert Corell, J Michael Hall, Carlo C Jaeger, Ian Lowe, James J McCarthy, et al. 2001. "Sustainability Science." *Science* 292 (5517): 641–42. doi:10.1126/science.1059386.

Turnhout, Esther, Bob Bloomfield, Mike Hulme, Johannes Vogel, and Brian Wynne. 2012. "Listen to the Voices of Experience." *Nature* 488 (7412): 454–55. doi:10.1038/488454a.

Yu, Siqing, Hana Vodeb, Joep Crompvoets, Trui Steen, Abbas Rajabiard, Jagannath Aryal, and Tina Jukić. 2025. "Measuring the Impact of Key Factors on Knowledge Co-Production Outcomes in Citizen Science." *Proceedings of the 58th Hawaii International Conference on System Sciences (HICSS-58).* https://hdl.handle.net/10125/109462

5

Using the Power of Geospatial Intelligence to Build Trust for the Renewable Energy Transition

Alice Kesminas

5.1 Introduction

The rapid expansion of renewable energy (RE) is urgent but must occur on already overstretched land, also needed for food, water, nature, and housing. We have a tremendous opportunity to harness geospatial data and tools to justly prioritise communities and nature and equip government and industry with the insights needed to plan, build, and operate RE infrastructure optimally – balancing competing land interests. Fair, transparent RE decisions depend on accessible geospatial tools built on diverse, high-quality data. While the technology exists to standardise, integrate, and trace data, barriers to data access remain. The geospatial sector must address this by advocating for open, high-quality data and building workforce capability to support trustworthy, artificial intelligence (AI)-informed decision-making.

As geospatial data volumes grow and are increasingly integrated into Multi-Criteria Decision Analysis (MCDA) with AI-driven computation, assuring geospatial quality becomes ever more critical. Relying on poor-quality data to justify decisions risks stalling the transition further. If done well, there is a clear opportunity to rebuild trust and strengthen social licence, enabling the RE transition to move forward with community support.

The RE transition requires the rapid deployment of infrastructure across diverse landscapes. This includes not only wind turbines and solar arrays but also supporting activities such as mining for lithium and steel, and the expansion of transmission networks. For this chapter, "RE infrastructure" encompasses all these interconnected components.

DOI: 10.1201/9781003638308-6

5.2 Community Engagement and Social Licence

Australia's RE targets are ambitious, and current approaches are not sufficient to meet them. Lack of trust between stakeholders continues to hinder the RE transition. Open communication, data-driven decision-making and shared benefits are all important for a just and effective shift to renewables. Unfortunately, the term "community" is too often used to imply an external stakeholder or "other". We all belong to multiple communities. Recognising this interconnectedness and undertaking early, meaningful, community involvement is fundamental to planning, especially in times where rapid change is needed, such as the RE transition.

Long and contested consultation processes for RE infrastructure projects in Australia are common, with many projects delayed or abandoned.

> Australia's current target is for 82 per cent of electricity to be generated by renewables by 2030, but, on current trends, it's an uphill battle to meet it. The way projects have been proposed, developed and managed in regional communities has caused divisions, and ultimately delays, for decades, with wind and transmission lines among those continuing to face degrees of opposition.

It's not uncommon to hear of nine-year consultation processes (Acciona, Mortlake, Victoria) and projects abandoned after six years of consultation (Ark Energy, Stanley, Tasmania), re-enforcing the Clean Energy Council (CEC) observation that "it's in the industry's interests to ensure proper community consultation occurs".

The CEC and other industry leaders recognise that successful projects balance community expectations from the start. A scrapped A$2 billion wind energy development in Australia shows the difficulties involved in winning community support. Research showed how, despite the proponents using a "best practice" mode of community engagement, the proposal caused much social conflict. Locals were concerned that the community consultation process was not impartial (McBain, 2021).

3D visualisation and immersive environments built on geospatial information can be used to visualise scenarios. Communities can view different options for RE infrastructure construction and assess the visual impact from different locations. By simultaneously giving stakeholders input and control while also enabling operators to design optimal locations for RE infrastructure – competing interests can be better balanced. But the effectiveness of such systems depends on more than just functionality; they must be trusted and accepted by the communities they affect. Requirements for social licence for "big data" apply equally to geospatial applications, especially the need for good governance and data management, clarity of system architecture, and the importance of high-quality data (McBain, 2021).

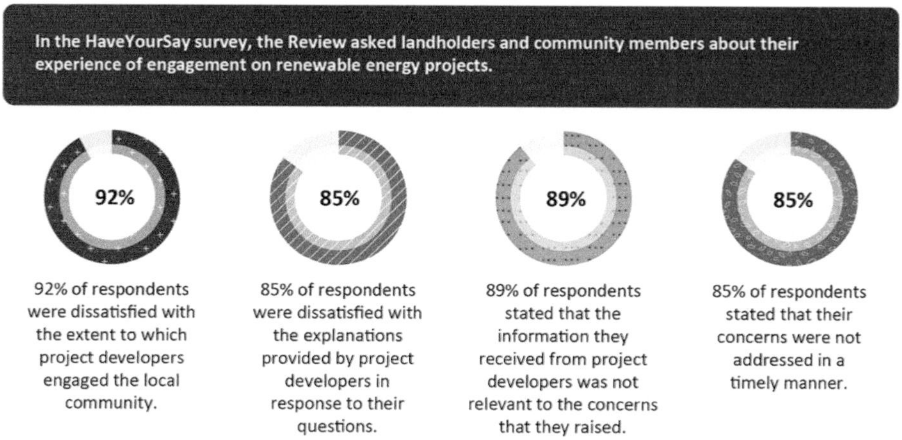

FIGURE 5.1
Landholders and community members were generally dissatisfied with the engagement that they received from project developers. (Adopted from Dyer (2023).)

A long history of mining projects, and the more recent global experience with RE developments, demonstrates that responsible land governance and genuine community consultation are essential to securing land for successful, land-based climate action and infrastructure delivery. The Food and Agriculture Organization (FAO) of the United Nations principles (FAO, 2022) which include "Consultation and participation; Holistic and sustainable approach; and Transparency, need to be central to integrating geospatial data for decision support".

Projects that engage communities from the outset, offer local ownership or employment, and communicate the benefits of RE are more likely to gain access to the necessary land and maintain support (Figure 5.1). RE infrastructure projects must go beyond technical planning. Geospatial tools should be paired with clear benefit-sharing models that communicate the social, economic, and environmental advantages of RE to affected communities. By offering local ownership, employment, or service opportunities, projects can ensure benefits flow to the land and people hosting the infrastructure, not just to distant cities. With the right intent, spatial technologies can help align land use, energy justice, and climate action.

To generate the quality data and transparent, accessible tools needed, we will need to train a new generation of professionals, across RE, government, the geospatial sector, IT, and AI, who can work collaboratively in interdisciplinary teams. Building a shared understanding will be essential for appreciating the full complexity of RE infrastructure challenges and embedding effective, inclusive, and sustainable energy transition planning from the ground up.

Geospatial intelligence can power more optimal RE infrastructure siting and enable circular economy outcomes. In practice, this promise can only be fulfilled if we prioritise community acceptance, data integrity, and the ethical use of AI. If we design our tools and processes with trust, transparency, and inclusion at the core, geospatial technologies can help deliver a RE transition that is not only fast, but fair.

5.3 Harnessing Geospatial Tools and AI for Informed Planning

Planning, constructing, and operating infrastructure is always complex, with layers of conflicting rights, restrictions, responsibilities, and priorities converging at potential sites. These range from environmental regulations and carbon commitments to planning overlays, mineral and energy licences, existing landowners, and the broader local community. The usual challenges are now intensified by the urgency of the RE transition.

Geospatial data and tools have long proven effective for strategic planning, site selection, and community engagement. They enable us to map and assess these overlapping factors, balance competing priorities, and visualise potential scenarios, such as trade-offs between protecting delicate ecosystems and increasing RE output. Hybrid solutions like agrovoltaics are emerging to address multiple priorities simultaneously. Projects that offer community equity and local employment demonstrate higher success in securing land access.

As RE deployment accelerates, tools capable of supporting regulatory compliance, ESG reporting, and circular economy assessments can benefit implementation. Integration with enterprise systems, particularly those used for insurance and risk management, can further enhance their value. While the technology is available, and data access and integration constraints can be addressed with open data and standards, now the challenge is to provide access to reliable tools. Data quality and governance must come first and must underpin every stage of planning, design, and implementation.

> Data quality refers to the usability and applicability of data used for an organization's priority use cases, including AI and machine learning initiatives. Data quality is usually one of the goals of effective data management and data governance. Yet too often organizations treat it like an afterthought.
>
> **(Gartner, 2020)**

Gartner research estimates that poor data quality costs organisations at least $12.9 million a year on average. Geospatial intelligence can support RE goals, but only if designed with community acceptance in mind. Biased AI, unclear models, and inaccessible tools risk undermining progress. Transparency and respect must be embedded in every stage.

Data and AI strategists such as Jorge Afonso (2025) articulate the importance of good data to AI. "Without data, there is no AI. And without good data, there is no useful AI". Afonso emphasises that data needs to be studied and mastered – "AI is the illusion of power. Data is the reality of it. Everyone wants to ride the rocket. Few want to build the launchpad". Good data is critical to AI in every field, but especially in the geospatial sector, where the complexity of location data and the risk of propagating spatial errors make data quality even more critical.

For some time, the geospatial data has been

> transforming the renewable energy sector by providing critical insights that help guide decisions for planning, development and in the operational phases of implementation. Data such as terrain models, including LiDAR, 3D building models and aerial photography can aid in site selection, infrastructure planning, operational efficiency, and environmental impact assessments.
>
> **(Hilton, 2024)**

Data products like Geoscape Solar, in Australia, give decision makers a valuable view of all installed rooftop solar, to assist utility in network planning and also inform community battery projects. These geospatial data and tools can also play an important role in community engagement. And we know that "the companies that achieve the best outcomes are those that are able to balance all community expectations from the start" CEC (Holmes, 2025).

Organisations like Spatial Media are using geospatial information to create online spaces to centralise project information and updates that are interactive and easy to use as the project progresses (Figure 5.2).

> The Virtual Engagement Room is set in the outdoors reflective of the proposed wind farm landscape and allows users to navigate through informative boards, review planning documents, engage with project experts and explore the interactive project map to discover the positive contributions of Wilan Wind Farm to the community".
>
> **(Spatial Media, 2024)**

FIGURE 5.2
Virtual Engagement Room for the Wilan Wind Farm. (Source: Spatial Media (2024).)

5.4 Summary

Geospatial data and tools are already being used to support RE infrastructure site selection, planning, design, and operations. These tools have even greater potential with the integration of emerging technologies and data, such as AI, 3D models, and near real-time environmental observations. Geospatial intelligence can enable better balancing of RE deployment targets with nature protection, community values, and land use priorities. However, many RE infrastructure projects are already facing social licence challenges and growing community opposition.

If poorly implemented, geospatial and AI-driven tools could worsen distrust, especially if

- They rely on low-quality, incomplete, or biased data,
- They are not transparent, making it hard for communities to understand or challenge decisions,
- Tools and outputs are not accessible or user-friendly, excluding non-expert stakeholders, and
- AI optimisation processes reinforce inequities or fail to reflect social and environmental complexity.

To avoid undermining the transition, we must ensure geospatial, and AI technologies are as follows:

- Traceable,
- Built on fair, high-quality data,
- Designed with transparency, accountability, and accessibility, and
- Used to genuinely support inclusive, place-based decision-making.

If done right, tools can help rebuild trust and accelerate a just and equitable energy transition.

References

Afonso, Jorge. 2025. *Post from AI for Management Postgraduate Course at ISCTE Business School.* https://www.linkedin.com/posts/jorgeafonso_data-analytics-ai-activity-7330140174137413633-g3-K?utm_source=share&utm_medium=member_desktop&rcm=ACoAAA5stVQBORqU6xfR9GnEqqLFTPJXRUsBe1g.

Andrew Dyer, Australian Energy Infrastructure Commissioner. 2023. *Community Engagement Review Report*, on behalf of the Department of Climate Change, Energy, the Environment and Water, Canberra, Australia, 2 February 2024. https://www.dcceew.gov.au/sites/default/files/documents/community-engagement-review-report-minister-climate-change-energy.pdf

FAO (Food and Agriculture Organization of the United Nations). 2022. *Voluntary Guidelines on the Responsible Governance of Tenure of Land, Fisheries and Forests in the Context of National Food Security*, Rome, Italy, 47 p. https://ppp.worldbank.org/sites/default/files/2024-09/i2801e.pdf

Gartner. 2020. *Data Quality: Best Practices for Accurate Insights*. https://www.gartner.com/en/data-analytics/topics/data-quality.

Gluckman, Peter. 2017. *Using Evidence to Inform Social Policy: The Role of Citizen-Based Analytics. A Discussion Paper.* Office of the Prime Minister's Chief Science Advisor. https://www.pmcsa.org.nz.

Hilton, Alice. 2024. "How Geospatial Data Is Transforming the Renewable Energy Sector and Guiding Strategic Decisions." *LinkedIn*. https://www.linkedin.com/pulse/how-geospatial-data-transforming-renewable-energy-sector-guiding-4ld8e/.

Holmes, Adam. 2025. "Australia's Regions Face 'Social Licence' Balancing Act as 2030 Renewable Energy Targets Come Closer." *ABC News*, April 10, 2025. https://www.abc.net.au/news/2025-04-10/regional-australia-grapples-with-renewable-energy-projects/105155280?utm_campaign=abc_news_web&utm_content=link&utm_medium=content_shared&utm_source=abc_news_web.

McBain, Bonnie. 2021. "Renewables Need Land – and Lots of It. That Poses Tricky Questions for Regional Australia." *The Conversation*, March 16, 2021. https://theconversation.com/renewables-need-land-and-lots-of-it-that-poses-tricky-questions-for-regional-australia-156031.

Spatial Media. 2024. *Our Solution for Wilan Wind Farm*. https://www.spatialmedia.io/our-work/wilan-wind-farm.

6

Machine Learning and Comparative Spatial Analysis for Mapping Indigenous Land Ownership in Urban and Rural Australia

Atie Asadikia

6.1 Introduction

Land ownership in Australia remains deeply embedded within broader conversations of equity, sustainability, and indigenous self-determination. For Aboriginal and Torres Strait Islander people, land is not simply a physical asset but a fundamental part of cultural identity, spiritual practice, and community continuity (Kingsley 2014). This enduring relationship with Country, which spans tens of thousands of years, has persisted despite the extensive disruptions caused by colonization. Recognition of this relationship is a moral imperative and a foundational step toward supporting indigenous communities in the stewardship and governance of their lands (CSIRO 2023).

When examined through geographic and demographic data, the spatial distribution of land ownership provides valuable insights into ongoing disparities in land tenure. This is particularly evident in the urban–rural divide, where the conditions surrounding indigenous land rights often differ significantly in both scale and nature (DAFF 2024). Mapping these patterns enables the identification of regions where indigenous land ownership remains underrepresented, offering a basis for more equitable policy interventions and stronger support for indigenous land management.

Recent technological advancements, particularly in data science, have introduced new methods for analyzing complex spatial and socio-economic phenomena. Machine learning, in particular, presents a promising approach for working with high-dimensional datasets to detect underlying trends and non-obvious relationships. Applied to land ownership data, these methods can illuminate how various factors, such as urbanization, demographic shifts, and historical policy contexts, influence indigenous land tenure across

DOI: 10.1201/9781003638308-7

different regions. This evidence can inform more precise and context-sensitive policy responses that advance both sustainability and indigenous rights.

This chapter applies a mixed-methods approach combining statistical modeling with machine learning tools, including logistic regression, interaction term analysis, and Geographically Weighted Regression (GWR). By incorporating temporal data, spatial classification, and local variation, the analysis identifies both broad national patterns and region-specific dynamics in indigenous land ownership. The goal is to contribute an evidence-based framework that supports the development of more inclusive, locally informed, and sustainable land policies.

6.2 Methodology

Our approach involves three key steps to explore the relationship between land ownership and socio-economic or geographic factors. These steps include data collection and preparation, logistic regression modeling, and GWR. We explain each step in the following sections.

6.2.1 Data Collection and Preparation

This study used data from the Australian Bureau of Statistics (ABS) (Australian Bureau of Statistics 2021) across four census years (2006, 2011, 2016, 2021), focusing on indigenous property ownership at the Local Government Area (LGA) level. The ownership ratio was calculated by dividing the number of indigenous property owners by the total population per LGA. To add spatial context, this data was merged with a custom urban–rural classification based on official state designations, "Cities" coded as urban (1) and "Shires," "Towns," or "Regional Councils" as rural (0). Data was processed using Python and libraries such as pandas, numpy, scikit-learn, matplotlib, seaborn, and GeoPandas. Rows with missing values in key variables were removed. A binary outcome variable, *high_indig_ownership*, was created based on whether an LGA's indigenous ownership ratio was above the national median. The *year* variable was scaled to [0, 1], and an interaction term between *urban_rural* and *year_scaled* was included to explore changes over time.

6.2.2 Logistic Regression Modeling

Logistic regression was used to estimate the likelihood of high indigenous ownership across LGAs. After data cleaning, predictors were prepared by scaling *year*, using *urban_rural* as a binary variable, and one-hot encoding *state* with interaction terms (e.g., *state_QUEENSLAND_year*) to assess jurisdictional trends. Three models were developed: one using the full dataset, one using only urban LGAs, and one using only rural LGAs. All models used an 80/20 train-test split and were evaluated using accuracy, precision, recall, and F1-score. Coefficients were interpreted to understand the impact of each predictor, and the urban–rural model split offered insights into regional differences.

6.2.3 Geographically Weighted Regression (GWR)

GWR, implemented using the *mgwr* Python package (Oshan et al. 2019), allowed spatial relationships to vary across LGAs using centroid coordinates and an optimal bandwidth for local fitting. It helped identify how predictors like non-indigenous ownership and urban/rural status influenced indigenous land ownership differently by region. Outputs included local R^2 values and spatially varying coefficients, visualized to highlight geographic disparities and regional clusters.

6.3 Result

The analysis produced both statistical outputs and spatial visualizations that shed light on patterns of indigenous land ownership. This section presents the key findings derived from mapping and modeling approaches.

Figure 6.1 shows choropleth maps of indigenous land ownership ratios across Australian states and territories at the LGA level, based on 2021 data. The maps use a purple gradient to indicate ownership levels, with darker shades representing higher ratios. Urban and rural LGAs are differentiated by boundary color, black for urban and green for rural.

The Northern Territory (NT) displays the highest indigenous land ownership ratios, particularly in the northern and central LGAs. Most of these areas approach or exceed a ratio of 1.0, consistent with the presence of Aboriginal Land Trusts and outcomes from historical land rights movements. In contrast, Darwin and other urban centers show notably lower ratios. Queensland (QLD) shows a mixed pattern. Remote northern regions such as Cape York and the Gulf Country have some of the highest ownership levels. However, urbanized areas like Brisbane and the Gold Coast have low ownership ratios, highlighting colonization and land development disparities. South Australia (SA) presents a clear north–south divide. Northern LGAs, including the APY Lands, display uniformly high indigenous ownership, while southern and coastal LGAs, including Adelaide, show much lower ratios. Western Australia (WA) reflects similar patterns, with high ratios in the Kimberley, Pilbara, and inland regions. Urban and coastal areas, such as Perth and its surroundings, show considerably lower levels of indigenous land tenure.

Tasmania (TAS) shows moderate and more evenly distributed ownership ratios. Higher values appear in remote areas like the Furneaux Group and

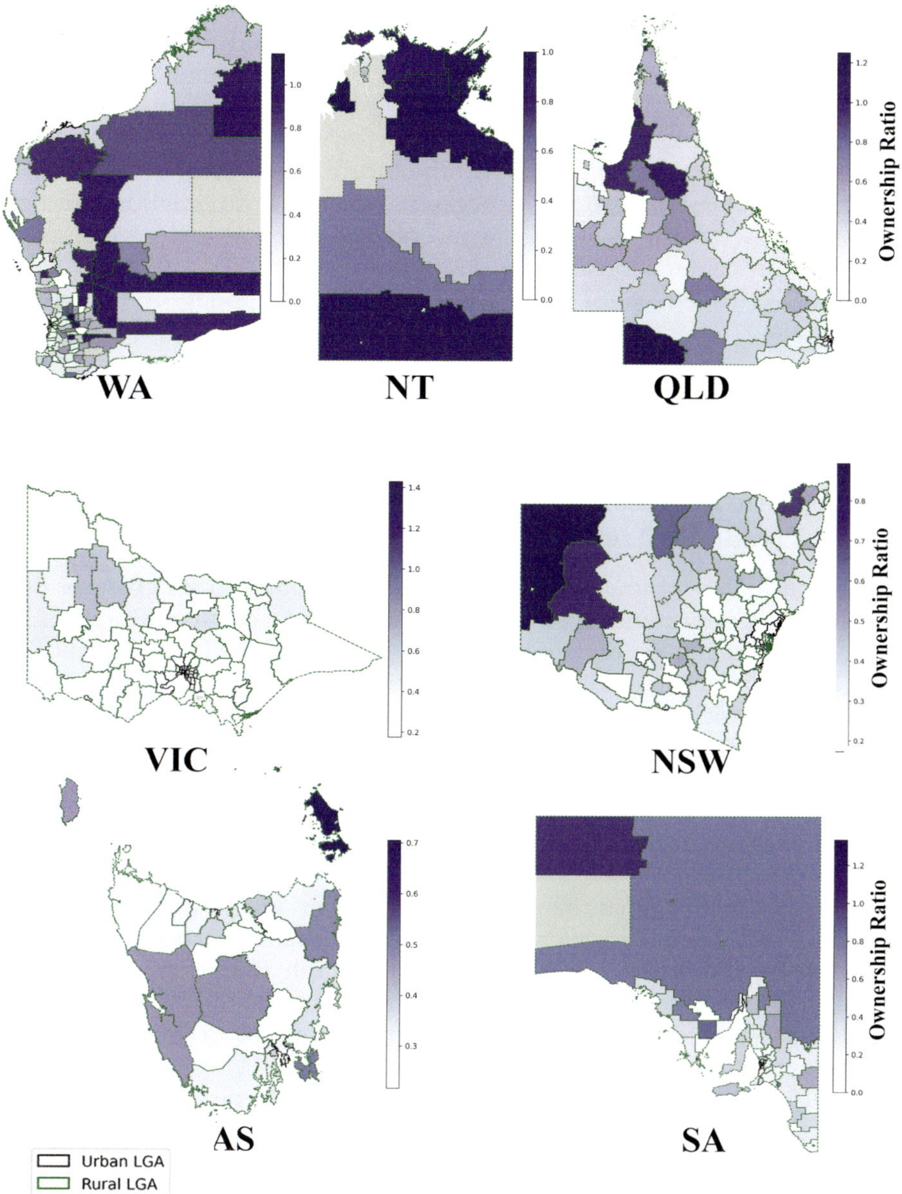

FIGURE 6.1
Australian States Indigenous Land Ownership Ratio by LGA (2021). (Visualization created by the author.)

the west coast, while Hobart and other urban centers show lower ratios. This reflects the limited and more recent formal land recognition in the state. Victoria (VIC) has the lowest indigenous land ownership ratios overall. Most LGAs are shaded in very light purple, indicating limited land return. Even the more rural areas in the northwest show only modest levels of indigenous tenure, reflecting both historical dispossession and limited recognition under native title processes. Across all jurisdictions, indigenous land ownership is strongly concentrated in rural and remote areas, while urban centers consistently show lower ratios. This pattern reflects the intersecting effects of colonization, land rights policy, and economic development.

6.4 Logistic Regression Findings: Urban, Rural, and National Trends

The main logistic regression model achieved an overall accuracy of 84%, with strong performance across both classes. It had a recall of 93% for LGAs with high indigenous ownership (Class 1), indicating effective identification, and a precision of 90% for low-ownership LGAs (Class 0), suggesting a low false positive rate. The most influential predictor was the non-indigenous ownership ratio ($\beta = 5.48$), showing a strong inverse relationship with indigenous ownership. The urban_rural variable also had a modest positive effect ($\beta = 0.38$), suggesting urban LGAs were slightly more likely to exceed the national median for indigenous ownership when controlling for other factors.

State-level interaction terms showed regional variation. Victoria, Tasmania, and South Australia had strongly negative coefficients, particularly Victoria ($\beta = -0.94$), indicating lower likelihoods of high indigenous ownership. Queensland showed a small positive trend over time ($\beta = 0.07$), while the NT's near-zero coefficient reflects limited variation due to consistently high indigenous ownership.

The urban-only model achieved 87% accuracy, with perfect precision and recall, confirming strong classification. Here, non-indigenous ownership remained the strongest predictor ($\beta = 4.29$), while the time variable (year_scaled) showed a negative effect ($\beta = -1.98$), indicating a slight decline in indigenous ownership over time in urban areas, likely due to development pressures and limited recognition in planning frameworks.

The rural model matched the main model's 84% accuracy and reinforced the inverse relationship between indigenous and non-indigenous ownership ($\beta = 4.91$). The year variable also had a negative effect ($\beta = -1.01$), suggesting a gradual decline in rural indigenous land ownership, possibly linked to changing land use, insufficient policy progress, or land sales.

6.5 Results and Interpretation of Geographically Weighted Regression (GWR)

GWR was employed to investigate how the relationship between key predictors and indigenous land ownership varies spatially across Australian LGAs. The selected bandwidth 933.0 suggests a relatively broad spatial kernel, capturing regional rather than highly localized effects. This degree of spatial smoothing indicates that the model is more sensitive to state or multi-LGA level variation in ownership patterns, consistent with the scale at which many historical and policy influences operate.

The mean local R^2 value of 0.28 indicates that, on average, the GWR model accounts for approximately 28% of the variation in indigenous land ownership across the LGAs analyzed. Although this explanatory power is lower than typically observed in global models such as logistic regression, it is not unexpected in socio-spatial studies. Tadayon and Saber (2022) noted that lower R^2 values are shared in spatial data analyses where local variation is substantial and key structural or historical determinants may be only partially observed or omitted entirely.

These findings reinforce the argument that highly context-specific factors shape indigenous land ownership. Relationships between variables, such as non-indigenous ownership ratios, urban–rural classification, and census year, exhibit considerable spatial heterogeneity, precluding the assumption of a single national pattern. Instead, the results highlight the importance of local historical, institutional, and demographic conditions in shaping land tenure outcomes.

The value of GWR lies in its capacity to uncover this spatial non-stationarity, offering a better understanding of how determinants of indigenous land ownership vary across space. Such insights are critical for informing land justice and policy responses. The evidence presented here supports the development of place-based strategies that account for regional complexity rather than relying on uniform, centralized policy prescriptions that may overlook specific communities' lived realities and governance structures.

6.6 Discussion and Conclusion

This study highlights the persistent and uneven spatial distribution of indigenous land ownership across Australian LGAs. By applying machine learning and spatial analysis techniques, specifically logistic regression and GWR, the research uncovers nuanced variations between states and between urban and rural contexts.

The logistic regression analysis revealed a strong inverse relationship between non-indigenous ownership and the probability of high indigenous land tenure.

While rural LGAs generally showed greater levels of indigenous ownership, urban areas demonstrated small but consistent declines. These patterns likely reflect increased development pressure in metropolitan regions and the continued presence of legal and bureaucratic obstacles that limit land restitution. At the state level, Victoria and Tasmania recorded the lowest probabilities of indigenous land ownership, suggesting slower progress in the formal recognition and return of land rights. GWR offered further insight into how these relationships vary spatially. The mean local R^2 value of 0.28 indicates that although the selected predictors are relevant, they account for only a portion of the variance in land tenure outcomes. This reinforces the importance of local dynamics, historical trajectories, institutional conditions, and political frameworks, all of which play a critical role in shaping ownership patterns (Altman and Markham 2015).

These findings have direct implications for sustainability and equity, particularly in relation to the SDGs. The urban–rural divide in indigenous land tenure points to longstanding structural inequality that must be addressed through place-based and responsive policy. Urban planning frameworks must evolve to better reflect indigenous worldviews and land rights, fostering cities that are inclusive, just, and culturally informed. At the same time, the preservation of life on land is closely tied to indigenous stewardship, which has consistently supported biodiversity and sustainable land management practices (Garnett et al. 2018). Effective prioritization of the SDGs must, therefore, consider local disparities, including the differing realities of urban and rural indigenous communities, to ensure sustainability agendas are both fair and representative (Asadikia et al. 2024).

Promoting indigenous land ownership is central to advancing a more equitable and sustainable future (Garnett et al. 2018). Land remains fundamental to cultural survival, economic resilience, and ecological integrity (Schultz et al. 2019). However, increasing tenure alone is insufficient. Long-term outcomes also depend on ensuring appropriate infrastructure, investment, and services, particularly in rural regions, where land ownership may be more common but institutional support often remains limited.

Finally, while the value of statistical and spatial modeling is evident, this study recognizes that data-driven methods must be used in partnership with indigenous communities. Approaches such as participatory mapping, community-led planning, and the integration of indigenous knowledge systems are essential for ensuring that research and policy are grounded in lived experience and guided by the priorities of First Nations peoples.

References

Altman, Jon, and Francis Markham. 2015. "Burgeoning Indigenous Land Ownership: Diverse Values and Strategic Potentialities." In *Native Title from Mabo to Akiba: A Vehicle for Change and Empowerment?* 125–150. Canberra: ANU Press.

Asadikia, Atie, Abbas Rajabifard, and Mohsen Kalantari. 2024. "Navigating Sustainability: Key Factors in Prioritising Sustainable Development Goals." *Sustainability Science* 19: 2041–2063. https://doi.org/10.1007/s11625-024-01561-y.

Australian Bureau of Statistics. 2021. Census of Population and Housing: Counting Persons, Place of Enumerationrc. Accessed May 16, 2025. https://www.abs.gov.au/census.

CSIRO. 2023. "Our Knowledge, Our Way: indigenous-Led Approaches to Strengthening and Sharing Knowledge for Land and Sea Management." CSIRO. https://www.csiro.au/en/research/indigenous-science/indigenous-knowledge/Our-Knowledge-Our-Way.

DAFF. 2024. "Australia's indigenous Land and Forest Estate (2024)." Department of Agriculture, Fisheries and Forestry. https://www.agriculture.gov.au/abares/forestsaustralia/forest-data-maps-and-tools/spatial-data/indigenous-land-and-forest.

Garnett, Stephen T., Neil D. Burgess, John E. Fa, Álvaro Fernández-Llamazares, Zsolt Molnár, Cathy J. Robinson, James E. M. Watson, et al. 2018. "A Spatial Overview of the Global Importance of Indigenous Lands for Conservation." *Nature Sustainability* 1 (7): 369–374. https://doi.org/10.1038/s41893-018-0100-6.

Kingsley, Jonathan. 2014. *If the Land Is Healthy… It Makes the People Healthy*. Melbourne: Deakin University.

Oshan, Taylor M., Ziqi Li, Wei Kang, Levi J. Wolf, and A. Stewart Fotheringham. 2019. "MGWR: A Python Implementation of Multiscale Geographically Weighted Regression for Investigating Process Spatial Heterogeneity and Scale." *ISPRS International Journal of Geo-Information* 8 (6): 269. https://doi.org/10.3390/ijgi8060269.

Schultz, Rosalie, Tammy Abbott, Jessica Yamaguchi, and Sheree Cairney. 2019. "Australian Indigenous Land Management, Ecological Knowledge and Languages for Conservation." *EcoHealth* 16 (1): 171–176. https://doi.org/10.1007/s10393-018-1380-z.

Social and Emotional Wellbeing. n.d. "Country, Culture and Spirituality." Australian indigenous HealthInfoNet. https://healthinfonet.ecu.edu.au/learn/health-topics/social-and-emotional-wellbeing/country-culture-spirituality/.

Tadayon, Vahid, and Mohammad Mehdi Saber. 2022. "A Spatial Logistic Regression Model Based on a Valid Skew-Gaussian Latent Field." *Journal of Agricultural, Biological and Environmental Statistics* 28 (1): 59–73. https://doi.org/10.1007/s13253-022-00512-3.

7

Integrating Global Parameters to Extend Land Administration Domain Model

Fatemeh Jahani, Abbas Rajabifard, Behnam Atazadeh, and Daniel Steudler

7.1 Introduction

A nation's sustainable development and economic resilience are fundamentally linked to the way land is governed and administered. Land Administration Systems (LASs) serve as institutional infrastructures for managing land tenure, value, and use, essential elements for achieving transparency, justice, and long-term planning. By recording rights, restrictions, and responsibilities (RRRs), LASs formalize the people-to-land relationship and contribute directly to the delivery of public services, disaster resilience, and the Sustainable Development Goals (SDGs). Yet, despite their critical role, land governance systems in many parts of the world are still underperforming. Advancing LASs is essential to address emerging governance expectations, ensure regular data validation, and enable integrated, fit-for-purpose, and equitable land management practices.

Advancing LASs has been and continues to be an ongoing process, which should be in line with the latest global initiatives, technological trends, and new community expectations. The need for advancement has become increasingly evident since international bodies and land professionals have been working hard to improve land administration practices by developing several LASs-related frameworks and models, such as SDGs (United Nations 2015), Framework for Effective Land Administration (FELA), and New Urban Agenda (NUA), all of which affect LASs directly or indirectly (Jahani Chehrehbargh et al. 2022). These global initiatives stress the importance of an effective, efficient, integrated LAS that is ongoing upgraded, validated, and that ensures consistency of data by highlighting key parameters essential for LAS advancement. Achieving data validation and integration according to global trends can be facilitated through a data model, a central element of the LAS data lifecycle.

 DOI: 10.1201/9781003638308-8

The Land Administration Domain Model (LADM), established as ISO 19152, is a globally recognized standard designed to structure and secure the complex legal, administrative, and spatial relationships inherent in land administration, covering both land and marine domains and supporting elements above and below the surface (ISO 2012). Since its first edition, LADM has undergone ongoing revisions to address operational gaps such as data acquisition and maintenance. The forthcoming LADM Edition II introduces significant modular extensions, guided by broad stakeholder input, including the UN-GGIM, ISO-TC 211, Open Geospatial Consortium, and other international bodies, to support emerging requirements such as valuation, spatial planning, marine regulation, and advanced 3D/4D land administration (ISO 2018). While Edition II is still under development, numerous studies have contributed to its evolution (Lemmen et al. 2021). The progression of LADM's development is illustrated in Figure 7.1.

Despite notable advancements in the development of the LADM, the comprehensive integration of global sustainability and equity-related parameters, such as gender equality, informal tenure, and valuation, remains insufficiently addressed in existing literature. Previous studies typically focus on isolated themes, for example, incorporating taxation or Indigenous land rights (Cagdas et al. 2016) or formalizing digital transition and SDG indicators (Kara et al. 2024), rather than presenting a holistic approach that embeds these dimensions collectively within land administration data structures. While such research demonstrates progress in aligning LADM with specific international frameworks, there is a persistent gap in the provision of unified models that capture the breadth of emerging global trends. This chapter addresses this gap by developing and validating an extended LADM that systematically integrates critical sustainability parameters, particularly those relevant for marginalized communities, thus enhancing the model's capacity to support inclusive, responsive, and future-oriented land governance systems in line with global development agendas. The proposed model extensions align with the book's overarching themes by promoting sustainability and social inclusion through smarter, data-driven land systems.

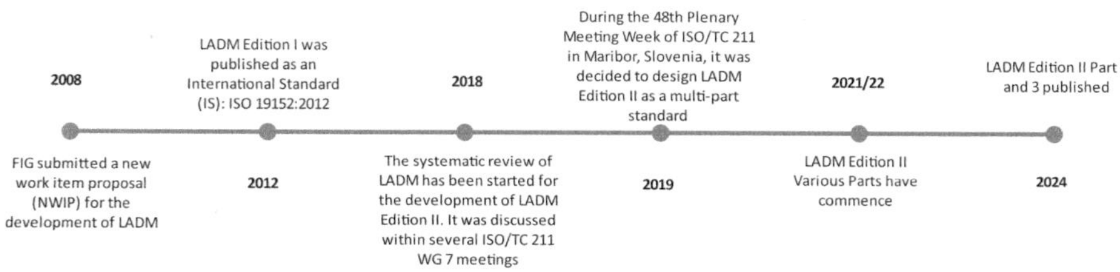

FIGURE 7.1
The development timeline of LADM (Jahani Chehrehbargh et al. 2025).

Furthermore, by implementing these extensions within spatial databases and proposing a modular framework, this work lays a foundation for digital and AI-enabled innovations in land governance, especially for developing countries. In doing so, the research bridges normative global goals with technical implementation, fostering more responsive and equitable land information infrastructures.

7.2 Methodology

This study applies a rigorous design science research approach to design, develop, and empirically validate an enhanced LADM tailored to address global governance imperatives. The methodological framework comprises three interconnected phases: parameter identification, model extension, and physical model implementation.

Phase 1: Global Parameter Identification

The research begins by systematically identifying key parameters influencing the evolution of LASs worldwide. A comprehensive document analysis was undertaken, targeting major international policy instruments such as the SDGs, the FELA, and the VGGT. The analytical process follows the three-tiered framework:

- **Initial Skimming** of high-level documents to filter and isolate land-related directives;
- **Focused Reading** to distill specific parameters associated with gender equality, Indigenous and informal tenure, valuation, and taxation;
- **Interpretive Mapping** to categorize these indicators against the Land Management Paradigm, thereby distinguishing elements of governance, operation, and sustainability.

This phase ensures that the foundation for LADM extension is firmly aligned with globally recognized challenges and opportunities for inclusive land governance.

Phase 2: Conceptual and Logical Model Extension

Building on the parameter identification, the next stage involves extending the LADM at both the conceptual and logical levels:

- **Conceptual Modeling**: Using Unified Modeling Language (UML), the base LADM was refined to incorporate new entities and attributes that reflect contemporary land governance needs, such as expanded tenure classifications, enhanced boundary definitions, and structured provenance metadata.
- **Logical Modeling**: The conceptual design was subsequently formalized into a logical data model using Enterprise Architect, which enabled explicit definition of data structures and inter-entity relationships. This logical layer was designed to be agnostic of implementation platform, supporting compatibility with relational database environments like PostgreSQL, Oracle, and SQL Server.

Phase 3: Physical Model Implementation and Empirical Validation

To ensure practical relevance, the enhanced LADM was instantiated as a physical database using PostgreSQL with the PostGIS spatial extension. The translation from logical model to physical schema involved:

- **Generating SQL schemas** with new fields to capture attributes such as gender, Indigenous status, and informal rights,
- **Populating the database** with simulated datasets that model diverse land tenure situations, particularly focusing on underrepresented groups,
- **Developing and executing targeted SQL queries** to validate key functionalities, including extraction of gender-disaggregated land ownership, identification of informal tenure, and spatial analysis of Indigenous parcels.

The empirical validation phase not only tests the robustness of the proposed extensions but also demonstrates their applicability in supporting real-world use cases.

7.3 Proposed Extension of LADM

Through comprehensive analysis of global initiatives, this study identifies key parameters essential for advancing LASs in line with international sustainability and equity goals. Figure 7.2 shows some of the identified parameters, mapped onto the well-established land management paradigm (Enemark 2006), which is crucial for contributing to the formulation

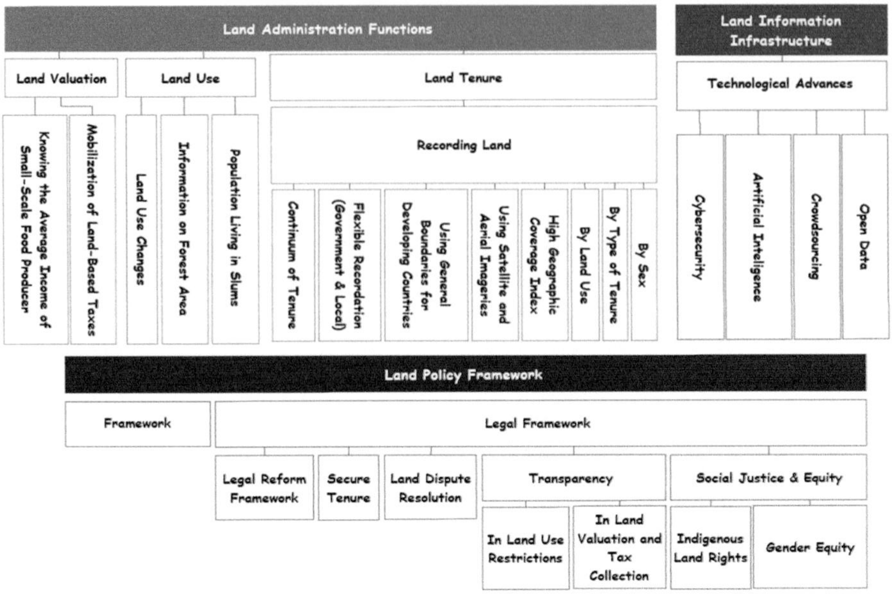

FIGURE 7.2
Some of the extracted global parameters for advancing LAS (Jahani Chehrehbargh et al. 2024b).

of a new framework for effective and advanced LAS (Jahani Chehrehbargh et al. 2024b).

This research focuses on incorporating previously overlooked or underrepresented global parameters, particularly those concerning gender, Indigenous populations, informality, and valuation, within the LADM framework. By anchoring these parameters within the broader land information infrastructure, the proposed extension enhances the capability of LASs to support sustainable development, particularly in developing contexts such as Indonesia.

7.4 Country-Specific Implementation for Indonesia

Indonesia, with its high volume of unregistered land parcels, serves as a compelling case study to demonstrate the practical application of the proposed model (Jahani Chehrehbargh et al. 2024a). National programs such as the World Bank-supported One Map Project aim to address fragmented and inconsistent land records by promoting digitalization. Integrating the extended LADM in Indonesia offers opportunities for environmental sustainability, disaster resilience, and improved service delivery.

Existing classes in the LADM, such as Person, Right, RegisterParcel, and SurveyDocument, already have some alignment with Indonesia's current system (Sucaya 2009). However, classes like Responsibility, Restriction, and ServingParcel are still largely underutilized. To bridge these gaps, this study adopts the core classes of LADM Edition II and extends them with attributes related to gender equality, Indigenous tenure, informality, valuation, and taxation.

7.4.1 Gender Equality

The initial step in securing women's land rights involves identifying the gender of parcel owners. The original LADM does not include a gender attribute in the *LA_Party* class. To make LAS gender-sensitive, this research proposes adding a gender attribute to the LADM, allowing for the collection of gender-specific data. Unger et al. (2023) introduced gender as an attribute in the *LA_Party* class, using the ISO/IEC sex type code list. To develop a comprehensive gender-sensitive system, it is necessary to recognize a broad spectrum of gender identities. By incorporating these identities under the attribute *LA_GenderType*, the system can accurately reflect an individual's gender. Figure 7.3 shows the proposed gender attribute and its code list, with green boxes indicating modified classes, yellow boxes highlighting new elements, and blue boxes for Indonesia's profile.

7.4.2 Indigenous Rights

Globally, Indigenous populations face historical marginalization in land tenure recognition. In Indonesia, millions identify as Indigenous, with at least 1300 ethnic groups officially recognized (IWGIA 2023). Despite partial legal recognition through instruments such as the Basic Agrarian Regulation and coastal or forestry laws, practical protection remains limited, and land conflicts persist. To improve recognition, this study introduces a binary *Indigenous_status* attribute under *LA_Party* and adds Indigenous group and group type attributes. Indigenous unit types are incorporated into *LA_BAUnitType*, and Indigenous rights are represented under *LA_RightType*. The model also proposes a dedicated classification for 31 of Indonesia's largest ethnic groups and integrates customary documentation into administrative data elements, aligning with practices such as registering *Ulayat* Land (i.e., customary communal land recognized under Indigenous law). These changes are shown in Figure 7.3.

7.4.3 Informal Rights

Informal tenure affects nearly one-quarter of urban populations globally yet remains largely undocumented in conventional LASs. Prior studies

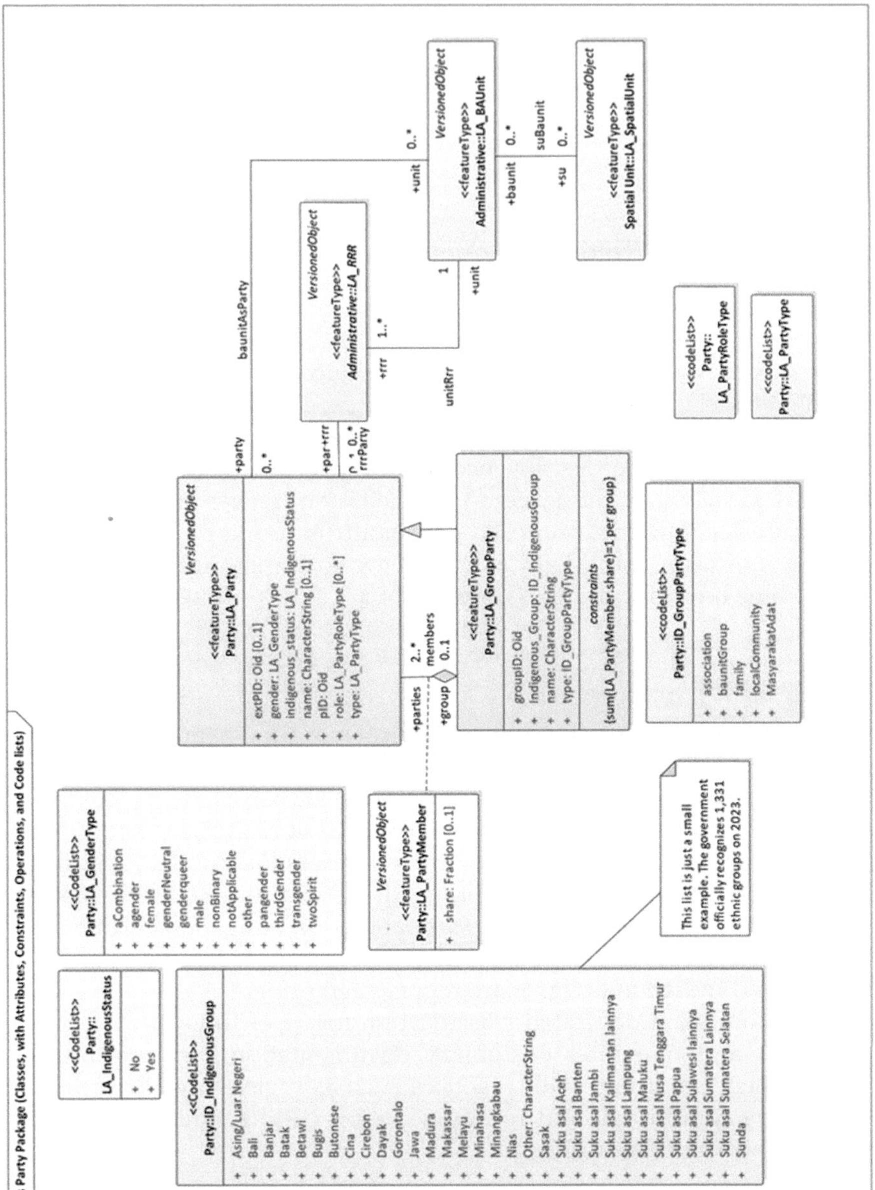

FIGURE 7.3
Extended LADM party package for Indonesia.

have significantly contributed to modeling informal tenure within the LADM. Despite substantial contributions (Paasch et al. 2013, Kalogianni et al. 2022), LADM has yet to fully encompass certain practical and contextual nuances required for comprehensive informal tenure documentation, especially aligned with global sustainability frameworks. This study specifically addresses these limitations by extending the *LA_Right* class to include a generalized category explicitly designed for informal rights. This includes attributes for settlement types, detailed population statistics, and documentation records, which facilitate a structured, hierarchical coding approach compatible with open-linked data methodologies. Furthermore, tailored Indonesia-specific documentation forms, such as management rights, designation forms, and physical control statements, have been introduced to support precise registration of customary and informal land scenarios. Certain necessary details from these forms need to be included in the spatial unit package of LADM. This includes information like the Land Identification Number (*NIB: Nomor Identifikasi Bidang*), Land Parcel Map (*PBT: Peta Bidang Tanah*), and land status, indicating whether the land is Indigenous or not.

7.4.4 Valuation and Taxation

Valuation and taxation are fundamental to sustainable land governance, and their integration within the LADM has evolved significantly in recent years (Cagdas et al. 2016).

As shown in Figure 7.4, this study advances the LADM by expanding its valuation and taxation modules to better support single and mass appraisal processes, property tax assessments, and the recording of diverse transaction types and sales statistics. New and refined classes within the model—such as *VM_Valuation*, *VM_SinglePropertyAppraisal*, *VM_MassAppraisal*, *VM_TransactionPrices*, and *VM_SalesStatistics*—enable detailed recording and analysis of market activity, appraisal methods, and property characteristics, while entities like *VM_ValuationUnit* and *VM_SpatialUnit* facilitate comprehensive fiscal registries linked to cadastral systems.

Further enhancements include the introduction of new enumerations and code lists that capture building features (such as number of rooms, quality indicators, and amenities), transaction specifics, and taxation details. The *TM_Taxation* class supports the management of property tax data, including fiscal year, assessed values, exemptions, payment histories, and appeals. By incorporating these features, the extended LADM enables a more robust and transparent fiscal management infrastructure, tailored for the complex requirements of countries like Indonesia.

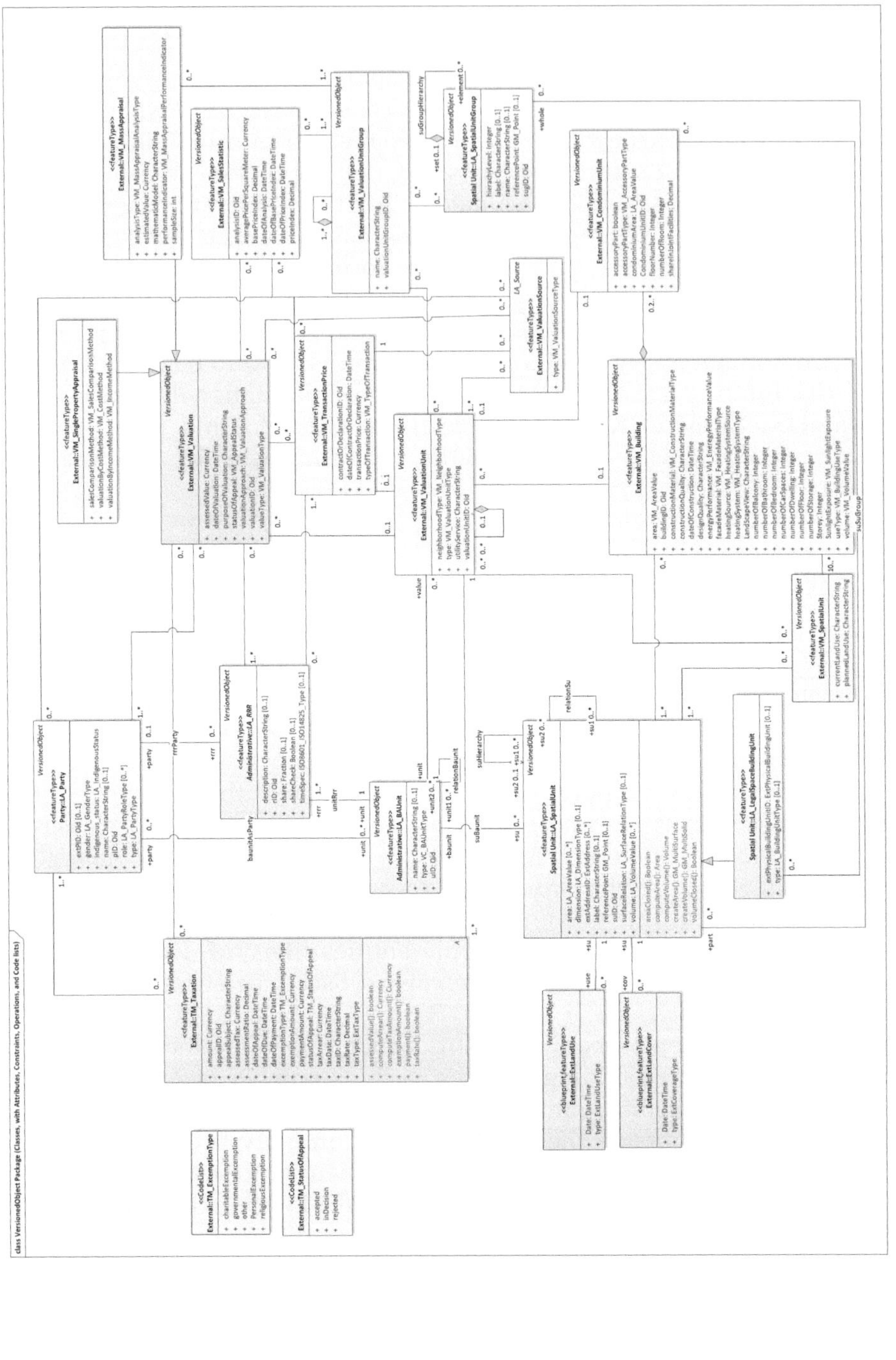

FIGURE 7.4
Extended valuation class within the versioned object package of LADM.

7.5 Physical Data Model Implementation

While the LADM provides a robust conceptual foundation, it does not offer direct instructions for practical implementation as a functional land administration database. This section details the process of translating the extended LADM, enhanced with parameters such as women's land rights, Indigenous tenure, informality, valuation, and taxation, into a working physical data model implemented in PostgreSQL with PostGIS. The workflow followed these sequential steps:

- **Logical Model as Input**: The extended logical schema developed in Enterprise Architect was used as the basis for conversion.
- **UML to DDL Conversion**: UML diagrams were translated into Data Definition Language (DDL) within EA, forming the foundational structure for the database.
- **Refining DDL**: Primary and foreign key constraints and class relationships were manually reviewed and updated to ensure accuracy and consistency.
- **SQL Generation and Editing**: The finalized DDL scripts were exported and converted to SQL format and subsequently refined in pgAdmin to address syntactic issues and support implementation requirements.
- **Table Creation and Linking**: Database tables were created using the refined SQL scripts, establishing correct relationships among entities as defined in the LADM structure.
- **Data Population**: Test data was inserted into the database, including gender attributes, Indigenous and informal classifications, and valuation metadata.
- **Spatial Query Integration and Visualization**: The PostgreSQL database was linked with QGIS to support spatial querying and visualization of land administration data.

To demonstrate the practical utility of the extended LADM, a real-world dataset comprising 156 Jakarta land parcels was prepared, with each parcel enriched to support advanced queries on gender, tenure, and Indigenous rights. Focusing on a scenario involving planned road construction, the system effectively identified and categorized parcels within a 15-meter buffer, highlighting those owned by Indigenous communities (Figure 7.5). Using the new *Indigenous_status* attribute, the model enabled spatial analysis and mapping in QGIS, allowing planners to visualize affected Indigenous parcels and facilitate informed decisions regarding legal compliance, compensation, or

FIGURE 7.5
Query Execution and Spatial Distribution of Parcels Affected by Planned Road Construction.

required consultations. This application illustrates the value of the extended LADM in supporting equitable land management and protecting community rights in development contexts (Jahani Chehrehbargh et al. 2025).

7.6 Conclusion

This chapter has demonstrated that extending the LADM to incorporate global parameters, including gender, Indigenous rights, informal tenure, valuation, and taxation, significantly enhances the inclusiveness and effectiveness of LASs. The research bridges critical gaps by unifying global sustainability objectives within a single data model. Through a structured methodology, the extended LADM was thoroughly analyzed, conceptually modeled, and implemented in a physical database environment, with validation confirming its practical utility, particularly in supporting vulnerable groups and improving land governance outcomes.

While the practical deployment of the extended model in PostgreSQL with PostGIS highlights its feasibility, the process also revealed technical challenges in translating conceptual models to robust database implementations. These challenges underscore the need for careful refinement and ongoing validation to maintain data integrity and support complex land administration requirements. Continued enhancement, including the adoption of advanced technologies and broader jurisdictional testing, will be essential to ensure that LASs remain relevant, equitable, and sustainable in a rapidly evolving global landscape.

References

Cagdas, V., A. Kara, P. Van Oosterom, C. Lemmen, Ü. Işıkdağ, R. Kathmann and E. Stubkjær (2016). "An initial design of ISO 19152: 2012 LADM based valuation and taxation data model." ISPRS Journal of Photogrammetry and Remote Sensing **4**: 145–154.

Enemark, S. (2006). Understanding the land management paradigm. *Proceedings: Innovative technology for land administration*, International Federation of Surveyors: 17–27.

ISO. (2012). ISO 19152: 2012. Geographic information-Land Administration Domain Model (LADM), International Organization for Standardization Geneva.

ISO. (2018). Result of voting on Systematic review ISO 19152:2012, Geographic information– Land Administration Domain Model (LADM), ISO/TC 211 N 4812.

IWGIA. (2023). "Indigenous Peoples in Indonesia." Retrieved 03 April, 2024, from https://www.iwgia.org/en/indonesia.html.

Jahani Chehrehbargh, F., A. Rajabifard, B. Atazadeh and D. Steudler (2024a). "Current challenges and strategic directions for land administration system modernisation in Indonesia." Journal of Spatial Science **69**: 1–33.

Jahani Chehrehbargh, F., A. Rajabifard, B. Atazadeh and D. Steudler (2024b). "Identifying global parameters for advancing Land Administration Systems." Land Use Policy **136**: 106973.

Jahani Chehrehbargh, F., A. Rajabifard, B. Atazadeh, D. Steudler, J. Shahidinejad and B. W. Nugraha (2025). "A feasibility assessment of an extended LADM: enhancing land administration systems through object-relational database management." Transactions in GIS **29**(4).

Jahani Chehrehbargh, F., A. Rajabifard and M. Kalantari (2022). Advancing Land Administration System-Implications of Recent International Trends. *FIG Congress 2022*. Warsaw, Poland, FIG.

Kalogianni, E., A. Kara, B. Anthony, J. M. Paasch, J. Zevenbergen, E. Dimopoulou, D. Kitsakis, P. van Oosterom and C. Lemmen (2022). Refining the Legal Land Administration-Related Aspects in LADM. 10th LAND ADMINISTRATION DOMAIN MODEL WORKSHOP, International Federation of Surveyors.

Kara, A., M. Chen, P. J. van Oosterom and C. Lemmen (2024). Monitoring Indicators of International Guidance Documents and Frameworks through LADM. 12th International FIG Land Administration Domain Model & 3D Land Administration Workshop, International Federation of Surveyors (FIG).

Lemmen, C., A. Alattas, A. Indrajit, E. Kalogianni, A. Kara, P. Oukes and P. J. van Oosterom (2021). The Foundation of Edition II of the Land Administration Domain Model. FIG e-Working Week 2021: Challenges in a new reality.

Paasch, J., P. van Oosterom, C. Lemmen and J. Paulsson (2013). Specialization of the LADM: Modelling of Non-formal RRR. LADM2013, 5th Land Administration Domain Model Workshop, Kuala Lumpur, Malaysia, 24–25 September, 2013, International Federation of Surveyors (FIG).

Sucaya, I. K. G. A. (2009). Application and validation the land administration domain model in real life situation: a case study in Indonesia, ITC.

Unger, E.-M., C. Lemmen and R. Bennett (2023). "Women's access to land and the Land Administration Domain Model (LADM): requirements, modelling and assessment." Land Use Policy **126**: 106538.

United Nations. (2015). "Sustainable Development Goals." Retrieved October, 2022, from https://www.un.org/sustainabledevelopment/poverty/

8

Urban Analytics in Planning Practice: Future Trends in Spatial Digital Technologies for More Resilient, Sustainable, and Livable Cities

Mark Allan

8.1 Introduction

The collaborative efforts of researchers, engineers, spatial design practitioners, and decision makers, supported by inclusive community engagement, are integral to addressing urban challenges effectively. Currently, over 55 percent of the global population lives in urban areas, a figure forecast to increase to 68 percent by 2050 (United Nations Department of Economic and Social Affairs (UNDESA) 2018). This rapid urbanization, coupled with unprecedented mobility and demographic changes, is accompanied by rising resource depletion, climate-change-induced weather extremes, and both natural and human disasters. City administrators face significant hurdles in sustainably managing growth, congestion, pollution, and disaster risks. Although urban areas occupy only 2–3 percent of the earth's land, they account for approximately 70 percent of carbon emissions associated with energy consumption and consume 60 percent of the world's freshwater resources (Grimm et al. 2008, United Nations (UN) 2012, Luqman et al. 2023). Cities serve as the planet's 'economic powerhouses' and 'safety valves,' generating approximately 80 percent of global gross domestic product and offering opportunities and the promise of a sustainable future for all (United Nations (UN) 2015, World Economic Forum 2022).

Global efforts to advance urban sustainability and resilience can significantly benefit from geospatial and digital infrastructures, including urban digital twins (UDTs). However, these technologies alone are insufficient to address current urban challenges and efforts to manage the environment and climate-change-related impacts. Increased collaboration and collective

DOI: 10.1201/9781003638308-9

action to reintegrate infrastructure engineering with spatial design are critical to the creation of sustainable cities of the future (Hooimeijer et al. 2022). This chapter contributes to the discourse on how spatial technologies can enhance the sustainability, resilience, and livability of cities as part of contemporary urban planning practice. It combines a review of literature and the synthesis of observations from stakeholders and the author as researcher and urban practitioner.

8.2 Sustainable Development and Livability

The concepts of livability and sustainability (and sustainable development), while distinct, contribute to social well-being. Sustainability encompasses a broader, long-term perspective on social wellbeing, whereas livability focuses on immediate needs and quality of life (Tennakoon & Kulatunga 2019). Both concepts are multifaceted and subjective, with no singularly accepted definitions of either. A long-standing definition of sustainable development is development that 'meets the needs of the present without compromising the ability of future generations to meet their own needs' (World Commission on Environment and Development 1987). Whereas livability can be defined as 'the life quality and satisfaction of people and communities' (Hill et al. 2021). In the context of cities, the United Nations (UN) Sustainable Development Goal (SDG) 11 articulates a framework to 'make cities and human settlements inclusive, safe, resilient and sustainable' (United Nations (UN) 2015). SDG 11 is one of the 17 SDGs, which together comprise the UN 2015 Agenda for Sustainable Development, outlining a pathway to a globally sustainable future by 2030. Although it is widely accepted that implementation of all SDGs will not now be achieved in all places by 2030, the 2015 agreement of all member states remains significant (Sachs et al. 2024). Digital technologies will have an increasingly important role in progressing the SDGs (Tzachor et al. 2022). The livability of urban environments is a crucial determinant of our wellbeing (Hill et al. 2021). There are several frameworks that measure livability, including the Economist Intelligence Unit's Global Livability Ranking. This annual survey ranks 173 cities across five weighted categories: stability, healthcare, culture and environment, education, and infrastructure (Economist Intelligence Unit (EIU) 2024). While the politics surrounding the ranking of cities may be debated, identifying key quality of life indicators can provide valuable support to urban policymakers. Strategies, such as the 'poly-centric city' and the '15-minute city,' hold promise for improving sustainability and livability by linking proximity and social equity (Allam et al. 2022).

8.3 Urban Resilience

The concept of urban resilience involves the preparedness of communities to maintain basic city functions, prosper, and bounce back from human and natural disasters (Meerow et al. 2016). Increasingly, digital and communication technologies are being applied to prepare and implement urban resilience plans. These plans are more likely to be successful when developed collaboratively with local communities involving public and private sector stakeholders (Agboola et al. 2023).

8.4 UDT Enhanced Urban Sustainability – Stakeholders' Views

An online survey of researchers, practitioners, and government officials was undertaken in 2024 at the University of Melbourne, Australia. A total of 79 individuals completed the survey and answered all questions including one asking participants to rate the contribution of UDT outputs to the overall sustainability of people-centric urban regeneration public space projects from a list of five outputs: (1) data management, (2) situational awareness, (3) simulation/prediction, (4) multidisciplinary collaboration, and (5) visualization/engagement. A large cohort of participants rates all five outputs highly. UDT-generated visualizations/engagement and simulation/predictions are considered likely to make the highest contribution. These observations are consistent with the literature that suggests UDT modeling can support inclusive planning and efforts to increase community preparedness and urban resilience (Vafaeva et al. 2024).

8.5 Importance of Spatial Digital Technologies

For spatial digital initiatives to significantly benefit the sustainability of cities and urban communities, they must prioritize human-centric implementation with a focus on integrating land, energy, and infrastructure (Steer 2024). Digitalization can serve communities by capturing public goods through land administration and significantly enhancing urban management and strategically aligning infrastructure provision with urban growth. The application of digital geospatial information systems (GIS), and increasingly UDTs and Geospatial Artificial Intelligence (AI) (GeoAI) is transforming the way urban planners visualize land use changes, engage with stakeholders, and track carbon emissions. Advances in these technologies are enabling more sophisticated monitoring and management of urban ecosystems, facilitating urban resilience and improved governance to promote urban sustainability (He & Chen 2024). Digital tools that simulate urban environments can facilitate collaboration and citizen engagement and enhance sustainable urban development and livability (Lim et al. 2021).

8.6 Urban Analytics

Conceptually, urban analytics is a quantitative approach that employs various data gathering and analysis techniques to address urban-related questions and challenges. It incorporates methods from statistics, geospatial, and computer science to derive insights into the complex and dynamic nature of urban environments (Boeing et al. 2021). Urban researcher Michael Batty posits that urban analytics goes beyond data collection to integrate spatial and temporal data to model urban scenarios, thereby facilitating informed data-driven decision-making (Batty 2019). Moreover, UDTs utilize data analytics and AI, connected to the Internet of Things (IoT), they replicate the real-time operation of physical cities, utilizing sophisticated automated sensors that generate data in near real time (Caldarelli et al. 2023). Given the inherent variability of cities and their continuous evolution, characterized by a vast array of diverse multi-dimensional socioeconomic and physical factors, creating accurate digital replicas is extremely difficult. However, analysis of large data sets can unlock intricate patterns of urban life, enabling exploration of complex demographics, mobility, land use, and environmental impacts to promote urban sustainability (Elena et al. 2024). Urban analytics can play a significant role in the future management of cities and multifaceted issues related to sustainability, resilience, and changing demographics.

8.7 Digital Technologies in Urban Planning Practice

Since the 1990s, urban planners have used GIS and other forms of technology in practice. This usage expanded during the 2010s, coinciding with faster computers, the IoT, increasingly detailed data sets, and the transition from analogue to digital applications. Throughout the 2020s, increasingly sophisticated platforms have evolved, such as virtual reality (VR), augmented reality (AR), GeoAI, and generative AI (Gen AI) (Christmann & Schinagl 2022). Internationally, professional planning associations, including the Planning Institute of Australia (PIA), are developing practice guidelines to proactively position the profession and society for the uncertainty and disruption associated with the increased use of evolving digital technologies (Planning Institute of Australia (PIA) & Daniel 2024). The PIA guidelines aim to respond to the increasing complexity of urban management to transform spatial planning and fully utilize the benefits of current and emerging digital technologies in planning more sustainable cities and places. Practical applications include online planning information, automating development approvals, speeding up report writing, considering Gen AI community submissions, and utilizing digital platforms and 3D visualizations to enhance engagement and effective process improvements. The PIA anticipates that digital technologies will continue to profoundly change the way urban practitioners work and the future skillsets they will require. In 2022, the PIA published PlanTech, a strategy containing ten guiding principles for digitization (Planning Institute of Australia (PIA) & Daniel 2024). While most planners do not have highly developed technological skills, their ability to engage and collaborate with stakeholders, including technical specialists, will become increasingly important in directing technology to benefit urban communities (Sabri & Witte 2023, Allan et al. 2024). The future skills and application settings of the 'digital urban planner' are summarized in Figure 8.1.

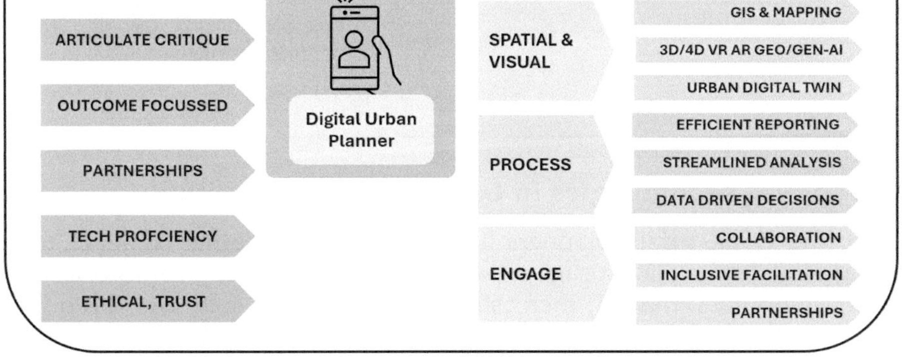

FIGURE 8.1
Digital urban planner. (Adapted from Planning Institute of Australia (PIA) & Daniel 2024.)

8.8 Trends in Generative AI Applied to Urban Design and Planning

Gen AI applications in urban design and planning are currently experimental; however, partnerships between humanity and intelligent machines are forecast to become the norm in the coming decades (Jiang et al. 2024). The application of these technologies will require collaboration between urban practitioners, data and computer scientists, as urban designers and planners transition from 'design focused' to 'decision centric' approaches (Jiang et al. 2024). While not intended to replace human creativity, Gen AI platforms can increase design efficiency, alleviate time-consuming considerations, including development approval applications, site conditions, regulations, and policies (Jiang et al. 2024). Successfully advancing these technologies is reliant upon accessible, quality open data sources and ethical governance that protects privacy. Planning and urban design are important applied social sciences that, in combination with scientific spatial platforms, present opportunities for a unique placed-based systems approach to urban resilience and sustainability (Batty & Yang 2022). The ongoing transformation of urban planning requires additional research into 'digital governance' to safeguard individual privacy and ensure ethical deployment of technology.

8.9 Conclusions

Rapid urbanization and challenges from population growth, climate change, and socioeconomic disparities require urban planners to integrate geospatial and digital technologies into their practice. Utilizing digital participatory planning approaches and emerging UDT and AI technologies can enhance community engagement and informed decision-making, critical for fostering livability, urban resilience, and sustainability. Furthermore, the creation of sustainable cities of the future relies upon innovation and bottom-up and top-down knowledge transfer among communities.

Disaster preparedness is essential for maintaining city operations under stress, emphasizing the need for collaborative resilience strategies that include diverse community input. Urban planners must utilize collective knowledge to advocate for inclusive, safe, resilient, and sustainable community development. Advancing SDG 11 relies upon the local community involvement and participatory urban planning processes to develop practical and actionable resilience strategies to effectively address complex urban challenges.

References

Agboola, Oluwagbemiga Paul, Henry Ojobo, and Anar Aliyev. 2023. "Ameliorating Climate Change Impacts on the Built Environment." Civil Engineering and Architecture 11 (3): 1324–36. https://doi.org/10.13189/cea.2023.110317.

Allam, Zaheer, Simon Elias Bibri, Didier Chabaud, and Carlos Moreno. 2022. "The '15-Minute City" Concept Can Shape a Net-Zero Urban Future." Humanities and Social Science Communications 9 (126). https://doi.org/10.1057/s41599-022-01145-0.

Allan, Mark, Abbas Rajabifard, and Greg Foliente. 2024. "Climate Resilient Urban Regeneration and SDG 11 – Stakeholders' View on Pathways and Digital Infrastructures." International Journal of Digital Earth 17 (1): 2385076. https://doi.org/10.1080/17538947.2024.2385076.

Batty, Michael. 2019, March. "Urban Analytics Defined." Environment and Planning B: Urban Analytics and City Science. https://doi.org/10.1177/2399808319839494.

Batty, Michael, and Wei Yang. 2022. "A Digital Future for Planning: Spatial Planning Reimagined." https://digital4planning.com/wp-content/uploads/2022/02/A-Digital-Future-for-Planning-Full-Report-Web.pdf.

Boeing, Geoff, Michael Batty, Shan Jiang, and Lisa Schweitzer. 2021, May. "Urban Analytics: History, Trajectory, and Critique." https://doi.org/10.2139/ssrn.3846508.

Caldarelli, Guido, Elsa Arcaute, Marc Barthelemy, Michael Batty, Carlos Gershenson, Dirk Helbing, Stefano Mancuso, et al. 2023. "The Role of Complexity for Digital Twins of Cities." Nature Computational Science 3 (5): 374–81. https://doi.org/10.1038/s43588-023-00431-4.

Christmann, Gabriela B., and Martin Schinagl. 2022. "Digital Urban Planning and Urban Planners' Mediatized Construction of Spaces." In Communicative Constructions and the Refiguration of Spaces, First. London: Routledge. https://www.taylorfrancis.com/chapters/oa-edit/10.4324/9780367817183-11/digital-urban-planning-urban-planners-mediatized-construction-spaces-gabriela-christmann-martin-schinagl.

Economist Intelligence Unit (EIU). 2024. "The Global Liveability Index 2024: The World's Most Liveable Cities." The Economist Intelligence Unit Limited. https://www.eiu.com/n/campaigns/global-liveability-index-2024/.

Elena, Vasilieva, Rajat Singh, Rajeev Sobti, Kshama Sharma, Rajat Sharma, and Pothireddy Surekha. 2024. "Leveraging Big Data Analytics for Urban Planning: A Study Using the Big Data Analytics Efficiency Test." BIO Web of Conferences 86: 01082. https://doi.org/10.1051/bioconf/20248601082.

Grimm, Nancy B., Stanley H. Faeth, Nancy E. Golubiewski, Charles L. Redman, Jianguo Wu, Xuemei Bai, and John M. Briggs. 2008. "Global Change and the Ecology of Cities." Science 319 (5864): 756–60. https://doi.org/10.1126/science.1150195.

He, Wei, and Mingze Chen. 2024. "Advancing Urban Life: A Systematic Review of Emerging Technologies and Artificial Intelligence in Urban Design and Planning." Buildings 14 (3): 835. https://doi.org/10.3390/buildings14030835.

Hill, Sarah, Zena Cumpston, Gabriela Quintana Vigiola, and Australian Government. 2021. "Livability | Australia State of the Environment 2021." https://soe.dcceew.gov.au/urban/environment/livability.

Hooimeijer, Fransje, Jeremy Bricker, Adam Pel, Adam Brand, Frans Van de Ven, and Amin Askarinejad. 2022. "Multi- and Interdisciplinary Design of Urban Infrastructure Development." Proceedings of the Institution of Civil Engineers – Urban Design and Planning 175 (4): 153–68. https://doi.org/10.1680/jurdp.21.00019.

Jiang, Feifeng, Jun Ma, Christopher John Webster, Alain J. F. Chiaradia, Yulun Zhou, Zhan Zhao, and Xiaohu Zhang. 2024. "Generative Urban Design: A Systematic Review on Problem Formulation, Design Generation, and Decision-Making." Progress in Planning 180 (February): 100795. https://doi.org/10.1016/j.progress.2023.100795.

Lim, Tian Kuay, Abbas Rajabifard, Victor Khoo, Soheil Sabri, and Yiqun Chen. 2021. "Chapter 3 – The Smart City in Singapore: How Environmental and Geospatial Innovation Lead to Urban Livability and Environmental Sustainability." In Smart Cities for Technological and Social Innovation, edited by Hyung Min Kim, Soheil Sabri, and Anthony Kent, 29–49. Academic Press. https://doi.org/10.1016/B978-0-12-818886-6.00003-4.

Luqman, Muhammad, Peter J. Rayner, and Kevin R. Gurney. 2023. "On the Impact of Urbanisation on CO_2 Emissions." npj Urban Sustainability 3 (1): 6. https://doi.org/10.1038/s42949-023-00084-2.

Meerow, Sara, Joshua P. Newell, and Melissa Stults. 2016. "Defining Urban Resilience: A Review." Landscape and Urban Planning 147 (March): 38–49. https://doi.org/10.1016/j.landurbplan.2015.11.011.

Planning Institute of Australia (PIA), and Claire Daniel. 2024. "PIA PlanTech Best Practice Guidelines Guide for Adopting Digital Technology in Planning Practice." https://www.planning.org.au/common/Uploaded%20files/PIA/Policy/PlanTech/PlanTechBestPracticeGuidelines

Sabri, Soheil, and Patrick Witte. 2023. "Digital Technologies in Urban Planning and Urban Management." Journal of Urban Management 12 (1): 1–3. https://doi.org/10.1016/j.jum.2023.02.003.

Sachs, Jeffrey, Guillaume Lafortune, and Grayson Fuller. 2024. The SDGs and the UN Summit of the Future. Sustainable Development Report 2024. Dublin: Dublin University Press. https://doi.org/10.25546/108572.

Steer, Andrew. 2024. "World Bank Land Conference 2024 – Opening Plenary." Text/HTML. World Bank. https://www.worldbank.org/en/news/video/2024/08/26/world-bank-land-conference-2024-opening-plenary.

Tennakoon, Thilina., and Udaya Kulatunga. 2019. "Understanding Liveability: Related Concepts and Definitions." In Proceedings of the 8th World Construction Symposium, 578–87. University of Moratuwa, Colombo Sri Lanka. https://doi.org/10.31705/WCS.2019.57.

Tzachor, Asaf, Soheil Sabri, Catherine E. Richards, Abbas Rajabifard, and Michele Acuto. 2022. "Potential and Limitations of Digital Twins to Achieve the Sustainable Development Goals | Nature Sustainability." Nature Sustainability 5: 822–29.

United Nations (UN). 2012. "Sustainable Cities: Building Cities for the Future | Department of Economic and Social Affairs." https://sdgs.un.org/publications/sustainable-cities-building-cities-future-17424.

United Nations (UN). 2015. "Transforming Our World: The 2030 Agenda for Sustainable Development Resolution Adopted by the General Assembly on 25 September 2015." https://www.un.org/ga/search/view_doc.asp?symbol=A/RES/70/1&Lang=E.

United Nations Department of Economic and Social Affairs (UNDESA). 2018. "World Urbanization Prospects." https://population.un.org/wup/faqs.

Vafaeva, Khristina Maksudovna, Manish Ghalwan, Pothireddy Surekha, Richa Nangia, and Deepak Bharadwaj. 2024. "Community Engagement in Smart Cities: A Social Network Analysis and Community Engagement Test." BIO Web of Conferences 86: 01073. https://doi.org/10.1051/bioconf/20248601073.

World Economic Forum, and China Academy of Information and Communication Technology. 2022. "Digital Twin Cities: Framework and Global Practices." World Economic Forum. https://www.weforum.org/publications/digital-twin-cities-framework-and-global-practices/.

World Commission on Environment and Development. 1987. Our Common Future. Oxford and New York: Oxford University Press.

9

Biodiversity Net Gain as a Pathway to Sustainable Urban Land Development

Salman Makhld M. Aldlbahi

9.1 Introduction

Urban land development today is increasingly evaluated not only by its economic and structural outcomes but also by its capacity to protect and regenerate natural systems. Within the geospatial ecosystem, this shift is underpinned by the growing recognition that land and geospatial information must capture ecological dimensions, positioning geospatial ecosystems in analogy to natural ecosystems—both being complex, adaptive systems shaped by interactions, dependencies, and flows of resources and information. Biodiversity Net Gain (BNG) has emerged as a leading international framework, ensuring that land development results in quantifiable and lasting improvements in biodiversity compared to pre-development conditions. For rapidly expanding desert cities such as Riyadh, integrating BNG is not merely an environmental enhancement measure but a strategic application of land intelligence, responding to climate challenges, land degradation, and urban expansion (Figure 9.1).

A key element in achieving meaningful outcomes is based on the development of intelligent geospatial tools and workflows for integrating ecological, land, and infrastructural datasets. These enable multi-scale habitat mapping, baseline ecological assessment, spatial-temporal connectivity analysis, and continuous monitoring through geospatial digital twins, which are virtual, data-driven replicas of land and natural systems that integrate real-time information to support intelligent location-based analysis and predictive decision-making. By connecting BNG with the geospatial ecosystem of interoperable data, standards, and models, decision-makers can integrate land, ecological, and infrastructural dimensions into one framework. This supports transparent valuation of ecosystem services, ensures compliance with legal and planning requirements, and strengthens resilience against climate risks.

DOI: 10.1201/9781003638308-10

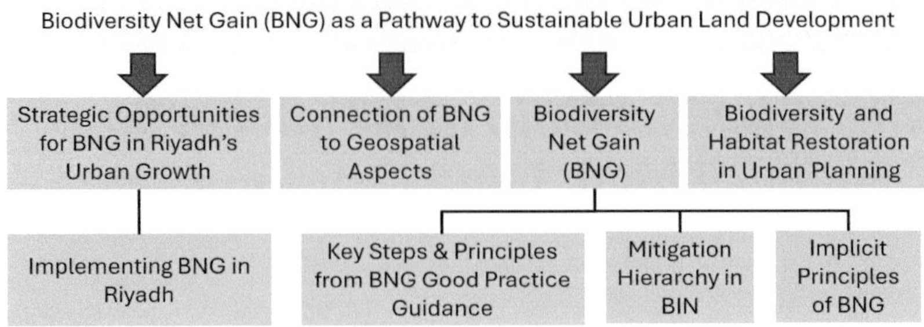

FIGURE 9.1
Methodology for BNG as a pathway to sustainable urban land development.

In the context of Saudi Vision 2030 and the Saudi Green Initiative, adopting BNG principles powered by geospatial planning and land intelligence represents more than an environmental safeguard. It redefines the relationship between urban growth and ecological resilience, demonstrating how future geospatial ecosystems can balance infrastructure expansion with natural regeneration, and ensuring that cities evolve as sustainable, data-driven landscapes.

9.2 Biodiversity and Habitat Restoration in Urban Planning

Restoring habitats in urban and peri-urban areas is essential for supporting biodiversity, particularly in human-altered environments (Figure 9.2). Studies show that such restoration enhances ecological connectivity and improves species survival potential (Breckheimer 2012). Urban biodiversity plays a growing role in supporting ecological resilience and climate adaptation. It also improves quality of life by providing key ecosystem services such as air and water purification, pollination, and mental and social well-being (Intergovernmental Panel on Climate Change 2022; Elmqvist et al. 2013). Habitat restoration is no longer solely an ecological goal but a planning necessity. In the face of increasing climate-related risks, it helps reconnect fragmented systems and supports species continuity within urban environments (Aronson et al. 2007). Integrating habitat restoration into urban planning enhances green infrastructure and requires a shift toward more holistic and sustainability-driven urban design, where biodiversity becomes a central planning element (Convention on Biological Diversity 2020). Embedding biodiversity into urban policy supports national and local climate goals, enhances urban adaptive capacity, and

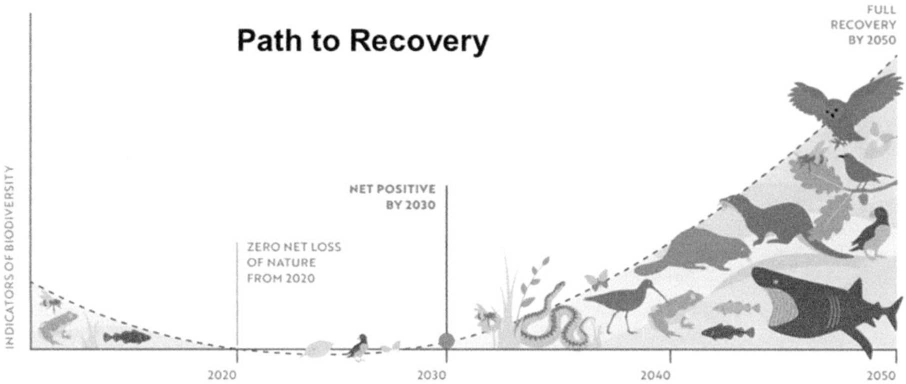

FIGURE 9.2
Thriving ecosystems: A green environment rich in biodiversity.

mitigates long-term risks. For rapidly growing cities like Riyadh, this represents a strategic path to balanced, sustainable development.

9.3 BNG

Although BNG is increasingly embedded in environmental and planning frameworks, its definition remains context-dependent, shaped by institutional and geographical variations. BNG policies have only been applied to a few local authorities in England, and only to major projects. Literature reflects diverse interpretations, from regulatory mechanisms to quantitative assessment tools to broader sustainability commitments. Key definitions to illustrate common principles and conceptual differences are:

- "Is intended to ensure developments are measurably positive for biodiversity. This involves creating or enhancing habitats to ensure that biodiversity is in a better state than it was before development and requires that gains are maintained for at least 30 years" (Parliamentary Office of Science and Technology 2023).
- "Ensures that the natural environment is in a measurably better state than beforehand by applying a structured approach to avoid, minimize and offset impacts, with the aim of delivering overall gains in biodiversity. It is measured using a biodiversity metric, and implemented through an evidence-based process" (WSP 2019).

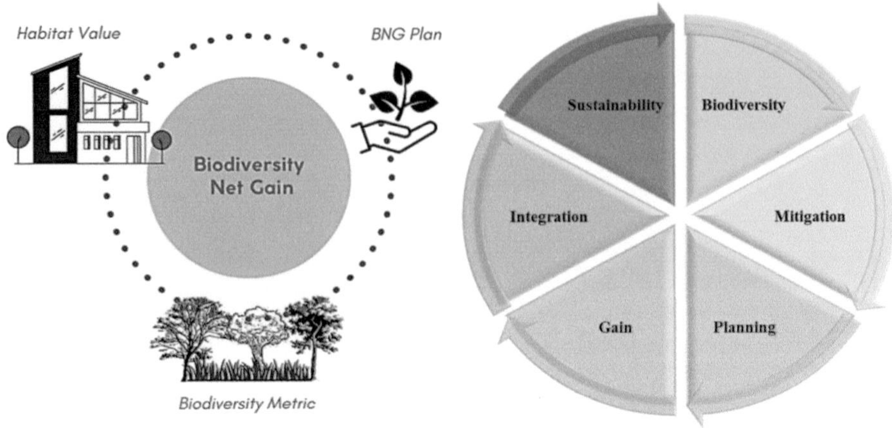

FIGURE 9.3
BNG definition.

- "Is the term used to describe an approach to development that leaves biodiversity in a better state than before, both in terms of species and habitats. It builds on existing policy and legislation, and relies on long-term commitment, ecological metrics, and good practice principles to achieve sustainable gains" (British Standards Institution 2022).

Through the synthesis of definitions found in relevant literature and policy frameworks, this study formulates a holistic definition of BNG that bridges theoretical insights with practical application. The definition is specifically adapted to suit the ecological and urban dynamics addressed in this research. BNG is a development approach that ensures biodiversity is measurably improved compared to the baseline state prior to development. It involves avoiding harm to irreplaceable habitats, applying the mitigation hierarchy, and delivering habitat improvements that exceed losses. This is achieved through long-term planning, monitoring, and the use of standardized metrics, and contributes to broader ecological and climate goals (Figure 9.3).

9.4 Mitigation Hierarchy in BNG

The mitigation hierarchy is a fundamental principle underpinning the effective implementation of BNG. It provides a structured, science-based framework for minimizing the adverse ecological impacts of development

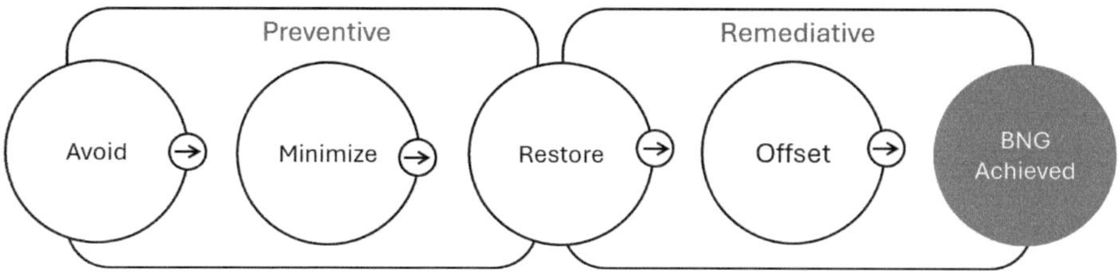

FIGURE 9.4
Four core stages of the mitigation hierarchy.

projects. The hierarchy follows four sequential steps: avoiding impacts entirely where possible, minimizing them when avoidance is not feasible, restoring affected habitats to their prior condition, and finally, offsetting any residual losses through compensatory measures that deliver equivalent or greater biodiversity value. This approach reflects a shift from reactive environmental management to proactive ecological integration within planning processes. Rather than addressing biodiversity as an afterthought, the mitigation hierarchy ensures that ecological considerations are embedded from the earliest stages of project design. It fosters a more responsible and forward-thinking model of urban development, aligned with global sustainability standards. The importance of this framework has been recognized and institutionalized in leading international guidance documents, such as the UK's BS 8683 standard and planning advisories by POST (Parliamentary Office of Science and Technology). These references emphasize the mitigation hierarchy as a practical and ethical pathway toward achieving a measurable net gain in biodiversity while balancing development needs. Figure 9.4 shows the four core stages of the mitigation hierarchy within the context of BNG implementation.

9.5 Key Steps and Principles from BNG Good Practice Guidance

Achieving BNG requires a systematic sequence of actions that aligns environmental planning with development objectives. This involves establishing a baseline, assessing impacts, applying the mitigation hierarchy, and ensuring long-term monitoring and community engagement. These principles and procedural steps are illustrated in Figure 9.5.

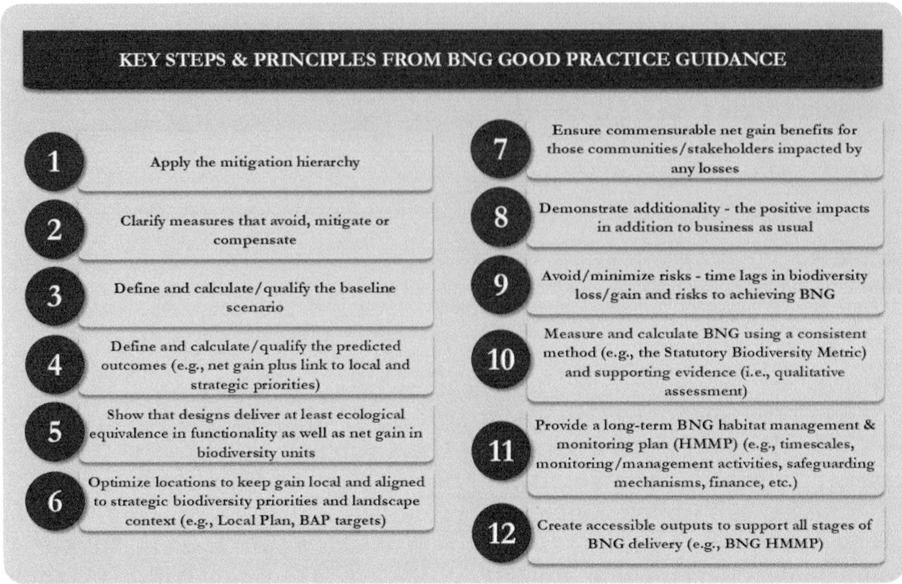

FIGURE 9.5
Key steps and principles from BNG good practice guidance (WSP 2021).

9.6 Implicit Principles of BNG

The effective delivery of BNG is underpinned by several core principles that, while not always stated explicitly in legislation, shape how the policy is applied in practice. These principles provide a framework for ensuring biodiversity outcomes are measurable, transparent, and integrated into development planning:

- Achieve Measurable Net Gain: "A biodiversity gain plan must demonstrate how at least a 10% gain in biodiversity units will be delivered over a 30-year period" (Wentworth 2024).
- Account for and Address Losses: The BNG process requires accounting for habitats lost on a development site and addressing these losses to ensure no net loss of biodiversity (Wentworth 2024).
- On-Site First, Then Off-Site: Gains are achieved by enhancing or creating the same types of habitats that will be lost, with a preference for on-site solutions, followed by a separate registered BNG site, and statutory credits as a last resort (Wentworth 2024).
- Consistent and Transparent Assessment: The BNG process relies on "biodiversity units" derived from data on habitat size, ecology,

and location, using metrics set out in legislation and provided in spreadsheets. This indicates a need for a standardized and repeatable approach to measurement (Wentworth 2024).

- Integrate Biodiversity into Planning: The introduction of BNG marks a significant shift in how nature is factored into local planning and development, transforming the natural environment from a passive responsibility to an active investment.
- Continuous Improvement and Learning: The policy includes opportunities for future legislative refinements, emphasizing information sharing and collaboration between larger and smaller developers to overcome.

9.7 Connection of BNG to Land Intelligence and Geospatial Ecosystem

Land intelligence and geospatial ecosystem play a pivotal role in the implementation of BNG, serving as essential tools throughout various stages of environmental assessment and planning. Within a geospatial ecosystem, Geographic Information System (GIS) data standards and interoperable models are a fundamental component of BNG evaluations, contributing to consistency and accuracy in the ecological, cadastral, and planning data used (EcoSpatial 2025). Additionally, specialized tools have been developed within GIS environments, such as the "Connect" tool on the ArcGIS 9.3 platform, to enable conservation practitioners to incorporate habitat connectivity considerations into land-use and property planning, environmental management, and decision-making processes (Breckheimer 2012).

From a land intelligence perspective, GIS is not only used to produce detailed mapping of vegetation types and habitats on development sites but also to integrate these ecological datasets with land parcels, property rights, and infrastructure models. This allows for an integrated analysis of ecological and geospatial characteristics in order to calculate biodiversity units with precision and link them to legal and planning frameworks (Wentworth 2024). Moreover, spatial analysis tools, such as "Connect Toolbox", help prioritize habitat restoration efforts within a geospatial ecosystem approach to land-use planning. Practical applications of this tool in areas such as Carvers Creek State Park have demonstrated how restoration efforts can be strategically directed to support ecological connectivity, while also aligning with land boundaries and urban development pressures, ensuring a balanced integration of natural and built environments (Breckheimer 2012)

9.8 Strategic Opportunities for BNG in Riyadh's Urban Growth

Urban development and expansion plans for Riyadh are largely informed by the Sustainable Planning Guidelines for Urban Growth in the Kingdom of Saudi Arabia (Ministry of Municipal and Rural Affairs and Housing 2024), which align with the broader objectives of Vision 2030. These guidelines emphasize energy efficiency, environmental sustainability, and livability (Ministry of Municipal and Rural Affairs and Housing 2024).

The Royal Commission for Riyadh City (RCRC) is among the key stakeholders contributing to the development of these planning standards (Sustainable Planning Guidelines for Urban Growth in KSA, 2024, Acknowledgment). As Saudi Arabia faces an accelerated pace of urbanization—with urban populations projected to reach 97.6% by 2030 and the total population expected to reach between 50 and 60 million—the need for extensive urban development and resource consumption is imminent (Ministry of Municipal and Rural Affairs and Housing 2024).

Vision 2030 aims to position at least three Saudi cities among the world's top 100 cities by enhancing quality of life, creating attractive living environments, and advancing environmental sustainability (Ministry of Municipal and Rural Affairs and Housing 2024). Early planning efforts in Riyadh highlight a focus on walkable public spaces, integrated urban development, mixed land use, and diversified housing types Historical data shows that a significant portion of Riyadh's urban expansion occurred on underutilized lands—36% of added built-up area from 1990 to 2000 and 25% from 2000 to 2013—indicating a strong tendency toward compact development (Ministry of Municipal and Rural Affairs and Housing 2024).

9.9 Implementing BNG in Riyadh

Although not explicitly mentioned, the principles of BNG can be strategically aligned with the environmental sustainability goals of Vision 2030 in Riyadh. In this context, the following aspects are critical:

- Leveraging Land Intelligence and Geospatial Ecosystem for Integrated Assessment: Given Riyadh's scale and rapid development, applying geospatial ecosystem tools and land intelligence workflows to map habitats, link them with land-use datasets, and assess their ecological quality and spatial connectivity will be critical to the successful implementation of BNG (EcoSpatial 2025; Breckheimer 2012).

- Integration into Urban Planning: BNG can be adopted as a mandatory component of major new developments or redevelopment projects in Riyadh. This would ensure that developments contribute positively to biodiversity rather than resulting in a net loss.
- Delivering Measurable Gains: Developers would be required to assess existing biodiversity units on-site and achieve measurable net gains (e.g., 10%) through habitat creation or enhancement (Wentworth 2024).
- On-Site Solutions: Creation or improvement of green spaces, urban parks, or natural habitats within new or redeveloped areas. Riyadh's emphasis on public space design and urban greening could benefit from BNG as a framework for quantifying the biodiversity benefits of such initiatives (Wentworth 2024; Ministry of Municipal and Rural Affairs and Housing 2024)
- Off-Site Solutions: Establishing registered BNG sites in less developed peripheral zones or designated conservation areas around Riyadh if on-site gains are unfeasible (Wentworth 2024).
- Restoring Degraded Areas: BNG could focus on rehabilitating brownfield sites or degraded urban land within Riyadh to restore and establish habitats. This aligns with the city's current strategy of promoting compact, infill development and could transform neglected areas into biodiversity assets (Nash et al. 2024; Ministry of Municipal and Rural Affairs and Housing 2024).
- Strategic Biodiversity Planning: BNG would support a citywide biodiversity strategy by identifying priority areas for habitat protection or enhancement, where interventions could yield the highest biodiversity returns (Wentworth 2024).
- Long-Term Management and Monitoring: BNG policies typically require long-term (e.g., 30-year) commitments to manage and monitor enhanced or created habitats, ensuring biodiversity gains are sustained within Riyadh's expanding urban footprint (Wentworth 2024).

9.10 Conclusion

Amid accelerating urban growth and rising environmental pressures, BNG serves as a strategic tool for restoring the balance between development and sustainability. Within a geospatial ecosystem, its integration into urban and land-use planning enables the integration of ecological data, land information, infrastructure networks, and socio-economic indicators into one interoperable framework. By applying land intelligence approaches, planners can not only mitigate habitat loss but salso simulate trade-offs, model connectivity across fragmented landscapes, and forecast long-term ecological outcomes under different development scenarios. This ensures that biodiversity considerations are embedded alongside land and property rights, zoning regulations, and infrastructure expansion.

Therefore, BNG becomes more than just a statutory necessity but also a land intelligence practice that benefits ecosystems, supports evidence-based decision-making, and boosts long-term urban resilience. Cities like Riyadh may connect development trajectories with sustainability goals by using continuous monitoring, geospatial analytics, and digital twins of land and ecological systems, guaranteeing that urban growth strengthens ecological resilience.

References

Aronson, Elliot, Timothy D. Wilson, and Robin M. Akert. 2007. *Social Psychology*. 6th ed. Pearson.

Breckheimer, Ian. 2012. *Mapping Habitat Quality in Conservation's Neglected Geography*. University of North Carolina.

British Standards Institution. 2022. *The Little Book of Biodiversity Net Gain*, BSI Knowledge

Convention on Biological Diversity. 2020. "Post-2020 Global Biodiversity Framework." Preprint.

EcoSpatial. 2025. *GIS Data Standards for Biodiversity Net Gain Assessments*.

Elmqvist, Thomas, Michail Fragkias, Julie Goodness, et al., eds. 2013. *Urbanization, Biodiversity and Ecosystem Services: Challenges and Opportunities: A Global Assessment*. Springer. http://link.springer.com/book/10.1007%2F978-94-007-7088-1.

Intergovernmental Panel on Climate Change. 2022. *Climate Change 2022: Impacts, Adaptation, and Vulnerability*. Cambridge University Press. https://www.ipcc.ch/report/ar6/wg2/.

Ministry of Municipal and Rural Affairs and Housing. 2024. *Sustainable Planning Guidelines for Urban Growth in KSA*. Kingdom of Saudi Arabia.

Nash, Caroline, Robert Vida, Mike Webb, and Stuart Connop. 2024. *Brownfield Habitats and Biodiversity Net Gain: Introduction and Context*. University of East London.

Parliamentary Office of Science, and Technology. 2023. *Biodiversity Net Gain (POST-PN-0728)*. UK Parliament.

Wentworth, James. 2024. *Delivering Biodiversity Net Gain: A Practical Overview*. UK Parliament.

WSP. 2019. *Biodiversity Net Gain White Paper: Building a Greener Future*. WSP.

WSP. 2021. *Brownfield Habitats and Biodiversity Net Gain: Introduction and Context*. Wsp.

10

Harnessing Land Information for Disaster Resilience and Recovery

Katie McDougall

10.1 Introduction: Land and Geospatial Intelligence as a Foundation for Disaster Resilience

Land administration systems and their supporting geospatial ecosystems are fundamental national resources, critical for a country's stability and economic growth by securing land and property rights (Williamson, 2002; Henssen, 1995). These systems, which include land registration, cadastral surveying and mapping, and land information systems, form the foundation for spatially enabling a society (Nasruddin and Rahman, 2006).

This chapter emphasizes the critical role of these systems in disaster risk management. Effective disaster management relies heavily on access to and utilization of land information, including cadastral data, parcel information, ownership details, land valuation, land-use information, and land development plans. The cadaster, as a core component, underpins land policies, management strategies, land markets, and disaster risk management practices (Steudler and Rajabifard, 2012).

Geospatial intelligence, integrated within land administration systems, provides essential details on land parcels, property rights, restrictions, and responsibilities, supporting activities that require accurate location information through technologies such as Geographic Information System and Spatial Data Infrastructure (FAO, 2002) (Figure 10.1).

The strategic integration of land administration and geospatial ecosystems is crucial for disaster resilience, enhancing pre- and post-disaster land management by facilitating secure land tenure, accurate land valuation, and effective land-use planning. This integration supports vulnerable groups by ensuring secure rights and access to land, which is vital for housing reconstruction, food security, and the recovery of production systems following a disaster (UN-HABITAT, 2010).

DOI: 10.1201/9781003638308-11

FIGURE 10.1
A Geospatial Intelligence System showing location information on high-risk bushfire areas in Victoria, Australia. (Image supplied by CSDILA.)

10.2 Enhancing Disaster Resilience Through Integrated Land and Geospatial Ecosystems

The integration of land administration systems with geospatial data significantly enhances disaster resilience, defined as the capacity of a system, community, or society to withstand, absorb, adapt to, and recover from hazards efficiently. In the context of land, resilience involves the ability of land and property, and the people-to-land relationships, to recover effectively, ensuring the continuity of land tenure, value, use, and development activities.

National land administration and geospatial data infrastructures are central to recording these people-to-land relationships. The stability of these systems and the tenure security they provide are essential for upholding land rights during and after disasters, protecting owners from land grabbing and ensuring their ability to return to their land (UNECE, 1996).

This integration offers benefits across the disaster management cycle:

- **Preparedness and Mitigation**: Comprehensive land data enables the identification of hazard-prone areas, risk assessment, and the development of mitigation strategies through informed land-use planning.
- **Early Warning and Response**: Access to up-to-date geospatial information during disasters supports effective emergency response by providing real-time data on affected areas, infrastructure damage, and population distribution to guide rescue operations, aid delivery, and evacuation planning (Mitchell et al., 2014a, 2014b).

- **Recovery and Reconstruction**: Accurate land records are crucial post-disaster for securing land tenure, facilitating aid and compensation, supporting infrastructure reconstruction, preventing land disputes, and enabling resilient "building back better" strategies.

10.3 Key Challenges to Land Resilience

Several challenges impede the development of land resilience. Inefficient land registration processes, outdated cadastres, a lack of trained professionals, and reliance on vulnerable paper-based records hinder effective land management and disaster preparedness. The absence of comprehensive, current, and accessible geospatial data, along with data fragmentation and incompatibility, limits effective disaster response. Weak governance, poor inter-agency coordination, and insufficient resources undermine land administration systems and information sharing. The lack of formal land tenure, particularly in developing regions, increases vulnerability to displacement and the loss of land rights following disasters. Furthermore, the vulnerability of land records, whether paper-based or digital, to physical destruction and cyber threats, poses a significant risk to land resilience.

10.4 Strategic Approaches to Enhance Land Resilience

To enhance land resilience, priority should be given to several strategic approaches. Investing in accurate and up-to-date cadastral systems is essential for recording land ownership and boundaries (Larsson, 1991). Formalizing land rights is necessary to strengthen land tenure security and ensure secure claims for all, including vulnerable groups. Modernizing land administration systems through investment in digital technologies is crucial for creating efficient, accessible, and resilient land information systems (UN-GGIM, 2023). This modernization includes digitizing existing records, establishing secure data storage and backup systems, developing interoperable systems for seamless data sharing between agencies, and implementing user-friendly interfaces to make land information easily accessible to stakeholders. Enhancing institutional capacity is also vital, which can be achieved through providing training and professional development for staff, improving governance and accountability, and fostering collaboration and coordination between different agencies involved in land management and disaster risk management. Promoting geospatial data sharing by establishing clear policies and protocols between government agencies, non-governmental organizations, and other stakeholders will ensure that information is available to those who need it during a disaster. Finally, ensuring the resilience of land records requires implementing measures to protect them from the physical impacts of disasters. This includes digitizing paper records and storing them in multiple secure locations, implementing robust data backup and recovery procedures for digital systems, and considering the use of distributed ledger technology, such as blockchain, to enhance the security and immutability of land records.

10.5 Global Frameworks Driving Land Resilience

Several global frameworks recognize the importance of land information in building resilience to disasters and promoting sustainable development. These include:

- **2030 Agenda for Sustainable Development**: The Sustainable Development Goals (SDGs) provide a global roadmap for achieving a more sustainable and equitable future. Several SDGs are directly related to land and disaster resilience, including:
  - SDG 1.5: By 2030, build the resilience of the poor and those in vulnerable situations and reduce their exposure and vulnerability to climate-related extreme events and other economic, social, and environmental shocks and disasters.
  - SDG 11.5: By 2030, significantly reduce the number of deaths and the number of people affected and substantially decrease the direct economic losses relative to global gross domestic product caused by disasters, including water-related disasters, with a focus on protecting the poor and people in vulnerable situations.
  - SDG 13: Take urgent action to combat climate change and its impacts.
  - SDG 15.3: By 2030, combat desertification, restore degraded land and soil, including land affected by desertification, drought, and floods, and strive to achieve a land degradation-neutral world.
- **Sendai Framework for Disaster Risk Reduction 2015–2030**: The Sendai Framework is a global agreement that aims to reduce disaster risk and losses. It emphasizes the importance of understanding disaster risk, strengthening disaster risk governance, investing in disaster risk reduction for resilience, and enhancing disaster preparedness for effective response and to "Build Back Better" in recovery, rehabilitation, and reconstruction.

These frameworks highlight the interconnectedness of land, disasters, and development and underscore the need for a holistic and integrated approach to building resilience.

10.6 Conclusion: The Imperative of Investing in Land Resilience

Land intelligence is a critical foundation for building resilience to disasters and ensuring sustainable development. By providing a clear understanding of who owns what, where, and how land is used, effective land administration systems and geospatial data enable societies to better prepare for, respond to, and recover from disasters.

Investing in land resilience is not simply a matter of technical improvement; it is a fundamental investment in the safety, security, and well-being of communities. By strengthening land tenure, modernizing land administration systems, and promoting the sharing of geospatial data, countries can significantly reduce their vulnerability to disasters and build a more resilient future for all.

References

FAO. (2002). *The role of land tenure in poverty alleviation and sustainable development.* Food and Agriculture Organization of the United Nations, Rome.

Henssen, J. (1995). "Basic principles of the main cadastral systems in the world." In *Proceedings of the one day seminar held during the Annual Meeting of Commission*, vol. 7.

Larsson, G. (1991). Land Registration and Cadastral Systems: Different Types and Purposes. *Land Use Policy*, 8, 3, 195–204.

Mitchell, D., Jacot des Combes, H., Myers, M., and McEvoy, D. (2014a). Addressing Land Issues in Disaster Risk Management in the Pacific Island Countries. *Land Tenure Journal*, 14, 1, 105–135. Retrieved from http://www.fao.org/3/a-i4712t.pdf

Mitchell, D., Myers, M., and Grant, D. (2014b), Land Valuation: A Key Tool for Disaster Risk Management. *Land Tenure Journal*, 14, 1, 37–69. Retrieved from https://www.researchgate.net/publication/279764940_Land_valuation_a_key_tool_for_disaster_risk_management

Nasruddin, A.A. and Rahman, A.A. (2006). Towards Developing a Sustainable Cadastral System in Malaysia. The FIG XXII Congress, April 16–21, Washington, DC, USA.

Steudler, D. and Rajabifard, A. (2012). Cadastral Systems and 3D Data in an Urban Environment. The FIG XXII Congress, April 16–21, Washington, DC, USA.

UN-GGIM. (2023). *United Nations Integrated Geospatial Information Framework – Part 1: Overarching Strategy*. 2023. https://ggim.un.org/UN-IGIF/documents/Part_1_UN-IGIF_Overarching_Strategy_Second_Edition_27Feb2023.pdf

UNECE. (1996). *Land administration guidelines: With special reference to countries in transition*. Economic Commission for Europe, United Nations, New York and Geneva.

UN-HABITAT. (2010). *Planning for Climate Change – A Strategic, Values-Based Approach for Urban Planners*. Cities and Climate Change Initiative. United Nations Human Settlements Programme.

Williamson, I.P. (2002). Best Practices for Land Administration Systems in Developed and Developing Countries. *Land Use Policy*, 19, 1, 55–71.

Part II

Digital Innovations Towards Geospatial Ecosystem and Land Intelligence

11

Ethical Considerations in Modernizing Land Administration Systems

Serene Ho

11.1 Introduction

This chapter aims to propose a basic framework that conceptualizes the links between digitalization and land administration practices, and how this may reconfigure epistemic and ontological assumptions – ways of knowing and doing – that may have ethical implications. Three basic universal ethical challenges are identified, being sovereignty, transparency and accountability. The intention is for this framework to stimulate further conversation and thinking about contending and responding to emerging ethical issues due to modernizing land administration.

11.2 Modernization Theory

Modernization in this chapter references those theories and ideas that attempt to understand how to improve development outcomes in poorer nations. Although these ideas took hold globally for policy-making following the Cold War, its fundamental principle is grounded in a binary archetype: traditional vs. modern societies (Gwynne 2009). Today, modernization theory is commonly attributed to Walt Rostow, who theorized that all nations inevitably move through five stages of development that start from agrarian-based 'traditional' economies and maturing through the various stages is characterized by increasing and prolific use of new technologies. This in turn produces a corollary shift in the nature of the economy: from a dominance of primary sectors towards a focus on tertiary and quaternary sectors that are service- and knowledge-based (Rostow 1991). Although Rostow's model is economic in its orientation, there are social, environmental and political implications of the economic transitions needed at each step.

DOI: 10.1201/9781003638308-13

11.3 The Modernizing Agenda in Land Administration

Such economy-wide transformations would undoubtedly impact land administration (alongside other public administration modernization initiatives) given the fundamental role of land in development. There are two aspects to this: ideological and practical. In land administration, 'modern' ideas around land and property rights are intrinsically linked to the concept of private ownership that have superseded traditional monarchy- or feudal-based systems of land rights (Sato 2018). Indeed, private property rights have only emerged over the last two to three centuries to become the dominant tenure construct in land administration (Kain and Baigent 1992). This shift has various motivations, including but not limited to:

- Colonial conquests from around the fifteenth century that dispossessed Indigenous territories and overrode longstanding local tenure systems to enable the amassing of territory and wealth. These 'settler models' (Blomley 2022) sought to territorialize land, i.e., treating land as a material object, through violent land acquisitions, and which operationalized 'regimes of exclusion' (Li 2014). These have had longstanding implications not just around how access to land is structured, governed and administered, but also the emplacement of racial, religious and economic hierarchies, many of which still endure (Bhandar 2018).
- Libertarian movements in the nineteenth century focused on the rights of individuals and their freedoms as the basis of modern societies (Vossen 2017). Property rights are a fundamental tenet of libertarianism although there are tensions around how to balance individual freedoms that private property offer with the inevitable restrictions this imposes on others' freedom to access the same space (e.g., see Peirce 2001). Where land and real property is concerned, this raises questions around what can and should be considered just and equitable under private ownership models when in pursuit of public interests and creating public value.
- The rise and dominance of new institutional economic thinking in the twentieth century. This argued and demonstrated that a model of private property rights that is effectively protected and enforced can significantly reduce conflicts over resources with macro-economic benefits to nations (North 1990). These ideas influenced international development organizations like the World Bank, who promoted 'law and justice' land reforms as part of structural modernization programmes of less-developed economies (Legal Vice Presidency 2003). This precipitated the introduction of private ownership models in many countries around the world.

- The rise of scientific and technological know-how, and the tendency for ideas to be exported from the west to 'the rest'. As developed countries mature faster and have transformed their economies to service- and knowledge-based sectors, these new sectors have inevitably found markets in less-developed countries, promulgating western science and technological expertise, business and service models, but also power and ideological biases that become established and perpetuated (Díaz Andrade and Urquhart 2012; Escobar 2011).

The *ideas* around land and property that are reflected in the above, alongside other contextual and political economic factors, have worked to crystallize land relations in *material* ways, overshadowing the social *relationships* that lie at the heart of land tenure systems. This material focus, centred on the relationship between a person and a property object (e.g., for revenue purposes), has had other material consequences primarily that the proliferation of the private ownership model has created new forms of land and property information and exponentially increased the scale of information being transacted. This has practical implications especially when paper has been the dominant material form of documentation.

The production, use and management of land and property information has changed in lockstep with broader technological developments around information management. Widespread internet penetration, proliferation of digital technologies and a general lowering of costs around technology and telecommunications infrastructure have had profound implications for everyday users, but also for governments and the task of governing (including land). The concept of e-government emerged in the early 2000s, recognizing that technology offered opportunities for improving the effectiveness and efficiency of service delivery, where 'modern' public services are imagined as 'integrated and seamless', potentially transforming state–society relations into one that is more collaborative rather than hierarchical (Silcock 2001, 88).

The modernization agenda in land administration sits at the intersection of these developments. On one hand, it is pulled by the currents of broader digital government agendas; on the other, it is pushed by the technical advancements from land-, survey-, and ICT-related disciplines. Modernizing land administration can arguably be conceived as an instance where globalization, capitalism and masculine technology sectors are intensely converging (Ho, Tanyag, and Scalise 2023). This may not be creating the types of public value that modernization is imagined to bring, particularly in non-western, less-developed settings (Ho, Choudhury, and Joshi 2024).

This chapter is interested in the ethical considerations of modernizing land administration because the rate of technological advancement is outpacing our ability to contend with these. It structures these considerations in a simple framework, as shown in Figure 11.1, organized by the three key areas that

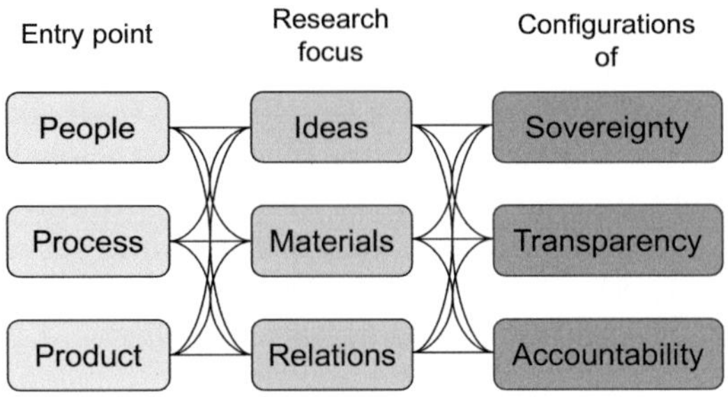

FIGURE 11.1
Proposed research framework for ethical modernization of land administration.

land administration is concerned with: **people**, and managing people–land relations; **process**, which is the primary modality for land administration and **product**, those textual and spatial records that constitutes the currency of land administration. These are used as the entry point for designing research to investigate how technology might be shaping **ideas, materials** and **relationships** in land administration. The framework concentrates initial ethical considerations as to how these translate into configurations of **sovereignty, transparency** and **accountability** being experienced on the ground. The framework is discussed briefly in the following section.

11.4 Ethical Concerns in Modernizing Land Administration

11.4.1 Focusing on People

Many modernization projects in land now emphasize inclusivity, especially communities whose rights to land (and land-related resources) have traditionally not been dealt with well – if at all – by state land administration systems. Groups such as informal/extralegal settlers, customary and Indigenous land holders, collective groups and even women, are some of the common beneficiaries of land projects today where new data models are better able to capture and reflect other ideas of land relations that are not privatized (e.g., communally held land). Indeed, the use of digital technologies is often framed as being in support of participatory or inclusion objectives, where a key value proposition could be that ensuring the inclusion of all owners in digital land records can reduce land fraud and disputes. For example, Cadasta is an organization involved in multiple land projects using digital technologies to improve inclusion of vulnerable communities in land documentation (www.cadasta.org).

Yet, there is emerging research showing how the design of these projects may not result in transparent datasets or accountable processes of data collection and production (Ho, Choudhury, and Joshi 2023). Further, with the production of large volumes of personal data in land administration, a primary ethical consideration must be digital sovereignty. Digital sovereignty refers to the right of individuals, organizations and states to be able to control not just digital data related to them, but also the technologies, services and infrastructure related to the production, management and use of the data and this needs to be articulated from the scale of the individual to the scale of the state.

11.4.2 Focusing on Process

Any discussion of digitalization – which is now synonymous with modernization – inevitably leads to a discussion about digital exclusion, which is treated as a detrimental phenomenon because it tends to be linked to inequality. For example, the (in)ability to participate in everyday activities that are increasingly transacted over the internet, which in turn requires appropriate technologies, connectivity and digital literacy (e.g., see commentary by Hernandez and Faith 2022). Digital exclusion can also impact the completeness (and hence, accuracy and quality) of datasets (i.e., the product). This reflects the normativity of the benefits of digital *inclusion*, mainly that we now think of inclusion as reflecting reduced inequality. Although often associated with the Global South, the link between digital inclusion/exclusion and inequality is a universal theme.

Digitalization enables integration of land information with other datasets for decision-making. To do this, digital infrastructures become essential. Today, these infrastructures present a material way of not just efficiently disseminating and enabling the use of data, they are also conduits for distributing ideas and cultural practices (Koch 2017). The modernization of land administration systems already has its (partial) epistemic origins in western ideas of private property rights which has been used to reconfigure land relations around the world with devastatingly violent consequences for some communities. The work of Science, Technology and Society (STS) scholars draw our attention to similar ethical issues related to the export of technologies from the west (e.g., see Jasanoff 2010) and there should also be critical attention paid to what ideas and assumptions are being diffused through hardware, software and other technologies being mainstreamed in digital land reforms. The ethical issues related to this have been a focus of STS (e.g., see Jasanoff 2010) and ICT4D scholars (e.g., see Heeks 2022); however, these have rarely been dealt with in land administration discourse where there is an ongoing lack of Global South – or rather, 'non-West' (Subramanian and Hassan 2024) – technologies penetrating the field. Further, the model of 'modernization' itself that is being upheld by donors and other funders of land reform projects needs to be critiqued.

11.4.3 Focusing on Products

Many land administration modernization projects often emphasize the production of new digital land data products that accurately reflect reality (who owns what), as well as being spatially accurate (where the holding is). Many land-related disputes stem from inaccurate land information (e.g., Daksh India (2016) found that two-thirds of the cases in the civil courts in India are land related) but as well, inaccurate land information inhibits the ability of states to properly plan for development. As mentioned above, inclusion is also now often a key objective.

More recently, scholars are starting to think about how such inclusion may be creating new inequalities for those data is captured within digital systems. Heeks (2022) has defined the concept of 'adverse digital incorporation' which refers to the phenomenon of some within a digital system being able to extract more value from digital resource than another group. In the context of land administration, a key example of adverse digital incorporation is the ethics around how digital land information is being used to inform carbon investments resulting in land grabbing at significant scales (GRAIN 2024; TMG Research and Robert Bosch Stiftung 2025). There are also longstanding issues around creating digital artefacts that are related to personal data and how this may be exposed, either intentionally or otherwise.

11.5 Conclusion

Land modernization projects are now underway in many parts of the world and there is now significant demand for digital land information and processes. This march towards 'modernization' is occurring rapidly and often, data, software, systems and new techniques are reconfiguring relationships and creating new or further embedding power asymmetries. Studies from fields such as science and technology studies and sociology of quantification have shown that digitalization comes with ethical challenges that have been dealt with a limited way in the field and discipline of land administration.

This chapter contributes to this gap by proposing an ethical framework to respond to three ethical challenges around sovereignty, transparency and accountability that likely has universal resonance. The framework is by no means a comprehensive one: it is intended to serve as a conversation starter and a basic research agenda to considering how and where attention should be focused to wrestle with the emerging ethical issues resulting from modernizing land administration.

References

Bhandar, Brenna. 2018. 'Colonial Lives of Property: Law, Land, and Racial Regimes of Ownership'. In *Colonial Lives of Property*. Duke University Press. https://doi.org/10.1515/9780822371571.

Blomley, Nicholas. 2022. 'Territory and Property'. In *The Routledge Handbook of Property, Law and Society*, edited by Nicole Graham, Margaret Davies, and Lee Godden, 1st ed., 339–48. London: Routledge. https://doi.org/10.4324/9781003139614-8.

Daksh India. 2016. 'Access to Justice Survey 2015–16'. https://dakshindia.org/wp-content/uploads/2016/05/Daksh-access-to-justice-survey.pdf.

Díaz Andrade, Antonio, and Cathy Urquhart. 2012. 'Unveiling the Modernity Bias: A Critical Examination of the Politics of ICT4D'. *Information Technology for Development* 18 (4): 281–92.

Escobar, Arturo. 2011. *Encountering Development: The Making and Unmaking of the Third World*. Princeton, New Jersey, USA: Princeton University Press.

GRAIN. 2024. 'From Land Grabbers to Carbon Cowboys: A New Scramble for Community Lands Takes Off'. https://grain.org/system/articles/pdfs/000/007/190/original/Carbon%20deals%2004.pdf?1726591548.

Gwynne, Robert N. 2009. 'Modernization Theory'. In *International Encyclopedia of Human Geography*, edited by Audrey Kobayashi, 2nd ed., 163–67. Oxford: Elsevier. https://doi.org/10.1016/B978-0-08-102295-5.10107-6.

Heeks, Richard. 2022. 'Digital Inequality beyond the Digital Divide: Conceptualizing Adverse Digital Incorporation in the Global South'. *Information Technology for Development* 28 (4): 688–704. https://doi.org/10.1080/02681102.2022.2068492.

Hernandez, Kevin, and Dr Becky Faith. 2022. 'Measuring Digital Exclusion'. *Digit*. https://digit-research.org/data_commentaries/measuring-digital-exclusion/.

Ho, Serene, Maria Tanyag, and Elisa Scalise. 2023. 'Women's Land Rights, Gendered Epistemic Tensions, and the Need for a Feminist Approach to Land Administration'. *Land Use Policy* 132 (September): 106841. https://doi.org/10.1016/j.landusepol.2023.106841.

Ho, Serene, Pranab R. Choudhury, and Richa Joshi. 2023. 'Community Participation for Inclusive Land Administration: A Case Study of the Odisha Urban Slum Formalization Project'. *Land Use Policy* 125 (February): 106457. https://doi.org/10.1016/j.landusepol.2022.106457.

Ho, Serene, Pranab Choudhury, and Richa Joshi. 2024. 'Drone Technology, Rapid Titling, and Value Creation: Early Insights from India's SVAMITVA Scheme'. In *Geospatial Science for Smart Land Management: An Asian Context*, edited by Walter De Vries, Iwan Rudiarto, and Milinda Piyasena. CRC Press. https://www.routledge.com/Geospatial-Science-for-Smart-Land-Management-An-Asian-Context/DeVries-Rudiarto-Piyasena/p/book/9781032393896.

Jasanoff, Sheila, ed. 2010. *States of Knowledge: The Co-Production of Science and Social Order*. Transferred to digital print. International Library of Sociology. London: Routledge.

Kain, Roger and Elizabeth Baigent. 1992. *The Cadastral Map in the Service of the State: A History of Property Mapping*. Chicago, IL: University of Chicago Press.

Koch, Gertraud. 2017. 'Ethnography of Digital Infrastructures'. In *Digitisation*, edited by Gertraud Koch. London: Routledge.

Legal Vice Presidency. 2003. 'Legal and Judicial Reform: Strategic Directions'. Text/HTML 26916. Washington, DC: The World Bank. https://documents.worldbank.org/en/publication/documents-reports/documentdetail/en/218071468779992785.

Li, Tania Murray. 2014. 'What Is Land? Assembling a Resource for Global Investment'. *Transactions of the Institute of British Geographers* 39 (4): 589–602. https://doi.org/10.1111/tran.12065.

North, Douglass. 1990. *Institutions, Institutional Change and Economic Performance*. Cambridge, MA: Cambridge University Press.

Peirce, Michael. 2001. 'Why Libertarians Should Reject Full Private Property Rights'. *The Philosophical Forum* 32 (1): 25–52. https://doi.org/10.1111/0031-806X.00050.

Rostow, Walter. 1991. *The Stages of Economic Growth: A Non-Communist Manifesto*. 3rd ed. Cambridge University Press. https://doi.org/10.1017/CBO9780511625824.

Sato, Hajime. 2018. 'The Emergence of "Modern" Ownership Rights Rather than Property Rights'. *Journal of Economic Issues* 52 (3): 676–93. https://doi.org/10.1080/00213624.2018.1495989.

Silcock, Rachel. 2001. 'What Is E-Government'. *Parliamentary Affairs* 54 (1): 88–101. https://doi.org/10.1093/pa/54.1.88.

Subramanian, Vidya, and Yousif Hassan. 2024. 'Peripheral Visions: STS and Digitalisation in the Non-West'. *Science, Technology and Society* 29 (4): 505–16. https://doi.org/10.1177/09717218241282531.

TMG Research and Robert Bosch Stiftung. 2025. 'Net Zero & Land Rights'. TMG. https://netzerolandrights.com/api/media/file/TMG_Climate%2BLandrights2025.pdf.

Vossen, Bas Van Der. 2017. 'Libertarianism'. In *Oxford Research Encyclopedia of Politics*. https://doi.org/10.1093/acrefore/9780190228637.013.86.

12

Copyright in Digital Land Administration: Challenges and Opportunities in Survey Data Management

Armin Agha Karimi, Glenn Campbell, and Zahra Gharineiat

12.1 Introduction

Surveyors currently use advanced technologies to collect and analyze massive volumes of data in order to provide their customers with services and products. Most of these services and products include digital survey data in various forms, such as Computer-Aided Design (CAD) files, point clouds, or 3D digital models. New technologies and the digital landscape have impacted land administration in different ways, which can be categorized into three broad categories: data collection, data modeling, and data management. Each category may have different implications from an intellectual property (IP) perspective, and this chapter will try to cover these perspectives.

Data-capturing technologies can play a significant role in creating accurate and up-to-date digital survey plans, enabling surveyors to collect, model, and represent data related to land parcels and legal boundaries. Digital survey data is captured by a variety of surveying and scanning methods. Although traditionally theodolites or total stations have been the main equipment to collect and set out cadastral survey data, laser scanning has drawn more attention recently because of its accuracy and capability in capturing details. However, this is more relevant to jurisdictions where property boundaries are defined by building walls rather than by monument-based cadastral systems, as is common in Australia. In regard to data modeling, various models are used to describe the geometry and topology of survey and land parcel data. More recently, the emergence of digital twins as well as augmented and virtual realities has created new opportunities and avenues for communicating and disseminating digital survey data for many stakeholders involved in digital survey lodgments (McDuff, Karimi, and Gharineiat 2025).

Data management is intertwined with data modeling in digital land administration, and as the volume of surveying data grows and technology

DOI: 10.1201/9781003638308-14

continues to evolve, new methods for processing and managing digital survey data are emerging. These methods have the potential to change the way cadastral survey data is managed and analyzed, and to improve the accuracy, efficiency, and speed of data processing. The emerging technologies of artificial intelligence (AI), blockchain, cloud computing, and Internet of Things (IoT) are transforming the way digital cadastral survey data is processed and managed. These technologies can improve the accuracy, efficiency, and speed of data processing, reduce the risk of fraud and disputes, and facilitate collaboration and data sharing among stakeholders. As these technologies continue to evolve, it is likely that they will play an important role in digital land administration.

Surveyors create IP in different forms, including survey plans. A common form of IP is copyright, which protects the work of surveyors in preparing survey plans. Surveyors hold the IP for both spatial data and legal data in survey plans in some jurisdictions, for instance, in Australia (Campbell, Kariyawasam, and Karimi 2024). With the ever-changing digital landscape and fast-evolving technologies in capturing, modeling, and managing survey data, the necessity of IP protection becomes more crucial in the digital land administration systems. Because of the technical simplicity with which digital survey plans, for example, may be copied, new mechanisms for preserving the interests and IP of surveyors should be developed. Kariyawasam, Palliyaarachchi, and Campbell (2024) have discussed these aspects in detail from different perspectives. For simplicity, the spatial data, models, and management methods will be referred to as survey data for the rest of this chapter. Additionally, the *Copyright Act* in this chapter mainly refers to the Australian Commonwealth's *Copyright Act 1968*. However, it can be extended to other jurisdictions broadly.

After this brief introduction, this chapter will address the global conventions and regulations in protecting the IP of different products and their applicability to survey data. It will then explore the digitalization and originality of survey maps as one of the end products and their implications in terms of copyright. This will be followed by scrutinizing the copyright of new data-capturing technologies and the AI factor. Survey and spatial data protection methods will be analyzed from a legal perspective, and new technologies for this purpose will be investigated briefly at the end.

12.2 Implications of Survey Data in International Copyright Conventions

The Berne Convention, the oldest international copyright treaty, provides comprehensive protection for literary and artistic works across various domains, including literary, scientific, and artistic productions (Ricketson and Ginsburg 2022). It states that 'literary and artistic works', which are protected by copyright, encompass a broad spectrum of expressions. Books, pamphlets, lectures, drawings, paintings, and photographs are only a few of these expressions. Under the Convention, both the original literary or artistic work and any derivative works are protected, provided that the modifications do not infringe upon the copyright of the original work. It extends copyright protection to authors' collections of literary and artistic works, which are considered intellectual creations due to their selection and arrangement. The Berne Convention does not explicitly reference survey plans or spatial data; however, survey plans and processed survey data meet the originality criteria by inference. For instance, a map or survey plan with distinctive design and originality would be entitled to copyright protection, granting the creator exclusive rights to authorize reproduction.

To address the challenges posed by the digital age, a protocol was added to the Berne Convention in the World Intellectual Property Organization (WIPO) Copyright Treaty in 1996 (Bontekoe and Wallot 2019). This treaty acknowledges the impact of technological advancements on the production and use of literary and artistic works. Article 5 of the WIPO Copyright Treaty stipulates that compilations of data or other material are protected if they constitute intellectual creations due to their selection or arrangement. Creators are endowed with exclusive rights to make their data publicly available, sell or transfer ownership, and communicate it to the public. However, this protection does not extend to the data or material itself. The Agreement on Trade-Related Aspects of Intellectual Property Rights (TRIPS) further clarifies that copyright protection applies to the expression of ideas, not the ideas themselves (Dhanjee and De Chazournes 1990).

Directive 96/9/EC of the European Parliament and the Council provides guidelines for protecting databases in Europe (SCOPE Chapter I 1996). It defines databases to include literary, artistic, and musical works, as well as other materials like texts, sounds, images, numbers, facts, and data organized systematically and accessible individually by electronic means. The Directive extends protection to both electronic and non-electronic databases and mandates that users avoid actions that harm the database's rightful owner.

In addition to these international conventions, different countries adopted a broad range of IP measures concerning the survey data. It has been shown that surveyors retain the IP of cadastral plans in Australia even if their creation is required by land registration legislation. In the case of Copyright Agency Ltd v State of New South Wales (CAL case), the Australian High Court determined that survey plans are classified as 'artistic works' under section 10(1) of the *Copyright Act*. This classification means that surveyors, as independent creators, are typically the first owners of the copyright in these artistic works. They hold exclusive rights to reproduce the survey plans in a material form and to communicate them to the public. This legal recognition underscores the importance of protecting the IP of surveyors, ensuring that surveyors' rights are upheld (Kariyawasam, Palliyaarachchi, and Campbell 2024).

In summary, copyright protection extends to creative and original expressions, not merely ideas or raw data. The digitalization of data presents significant challenges in protecting the IP of spatial databases, necessitating a clear determination of the originality required for their protection.

12.3 Artistic Work and Originality in New Data Collection, Processing and Presentation Methods

For a work to qualify for copyright protection, it must be original, meaning it has been created with the author's skill, labor, and intellectual effort and be fixed in some material form. The level of skill and labor required for a survey plan to be considered original depends on the specifics of each case. Generally, design plans, subdivision plans, reports, letters, opinions, and field notes are protected by copyright. Literary works include a map, chart, plan, table, and compilation, thus, qualify for copyright protection. Derivative works, which reproduce or adapt existing works, are protected only for the new material contributed by the author, not the pre-existing material. This ensures that survey plans, as original artistic or literary works, are protected under copyright law, safeguarding the IP of surveyors in the evolving landscape of spatial data and land administration (Kariyawasam, Palliyaarachchi, and Campbell 2024).

With the rise of digital technologies, the use of satellites and Unmanned Aerial Vehicles (UAVs) to capture spatial imagery has become widespread. Unlike traditional photography, which requires manual adjustment of camera settings, these technologies can autonomously capture high-resolution images (Cienciała, Sobolewska-Mikulska, and Sobura 2021). This raises two legal questions, the first whether such imagery qualifies as 'photographs' under copyright law, and second, who, if anyone, is the rightful author under the *Copyright Act*, in Australia for example, a photograph includes any image produced by photography or similar processes. This definition encompasses satellite and UAV-captured spatial images, as they involve a photographic process and result in visual outputs. These images are thus considered 'photographs' and may be protected as artistic works under various copyright laws. To be copyrightable, however, a photograph must originate from an author. This encompasses satellite and UAV-derived outputs, as they result from a photographic process that produces visual imagery. Even though these images are captured autonomously, they may still be regarded as artistic works eligible for copyright protection, provided there is evidence of human authorship. In the context of photogrammetry, this authorship is typically attributed to the spatial professional or surveyor who is responsible for mission planning, parameter configuration, ground control points setup, and data quality assurance.

Courts have affirmed that copyright can apply to digital formats, meaning orthomosaics, point clouds, and derived models are potentially protected IP. The creator retains rights over the digital files, even where physical copies or rendered outputs are shared with clients. However, legal ambiguity may arise regarding whether outputs produced through highly automated workflows meet the threshold of originality. In photogrammetric practice, originality, and authorship are often evident in two key stages: (1) Pre-capture planning, which includes selecting the area of interest, setting flight paths and overlap percentages, determining ground sampling distance (GSD), and configuring image and sensor settings; (2) Post-processing, where professionals apply judgment in aerial triangulation, generating 3D point clouds, orthophotos, textured meshes, and digital terrain models (DTMs), often followed by cartographic enhancement or data integration. These stages demonstrate that, despite automation, significant intellectual skill and discretion are still exercised, supporting a case for copyrightability.

Similar principles apply to Light Detection and Ranging (LiDAR) systems, which, although not producing conventional photographic images, generate detailed 3D spatial representations. Outputs such as point clouds, digital surface models (DSMs), and mesh files in CAD or Building Information Model (BIM) formats are the result of deliberate decisions in scan configuration, spatial resolution, and post-processing workflows (He and Li 2020). The process of configuring scan settings, determining scan coverage, and post-processing outputs involves skill and judgment. These activities potentially satisfy the originality requirement under copyright law. Therefore, the outputs of laser scanning and LiDAR systems, such as 3D models or processed point clouds (e.g., in mesh, CAD, or BIM-compatible formats), may similarly be protected, provided that human agents have exercised sufficient creative control in their generation and refinement. Like with satellite or UAV imagery, the copyright in these outputs would reside with the individual or entity responsible for the intellectual input by the surveyor or geospatial professional.

Digital maps, which visually represent spatial data on electronic screens, raise questions about their eligibility for copyright protection as artistic works (Menell 2002). The primary consideration is whether digital maps qualify as a type of work protected under the *Copyright Act*. The focus here is on the digital map itself, not the underlying code that generates it, making the 'artistic work' category relevant. Digital maps generally fit into the category of 'drawing' in this context. The challenge lies in determining whether digital maps, like those displayed on Google Maps, qualify as artistic works. Images on electronic screens can be considered artistic works, however, digital maps present additional complications due to their dynamic nature and the need to separate the output from the computer program. Courts require 'fixation' in some material form to recognize artistic work. The ephemeral nature of screen displays poses a challenge for copyright protection. It may not be problematic for static digital maps like JPGs or PDFs, but it raises issues for real-time or interactive maps, which are constantly updated. Another issue is whether a digital map, as the output of a computer program, can be separated from the program that generates it. This involves distinguishing between the underlying code and the output image. Digital maps are based on electronic databases containing geographical coordinates. Spatial databases fall within the scope of 'compilations', 'databases', and computer programs which are recognized as literary works under section 10 of the *Copyright Act*, in the case of Australia. Computer programs are protected as literary works, including subsets of instructions and statements. Whether a digital map constitutes artistic or literary work is a complex legal issue, dependent on factors such as creativity, purpose, and originality. This complexity underscores the importance of clear guidelines and protections for digital maps and spatial data in the evolving landscape of land administration.

Copyright protection in Australia, and similarly in jurisdictions like the US and EU, applies only to works demonstrating originality and human authorship. This poses challenges for spatial data products generated through AI or automated digital platforms, where direct human input may be minimal or absent. Under Australian law, 'original works' under Part III of the *Copyright Act* require evidence of independent intellectual effort by a human author, while Part IV addresses products resulting from investment or effort that may lack creativity. For spatial data, particularly those derived from AI-driven or automated survey systems, establishing human authorship is complex. Courts have emphasized that the key determinant of copyrightability is not the process, but the intellectual effort involved in the selection, organization, and presentation of data. This distinction is critical for AI-generated spatial data: while the surveyor or developer may design the system, the system's independent operation raises questions about authorship of the resulting output. Nonetheless, if surveyors actively determine key parameters similar to UAV photogrammetry, they may demonstrate the required level of intellectual skill and judgment. In such cases, surveyors are not merely passive operators but apply professional knowledge in determining where and how to collect spatial data, thereby contributing creatively to the final work. This distinction is crucial when dealing with cadastral maps or survey plans. The spatial data points themselves may be seen as raw data. However, the surveyor's decision-making process reflects professional expertise and judgment. These activities may justify recognition of the surveyor as the author under copyright law. In conclusion, for AI-assisted or automated spatial data products to be copyrightable, courts will consider whether the human contributor exercised sufficient skill, labor, and judgment in the creation or arrangement of the data. While automated systems may collect or process data, copyright protection may still apply where surveyors make substantive, creative contributions in planning, input, or interpretation (Kariyawasam, Palliyaarachchi, and Campbell 2024).

In summary, survey data produced by new technologies in collection, processing, and presentation can be eligible for copyright protection. The critical issue is not the automation itself, but whether the human operator has exercised professional skill and intellectual effort in guiding or interpreting the data collection and output process.

12.4 Protecting Survey Data in the Digital Era: Legal and Technical Perspectives

While IP laws, particularly copyright, are frequently used to protect survey data, they offer limited coverage. Copyright does not extend to raw data or unoriginal compilations and only protects spatial products that show sufficient originality and human intellectual effort. Given that modern survey data now often involves automated processes and may include AI in the near future, they frequently fall outside the scope of this protection. To address this gap, unfair competition law can serve as an alternative legal framework (Henning-Bodewig 2006). It targets commercial behaviors that violate ethical standards, such as unauthorized use or replication of valuable spatial datasets, by preventing entities from unfairly capitalizing on another's investments. In the digital age, the ability to rapidly copy and redistribute survey data without compensation undermines the incentive for creators to invest in data generation and maintenance. Unfair competition principles, therefore, can be used to uphold fairness by discouraging free-riding and protecting the commercial interests of those who invest in costly survey data acquisition, processing, and presentation. Survey data are often expensive and technically demanding to collect and process. These datasets support essential applications such as land administration and infrastructure planning. As such, protecting their competitive value is critical. Unfair competition law can deter unauthorized commercial exploitation and preserve the economic value of the data creators' efforts, even when traditional copyright cannot. Another legal avenue is protection through trade secrets or confidential information. Where copyright is uncertain, surveyors may protect their work by keeping it confidential. They can therefore consider contractual mechanisms, such as non-disclosure agreements (NDAs) or confidentiality clauses, when sharing spatial data with third parties (Tayyeb 2021). These legal instruments allow surveyors to define who owns the data, restricts redistribution, and clarifies the obligation to keep the information confidential. However, courts have also made it clear that once spatial data enters the public domain, either through government submissions or statutory obligations, its confidentiality is lost.

Blockchain, smart contracts, and technical protection measures (TPMs) offer emerging opportunities for surveyors and geospatial professionals to protect, license, and manage spatial data more effectively (Rawal, Amaduzzi, and Seedorf 2024). As spatial data becomes more digitized and widely shared, traditional IP frameworks, such as copyright, often fall short, as mentioned. Blockchain provides a decentralized ledger system that can authenticate authorship through timestamping, support the tokenization of spatial outputs, and track the origin and use of datasets in an immutable and transparent manner. This is particularly useful in evidencing the intellectual contribution of surveyors, especially where originality is disputed or where authorship must be established in collaborative or AI-assisted workflows. Blockchain can also act as a governance tool for copyright, enhancing digital rights management (DRM) by enabling individualized licensing, access control, and real-time monitoring of spatial data usage. These capabilities support the enforcement of licensing terms even in contexts where conventional copyright protection is unavailable or uncertain, such as with raw coordinates or point cloud data. Smart contracts further extend this functionality by allowing automated, self-executing agreements that facilitate licensing, access rights, and royalty payments without manual intervention. For surveyors, this provides clarity in ownership and licensing arrangements, helping to reduce administrative burden and protect against unauthorized use. These measures are particularly relevant when data is stored on third-party cloud platforms, where the risk of unauthorized resale or data leakage is high. More structured approaches, such as DRM models, help maintain data confidentiality and traceability, ensuring that spatial datasets remain identifiable and tied to their rightful authors (Ma 2017). Despite these advantages, blockchain and smart contracts carry certain risks. The decentralized nature of blockchain can obscure authorship if metadata is poorly managed, and cross-jurisdictional enforcement may be difficult. Smart contracts, while enforceable in theory, may contain terms that are illegal or unenforceable under local law. Furthermore, cryptocurrency-based transactions pose challenges due to volatility and regulatory uncertainty (Kariyawasam, Palliyaarachchi, and Campbell 2024).

To balance innovation with legal reliability, surveyors should adopt these technologies strategically. Blockchain can complement existing legal frameworks by reinforcing claims to authorship and ownership. Smart contracts can simplify licensing and ensure fair compensation, and TPMs can enhance the security of data storage and access. Together, these tools offer a robust, technology-driven framework for protecting spatial data in an increasingly digital and decentralized environment.

References

Bontekoe, Frits E., and Max Wallot. 2019. "World Intellectual Property Organization (WIPO)." In *Research Handbook on the European Union and International Organizations*, 275–292. Edward Elgar Publishing.

Campbell, Glenn, Kanchana Kariyawasam, and Armin Agha Karimi. 2024. "What Effect Will Electronic Cadastral Survey Plans in Australia Have on Surveyor's Intellectual Property?" In *FIG Working Week*. Accra, Ghana. https://www.fig.net/resources/proceedings/fig_proceedings/fig2024/papers/ts10g/TS10G_campbell_kariyawasam_et_al_12398.pdf

Cienciała, Agnieszka, Katarzyna Sobolewska-Mikulska, and Szymon Sobura. 2021. "Credibility of the cadastral data on land use and the methodology for their verification and update." *Land Use Policy* 102: 105204.

Dhanjee, Rajan, and Laurence Boisson De Chazournes. 1990. "Trade-Related Aspects of Intellectual Property Rights (TRIPs): Objectives, approaches and basic principles of the GATT and of intellectual property conventions." *Journal of World Trade* 24 (5).

He, Guang-bo, and Lison Li. 2020. "Research and application of LIDAR technology in cadastral surveying and mapping." *The International Archives of the Photogrammetry, Remote Sensing and Spatial Information Sciences* 43: 33–37.

Henning-Bodewig, Frauke. 2006. *Unfair Competition Law: European Union and Member States*. Vol. 18. Kluwer Law International BV.

Kariyawasam, Kanchana, Rangika Palliyaarachchi, and Glenn Campbell. 2024. *The Copyright Law of Spatial Data*. Taylor & Francis.

Ma, Zhaofeng. 2017. "Digital rights management: Model, technology and application." *China Communications* 14 (6): 156–167.

McDuff, Joseph Kevin, Armin Agha Karimi, and Zahra Gharineiat. 2025. "Enhancing spatial awareness and collaboration: A guide to VR-ready survey data transformation." *ISPRS International Journal of Geo-Information* 14 (2): 59. https://www.mdpi.com/2220-9964/14/2/59

Menell, Peter S. 2002. "Envisioning copyright law's digital future." *New York Law School Law Review* 46: 63.

Rawal, Darshana, Salvatore Amaduzzi, and Jan Seedorf. 2024. "Crypto-spatial: A new direction in geospatial data." *The International Archives of the Photogrammetry, Remote Sensing and Spatial Information Sciences* 48: 89–96.

Ricketson, Sam, and Jane Ginsburg. 2022. *International Copyright and Neighbouring Rights: The Berne Convention and beyond*. Oxford University Press.

SCOPE Chapter I. 1996. "Directive 96/9/EC of the European Parliament and of the Council, on the legal protection of databases." *Official Journal L* 77 (27/03): 20–28.

Tayyeb, Hasan. 2021. "Contract Design and Related Agreements." In *Innovative and Agile Contracting for Digital Transformation and Industry 4.0*, 40–61. IGI Global.

13

AI-Driven Transformations for Land Administration from Data to Decisions

Hamid Hosseini, Behnam Atazadeh, and Abbas Rajabifard

13.1 Introduction

Land and property data can be viewed through diverse analytical lenses for various purposes in different domains. However, governing land is a fundamental perspective. In this regard, land administration systems (LASs) are responsible for providing an operational environment for implementing land policies and strategies. They have four core functions, including ownership, value, use, and development (Williamson et al. 2010, Rajabifard et al. 2019), aiming at working together to achieve sustainable and resilient land governance. Basically, effective management of land and property data plays a foundational role in the functionality of LASs. More effective data management during its lifecycle (Saeidian et al. 2023) results in more informed decisions within LASs. Hence, land information infrastructure (LII) constitutes the central core of LASs, which can be equipped with information technologies to be advanced.

However, inadequacies and inefficiencies of current LASs are evident, especially in complex metropolitan areas. Since completeness and digitalisation of LASs are the top priorities in many countries and jurisdictions, these systems currently focus only on data maintenance (i.e., land records), particularly in digital formats like LandXML. Nevertheless, registration of most of land records in many countries, exemplified by Kenya, has not yet been done because of ineffective land and property management (Bennett et al. 2019). Several real-world instances illustrate the shortcomings of current LASs. Land disputes due to out-of-date mutation records in India (Mishra and Suhag 2017, Thakur et al. 2020), many uncertified land parcels in Indonesia (Rahim et al. 2024), and delays in resolving land disputes in Pakistan due to inefficiencies and complexity of their LASs (Ullah and Hussain 2023) are all examples of poor LASs worldwide. Furthermore, yet many land administration procedures rely on manual or semi-automated

 DOI: 10.1201/9781003638308-15

methods, resulting in significant delays. For instance, as of 2018, digitising one single paper/PDF-based survey plan in the state of Victoria, Australia, needed an average processing time of two weeks (Olfat et al. 2018). Ultimately, these issues pinpoint the need for advanced automated, real-time, and predictive solutions in LASs to support more informed decision-making.

Although LASs have employed various information technologies, such as blockchain and geospatial information systems (GIS), they remain inflexible in terms of responsivity. This is because rule-based processes and static databases are typically used in current systems. However, modern land administration is accompanied by dynamic and complex difficulties. Land and property data is classifiable as big data due to its minute-to-minute production during various processes such as document verification and leasing transactions (Junaid et al. 2024). Many processes in land administration are labour-intensive and time-consuming. For example, tracing rights, restrictions, and responsibilities (RRR) in 3D urban environments requires substantial effort, as relevant data is highly fragmented among several documents and databases, making it difficult to easily retrieve and query. Therefore, more responsive LASs are necessary, which must extend beyond basic digital record-keeping to encompass reasoning, adaptive learning, and predictive capabilities to facilitate decision-making. Consequently, intelligent approaches are required for shifting LASs from passive to active ones.

13.2 AI for Land Administration and Geospatial Science

One of the most powerful operational drivers advancing LASs is artificial intelligence (AI) (UNECE 2021, Chehrehbargh et al. 2024, Sagandykova et al. 2024). While there is not a single universally recognised definition of AI, it often refers to the ability of machines to replicate certain aspects of human intelligence, adapt to new situations, and carry out tasks that resemble those of humans (Russell and Norvig 2016, Duan et al. 2019). Different facets of human intelligence are replicated using a variety of methods that can be broadly divided into rule-based and data-driven paradigms.

- Reasoning, which is the process of logical thinking through something, aiming at deciding using evidence. To enable machines to reason like humans, explicit knowledge is represented in the form of encoded if-then rules by human experts. Expert systems are one of the examples of machines with the ability to reason. The reasoning aspect of AI increases transparency but decreases adaptability.

- Learning, which is the process of gaining knowledge by experience and situational adaptation, hence making predictions. To enable machines to learn like humans, experience data is fed into learning algorithms. Once having trained models, they can be applied to unseen data and make predictions. There are a wide range of techniques for machine learning that can be categorised into supervised (e.g., decision trees) and unsupervised learning (e.g., k-means). A subset of machine learning called deep learning uses a multi-layer neural network to automatically extract deep patterns from data while simulating the structure of the human brain (Bengio et al. 2017, Guangqi 2024). The learning aspect of AI is less transparent but more flexible than reasoning.
- Cognition, which is the process of acquiring and processing information from the environment in the form of vision and speech. Computer vision techniques such as optical character recognition (OCR) enable machines to perceive and comprehend visual information such as textual descriptions in survey plans. On the other hand, natural language processing (NLP) techniques enable computers to understand and generate linguistic information, such as extracted textual descriptions from survey plans (Nishant et al. 2020). Combining cognition aspects with reasoning and learning has led to the emergence of more robust tools such as multimodal large language models (MLLM) such as BERT and GPT, which can powerfully understand and generate data in different modalities such as text and image.

Geospatial science is one of the well-known domains in which AI has had revolutionary effects, leading to a new field known as geospatial AI (GeoAI) (Li and Hsu 2022) with the ability to resolve issues of geographic nature in an intelligent way. Advanced insights into geographic phenomena are made possible by GeoAI using its capabilities such as automation, real-time functionality and prediction, and intelligent decision-making. As a result, it provides an intelligent environment for analysing and understanding geospatial data for a variety of useful applications. However, adopting AI in LASs, as a subset of geospatial science, needs specific consideration and is challenging technically, legally, and institutionally.

Streamlining workflows in land development projects, managing vast amounts of land and property data, and decreasing the significant amount of cognitive load required for linking fragmented data are all examples of positive effects that AI can bring into LASs. As a specific example, using LLMs not only enhances efficiency in terms of speed but also eliminates needing to know specialised knowledge about another stakeholder's technical language and standards, resulting in easier interaction with LASs for all stakeholders. Overall, adopting AI into land administration offers revolutionary ways to improve the effectiveness of LAS functions, especially when considering the continuous management of land and property information.

13.3 A Conceptual Framework for Intelligent Land Administration Systems (ILASs)

To adopt AI in LASs, a conceptual framework has been proposed, which is illustrated in Figure 13.1. This framework is in line with the land management paradigm in which the four core functions are operated holistically. It has two layers in which different AI technologies in different aspects of AI are leveraged in the lifecycle of land and property data, leading to more effective data management. This effective land and property data management enables functions in the second layer to be operated in an intelligent environment by providing capabilities such as automation and real-time predictions. Consequently, the decision-making process in the functions would be more intelligent. The intelligent LII layer can potentially resolve issues regarding land and property data, such as fragmentation and out-of-date. Hence, the efficiency of the functions of the upper layer would be enhanced, which would lead streamlining various land administration practices such as land registration, land transfer, land allocation, and planning and building permits. As a result, it helps prevent delays and further costs in land administration projects.

Land administration has a wide range of stakeholders, such as government agencies, private surveyors, land developers, and legal institutions, each having its own data formats, standards, and procedures. The diversity of stakeholders leads to challenges in data sharing and query. Although some solutions have been proposed based on ontology and data integration techniques to provide an interoperable environment, enquiry is still not very efficient. Using natural language as a common language among the stakeholders not only enhances efficiency in terms of speed but also eliminates the need-to-know specialised knowledge about another stakeholder's technical language and standards. Instead of relying on understanding each other's data standards or procedures, stakeholders can use natural language queries to interact with LASs. It can efficiently and effectively eliminate the need to search manually through cadastral databases. NLP techniques can be used for semantic similarities in terms and definitions between data models in land administration (Vaněk et al. 2024). AI chatbots allow faster communication, greater accessibility, better data integration, streamlining processes, and making it easier for stakeholders to collaborate without needing specialised technical knowledge.

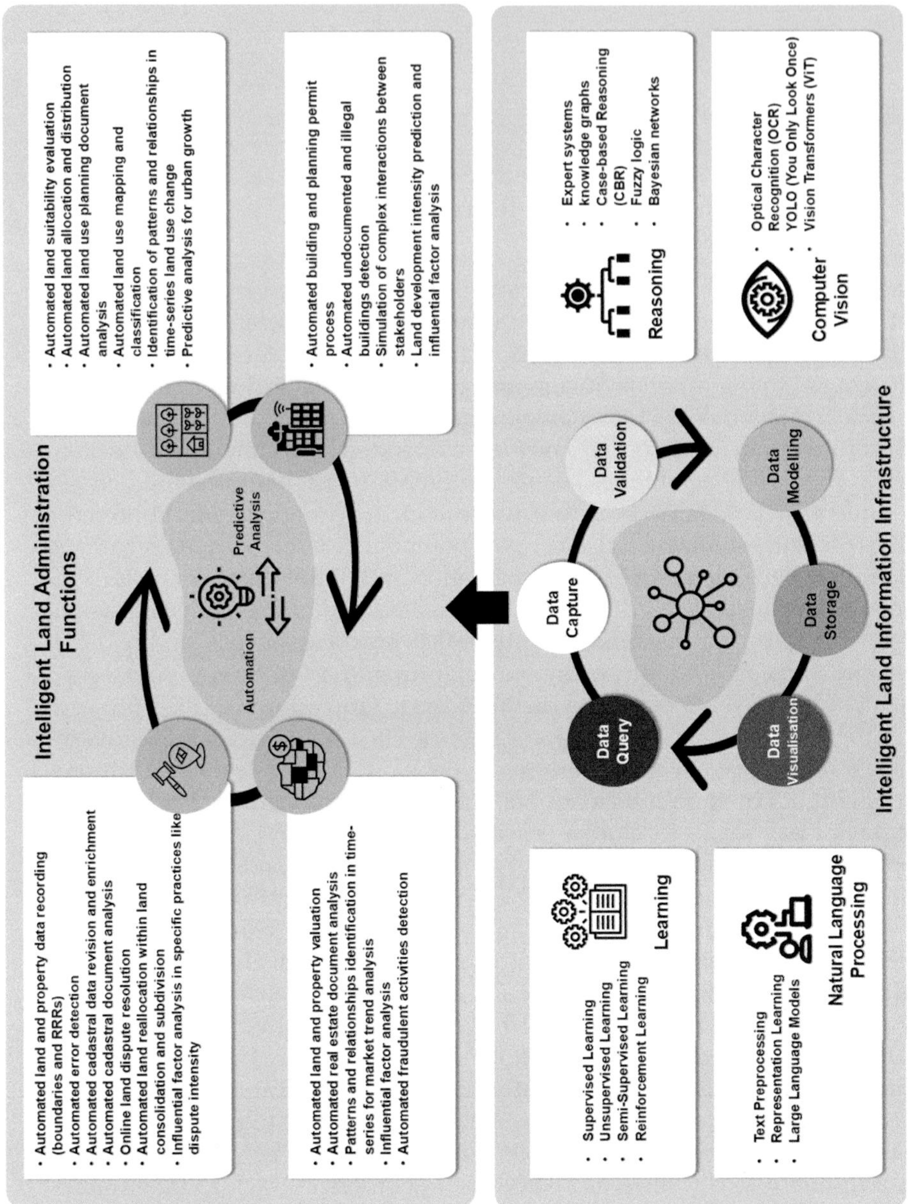

FIGURE 13.1
A conceptual framework for intelligent land administration systems (ILASs).

13.4 Discussion on Challenges of Implementing AI-Driven ILASs

To successfully shift legacy LASs to intelligent ones, the establishment of core infrastructures such as servers equipped by GPUs for a digital environment is a prerequisite. Following that, depending on the specific task that is going to be done by AI, AI models are trained from scratch or refined using prepared datasets to be suited to the task. Once models are developed, they are deployed in the existing systems in which they must be interoperable with other systems. Finally, the performance of the models is tracked and enhanced gradually over time. However, several technical, legal, and institutional issues may be raised when implementing AI-driven ILASs, which are discussed in the sections that follow.

13.4.1 Technical Issues

Since AI is shifting from rule-based to data-driven, issues regarding land and property data must be given special consideration. Land administration data in many jurisdictions and countries is often fragmented and outdated. Moreover, much of the data is frequently maintained in paper form, which needs to be digitised first. These issues may cause difficulties when training machine learning and deep learning models, as they need vast amounts of up-to-date data to be well-operated. On the other hand, training on biassed or incomplete datasets leads to inaccurate results, which can potentially have negative impacts on the decision-making process in property rights and land values. Additionally, cleaning training data is critically important to make sure it does not contain any error and anomaly. Overall, sufficient and high-quality data as a benchmark data is essential for training AI models.

Model scalability, explainability, and sustainability are other technical issues that must be considered carefully. Land administration terminologies, rules, and regulations in both surveying and planning processes differ between jurisdictions. For instance, the name and number of primary parcels vary across Australia and New Zealand, as each government has implemented cadastral systems in a distinct way (Atazadeh et al. 2023). Hence, small-scale models developed on small datasets could not work in other jurisdictions. Moreover, while machine learning and deep learning models can process vast amounts of data, it is not often transparent how they work due to the black-box nature of the models. This can impact the validity of the decisions made by AI, leading to diminishing stakeholder trust. To address this, explainable AI (XAI) is essential for understanding how factors affect predictions, like property values, since users might question results in the absence of transparency. Additionally, AI models in the land administration domain must be in line with sustainability principles. To clarify, AI models must generate reliable and repeatable results and must be adaptable and flexible in different circumstances due to the dynamic character of the landscape of land administration.

13.4.2 Legal Issues

Ownership rights associated with AI models is one of the legal challenges. When AI models are enhanced due to changing requirements, the stakeholders (e.g., developers, owners, and users) may have concerns about benefit distribution and intellectual property (Battah et al. 2022). Hence, the necessity of a clear framework supported by technologies like blockchain is crucial. Furthermore, in situations where models unintentionally introduce bias or unfairness, identifying responsibility is challenging. This is because several factors may have led to the AI failures, such as poor training data or programming flaws, in which pinpointing the exact root cause is often difficult (Novelli et al. 2024).

If policymakers establish clear guidelines for the pre-deployment assessment and continuing monitoring of AI models embedded in LASs, legal disputes would be significantly decreased. For example, specific restrictions can be established regarding the usage of MLLMs, such as using on-device models like Llama instead of cloud-based ones like ChatGPT and making sure that access to data is authorised. Finally, even if accountability of AI models operating in LASs is determined clearly and biases are minimised adequately, the generated results must be formally accepted for final decision-making. To make it possible, human oversight is advised to guarantee confidence and compliance.

13.4.3 Institutional Issues

The high level of uncertainty of adopting information technologies (Strauss et al. 2024) due to institutional factors is another challenge when implementing AI for organisations. AI developers would be added to the landscape of land administration as a new stakeholder, which demands new skills and technical expertise. Moreover, current policies within organisations might not be ready for letting AI models be operated due to their complexities. In addition, the adoption of AI in LASs also necessitates large expenditures for personnel training and digital infrastructures, in addition to continuous maintenance costs. Because of these issues, the transition from traditional LASs to AI-driven ones might face reluctance and resistance from institutional perspectives.

However, the adoption of AI for LASs may be improved if the main concerns of professionals with organisations, such as unfamiliarity with new information technology, fears of potential errors, and job displacement, are sufficiently addressed. In this regard, sufficient training, ongoing support, and the development of an innovation-friendly culture can all be highly beneficial to alleviate these concerns, resulting in better implementation of AI-driven LASs.

13.5 Conclusion

Due to the dynamic and complex nature of LASs, these systems must go beyond data maintenance, which requires a shift from legacy and static systems to more responsive and adaptive ones. Limitations of traditional LASs in terms of data fragmentation and manual processes highlight the need for an intelligent environment for land administration functions to be operated. Considering the high-level goal of land administration, which is achieving sustainable and resilient development, current AI usage in land administration is still in its infancy. By introducing a conceptual framework for AI-driven intelligent LASs (ILASs), this chapter illustrates how the integration of AI technologies in different aspects, including reasoning, learning, and cognition, can enhance the management of land and property data and enable automation, real-time functionality and predictions, and intelligent decision-making within LASs.

Despite advantages, implementing ILASs is not without challenges. Technical issues include data availability and quality and model scalability and explainability. On the other hand, legal and institutional issues include accountability, formal recognition of AI outputs, and institutional resistance due to unfamiliarity, infrastructure demands, and organisational inertia. Overcoming these challenges thoughtfully can significantly help better implementation of AI-driven ILASs, which, of course, needs an interdisciplinary strategy involving effective collaboration among stakeholders, revising regulations, and investments in digital infrastructure within organisations. Upon successful adoption, LASs might facilitate land development in a more sustainable and resilient way.

References

Atazadeh, B., A. Rajabifard and H. Olfat (2023). "Proposing a multi-jurisdictional framework for 3D digital cadastre in Australia and New Zealand." *Land Use Policy* **131**(C).

Battah, A., M. Madine, I. Yaqoob, K. Salah, H. R. Hasan and R. Jayaraman (2022). "Blockchain and NFTs for trusted ownership, trading, and access of AI models." *IEEE Access* **10**: 112230–112249.

Bengio, Y., I. Goodfellow and A. Courville (2017). *Deep learning*. MIT Press, Cambridge, MA.

Bennett, R. M., M. Pickering and J. Sargent (2019). "Transformations, transitions, or tall tales? A global review of the uptake and impact of NoSQL, blockchain, and big data analytics on the land administration sector." *Land Use Policy* **83**: 435–448.

Chehrehbargh, F. J., A. Rajabifard, B. Atazadeh and D. Steudler (2024). "Identifying global parameters for advancing Land Administration Systems." *Land Use Policy* **136**: 106973.

Duan, Y., J. S. Edwards and Y. K. Dwivedi (2019). "Artificial intelligence for decision making in the era of Big Data–evolution, challenges and research agenda." *International journal of information management* **48**: 63–71.

Guangqi, C. (2024). "Prospects, challenges and guidelines for practical deep learning in geoengineering." *Intelligent Geoengineering* **1**.

Junaid, L., K. Bilal, J. Shuja, A. O. Balogun and J. J. Rodrigues (2024). "Blockchain-enabled framework for transparent land lease and mortgage management." *IEEE Access* **12**.

Li, W. and C.-Y. Hsu (2022). "GeoAI for large-scale image analysis and machine vision: recent progress of artificial intelligence in geography." *ISPRS International Journal of Geo-Information* **11**(7): 385.

Mishra, P. and R. Suhag (2017). "Land records and titles in India." *Ideas for India* **20**: 24.

Nishant, R., M. Kennedy and J. Corbett (2020). "Artificial intelligence for sustainability: Challenges, opportunities, and a research agenda." *International Journal of Information Management* **53**: 102104.

Novelli, C., M. Taddeo and L. Floridi (2024). "Accountability in artificial intelligence: what it is and how it works." *Ai & Society* **39**(4): 1871–1882.

Olfat, H., D. Shojaei, M. Briffa, S. Maley and A. Rajabifard (2018). "Strategic actions for increasing the submission of digital cadastral data by the surveying industry based on lessons learned from Victoria, Australia." *ISPRS International Journal of Geo-Information* **7**(2): 47.

Rahim, S. N., L. Husni and A. Arba (2024). "Legal implications of the use of digitalization of online land registration." *Path of Science* **10**(5): 2011–2018.

Rajabifard, A., B. Atazadeh and M. Kalantari (2019). *BIM and urban land administration*. CRC Press.

Russell, S. J. and P. Norvig (2016). *Artificial intelligence: A modern approach*. Pearson.

Saeidian, B., A. Rajabifard, B. Atazadeh and M. Kalantari (2023). "Data lifecycle of underground land administration: A systematic literature review." *Survey Review* **55**(392): 396–415.

Sagandykova, D., G. Ussipbayev, E. Khassamdinova, A. Omarbekova and N. Jangarasheva (2024). "Scientific foundations of application of new effective technologies in land surveying studies (on the example of Talgar District, Almaty Region)." *Instrumentation, Mesure, Metrologie* **23**(3): 183.

Strauss, L. M., A. Z. Klein and E. Scornavacca (2024). "Adopting emerging information technology: A new affordances process framework." *International Journal of Information Management* **76**: 102772.

Thakur, V., M. Doja, Y. K. Dwivedi, T. Ahmad and G. Khadanga (2020). "Land records on blockchain for implementation of land titling in India." *International Journal of Information Management* **52**: 101940.

Ullah, I. and S. Hussain (2023). "Impact of early access to land record information through digitization: Evidence from Alternate Dispute Resolution Data in Punjab, Pakistan." *Land use policy* **134**: 106917.

UNECE (2021). *Scenario study on future land administration in the UNECE region*. United Nations, United Nations Economic Commission for Europe.

Vaněk, Martin, Karel Janečka and Otakar Čerba (2024). "Determining the semantic similarity of definitions by artificial intelligence for the needs of 3D Land Administration: A case of building units." 12th International FIG Land Administration Domain Model & 3D Land Administration Workshop 24-26 September 2024, Kuching, Malaysia. https://www.gdmc.nl/3DCadastres/workshop2024/programme/3DLA2024_paper_V.pdf

Williamson, I., S. Enemark, J. Wallace and A. Rajabifard (2010). *Land administration for sustainable development*. ESRI Press Academic Redlands.

14

Land Intelligence: Innovations in Online Dispute Resolution for Land and Property Disputes

Fahimeh Abedi

14.1 Introduction

Land has always been essential to human life, supporting nearly all activities. While traditionally seen as a resource for agriculture, it is now increasingly viewed as a strategic asset for industry, housing, and urban development (Munir et al. 2023). As the population grows, the demand for land also rises, as humans seek space for both living and livelihood, leading to competition for control and ownership rights (Kartika 2021). However, access to legal rights remains unequal. The United Nations Commission on Legal Empowerment reports that, in the early 21st century, 4 billion people worldwide still live without legal protection (United Nations Commission on Legal Empowerment of the Poor 2008).

Land conflict involves interactions between two or more parties, each asserting their interests over the same land or related resources such as water, vegetation, minerals, or the airspace above the land. At the micro level, these conflicts often arise due to cultural value differences, varying interpretations of technical data or local land conditions, or economic disparities reflected in unequal land ownership and control. Boundary disputes, in particular, frequently arise from unregistered land, overlapping claims by third parties, or delays and inefficiencies in dispute resolution processes. When settlements are poorly managed, aggrieved parties may pursue legal action. However, many individuals are reluctant to engage in court proceedings due to high costs, lengthy durations, and procedural complexity. Consequently, communities tend to favor alternative, non-judicial approaches when addressing land disputes (Munir et al. 2023).

Recent innovations have transformed various sectors, including law, leading to the development of Online Dispute Resolution (ODR) systems. These platforms offer an efficient, cost-effective way to resolve disputes, much like the growing preference for alternative, non-judicial methods in land dispute

DOI: 10.1201/9781003638308-16

resolution. By eliminating the need for physical presence, ODR enhances access to justice and is increasingly being adopted worldwide. With over 100 courts embracing this approach, ODR is extending its reach, providing a more accessible and flexible means of addressing conflicts, especially in regions where traditional legal processes may be less accessible.

This chapter examines how ODR can offer innovative solutions to land disputes by exploring its challenges and identifying strategies to address them in the context of land rights.

14.2 The Rise of ODR in Land Dispute Resolution

Land disputes can arise from a range of factors, including poor or incomplete documentation, lack of awareness, insufficient knowledge of family history, fraudulent practices by land sellers to deceive buyers, repeated sales of the same land, illegal occupation, and inefficiency in the land administration system (Boafo-Anang et al. 2021). These disputes arise in various forms, from conflicts between individuals, such as neighbors disputing land boundaries, siblings contesting inheritance divisions, and disagreements over sales and usage rights, to more complex conflicts involving multiple parties, with varying degrees of impact at the community level (Wehrmann 2008). To bridge these challenges, it is essential to explore how modern technological solutions, such as ODR systems, can provide a more efficient and accessible approach to resolving land disputes.

ODR uses digital platforms to facilitate communication and resolution processes, removing the need for physical presence in court. This model not only reduces the time and cost associated with dispute resolution but also increases accessibility, especially for individuals in remote areas or those with limited resources (Abedi and Zeleznikow 2014). The adoption of ODR has expanded across various types of disputes, from consumer issues to family matters, and more recently, land-related conflicts. In addition, ODR provides additional practical advantages beyond the efficient and fair resolution of individual cases. It enables Dispute Resolution Professionals (DRPs) to manage cases without being limited by the geographic locations of the parties involved (Ojelabi et al. 2023a, 2023b).

Building on this evolution, recent advancements have seen the integration of Artificial Intelligence (AI) into ODR systems. AI-enabled ODR platforms utilize large datasets to develop algorithms capable of predicting dispute outcomes and supporting real-time decision-making. In the context of land disputes, often complex and emotionally charged AI can enhance consistency and speed. ODR platforms can be government-operated, court-affiliated, or privately managed, each offering different levels of integration. Court-annexed ODR systems are designed to reduce pressure on overburdened judicial systems by embedding technology into traditional alternative dispute resolution (ADR) mechanisms. For instance, the Civil Resolution Tribunal in British Columbia (Civil Resolution Tribunal, n.d.) serves as a leading example of a court-connected model that delivers a comprehensive digital pathway for resolving disputes, including those related to land and housing efficiently and affordably.

14.3 Challenges and Considerations in Implementing ODR for Land Disputes

The integration of ODR mechanisms into land dispute contexts presents several challenges that must be addressed to ensure their effectiveness and fairness.

14.3.1 Complexity of Land Dispute

Land-related conflicts are intrinsically complicated, often characterized by overlapping claims, long-standing historical tensions, and deeply rooted cultural factors, all of which pose challenges to uniform resolution methods. Such disputes commonly arise from unclear land tenure arrangements and inconsistent land use practices, which result in competing interests grounded in historical injustices and past ownership patterns (Rubakula et al. 2019). The complexity of land disputes is further intensified by insufficient documentation and ambiguous legal titles, which often extend the duration of conflicts and impede their resolution (The Legal Lo 2024). Given these complexities, ODR systems must be designed with flexibility to address the specific needs, values, and expectations of the communities they serve. Failure to do so may result in processes that are perceived as illegitimate or disconnected from local realities.

14.3.2 Issues of Fairness and Trust in ODR Systems

For ODR systems to be effective, especially in the context of land disputes, the outcomes must be legally recognized and enforceable within existing legal frameworks. Although these technologies offer numerous social and economic benefits, they also raise several concerning issues (Abedi et al. 2025). This is crucial for ensuring that the decisions reached through ODR are not merely advisory but have the force of law, thereby promoting compliance and respect from all parties involved (Abedi and Yusoff 2011). Ensuring both procedural fairness and user trust is critical for the success of ODR in land disputes. Procedural fairness requires that all parties perceive the process as unbiased, transparent, and consistent (Abedi et al. 2019a). Establishment of clear regulatory standards for fairness is essential to ensure that ODR outcomes are fully integrated into the judicial system and have legal standing. Without such recognition, the ability of ODR platforms to resolve land disputes may be severely limited, as parties may not view the results as legitimate or binding (Abedi et al. 2019a).

Trust is an important element for ODR users and is especially vital in digital environments where face-to-face interactions are absent. It can be reinforced through both technical and organizational safeguards (Abedi and Miller 2024). Auditability, the ability to review and verify each stage of the ODR process, plays a key role in building trust by ensuring transparency and offering clear information about how decisions are reached (Abedi et al. 2019b). In addition, secure data handling and strong privacy protections help ensure that sensitive land records are treated responsibly and not misused, thereby increasing trust (Abedi et al. 2019b). In culturally diverse land dispute settings, embedding local norms and providing multilingual support also foster trust by demonstrating respect for community practices and ensuring all parties can effectively engage (Abedi et al. 2019c). Collectively, these elements of fairness and trust help create an environment in which technology-driven ODR can fulfill its promise of accessible, equitable, and enduring land dispute resolution.

14.3.3 Digital Literacy and Accessibility

The effective implementation of ODR for land-related conflicts is heavily influenced by both digital literacy and access to technology. Although ODR platforms are frequently promoted as mechanisms to enhance access to justice especially for individuals in remote or underserved areas, their potential is fundamentally limited by users' ability to interact competently and consistently with digital systems. ODR platforms require participants to have not only reliable internet connectivity but also the digital competence to navigate and engage with online systems effectively (Thomas et al. 2018).

In many regions where land disputes are prevalent such as rural communities, informal settlements, or post-conflict zones, there are significant barriers to digital access. As an example in Australia, a significant portion of the population remains digitally excluded due to limited access to digital technologies and insufficient digital skills (Thomas et al. 2018). Practical barriers to digital access include coverage gaps and unreliable broadband, particularly in rural and regional areas of Australia, website accessibility issues, the cost of internet services, and the restricted hours and limited resources of free internet providers like libraries (Cashman and Ginnivan 2019). Without addressing these issues, ODR may risk replicating or even exacerbating structural injustices, rather than offering a more inclusive path to resolution.

14.3.4 Cultural Sensitivity

Land disputes are often more than disagreements and deeply embedded in local customs and societal norms. ODR systems must be designed with cultural sensitivity to address the specific needs and expectations of the communities they serve (Yusoff et al. 2013). Failure to do so may result in processes that are perceived as illegitimate or disconnected from local realities. In many regions, particularly among Indigenous populations and traditional communities, land is not just a physical asset but a central part of identity, heritage, and livelihood (Figueroa 2011). These connections mean that disputes over land are frequently influenced by customary practices, histories, and collective memory, which may not align with formal legal frameworks. ODR systems that fail to account for cultural factors risk overlooking critical dimensions of disputes, potentially excluding users and producing outcomes that lack legitimacy in the eyes of the community. For instance, decision-making processes that reflect local values or traditional dispute resolution mechanisms may be perceived as unjust, even if they conform to legal standards (Yusoff et al. 2013). Therefore, ODR platforms need to be thoughtfully designed with cultural sensitivity at their core. This involves integrating local dispute resolution practices, engaging community elders or traditional leaders when suitable, and ensuring that the platform's language, interface, and communication methods are both accessible and respectful of the cultural backgrounds of its users.

14.4 Conclusion

As land remains a crucial yet contested resource, the need for fair, accessible, and efficient dispute resolution mechanisms has become more pressing than ever. This chapter has illustrated how ODR systems can transform land conflict resolution by offering new ways to deliver justice where traditional systems fall short. ODR presents a promising alternative to court-based models by lowering costs, increasing accessibility, and reducing delays; benefits are especially critical for marginalized, rural, and Indigenous populations, often affected by land disputes. Its digital nature allows scalability across diverse socio-legal contexts, and AI integration enhances efficiency through predictive analytics, real-time feedback, and more consistent decision-making. Examples like British Columbia's Civil Resolution Tribunal demonstrate how court-connected ODR can ease judicial burdens while empowering users to resolve disputes independently and fairly. However, the effectiveness and legitimacy of ODR depend on key factors: procedural fairness, legal recognition, cultural sensitivity, user trust, and inclusive design. Land disputes are deeply tied to historical grievances and cultural identities. Digital literacy and equitable access to technology remain significant barriers. Without investment in infrastructure, training, and outreach, the benefits of ODR may remain inaccessible to many.

Building and maintaining trust in both technology and institutions is essential. Trust goes beyond technical features—it requires transparency, enforceable outcomes, user consent, auditability, and cultural sensitivity. Collaboration among legal experts, technologists, policymakers, and communities is key to creating systems that are efficient, fair, and inclusive. When done right, these innovations can foster a just, transparent land governance system where rights are respected, disputes are resolved peacefully, and trust in institutions is restored.

References

Abedi, Fahimeh, and Tim Miller. 2024. "Closing the Gaps in Law around Emerging Technology: Recommendations for the Legal Profession." *Precedent (Sydney, NSW)* 183: 36–39.

Abedi, Fahimeh, Tim Miller, and Atif Ahmad. 2025. "Legal, Ethical and Technological Challenges of New Emerging Technologies for Lawyers: A Mixed Methods Approach." *Legal Information Management* 25 (1): 48–58.

Abedi, Fahimeh, and Sakina Shaik Ahmad Yusoff. 2011. "Consumer Dispute Resolution: The Way Forward." *Journal of Global Management* 2 (2): 204–15.

Abedi, Fahimeh, and John Zeleznikow. 2014. "The Provision of Trustworthy Online Dispute Resolution for Business to Consumer Electronic Disputes." In *Proceedings of 7th Asia-Pacific Business Research Conference, Bayview Hotel, Singapore, August 25–26, 2014*.

Abedi, Fahimeh, John Zeleznikow, and Chris Brien. 2019a. "Universal Standards for the Concept of Fairness in Online Dispute Resolution in B2C E-Disputes." *OSJDR Ohio State Journal on Dispute Resolution* 34: 357.

Abedi, Fahimeh, John Zeleznikow, and Chris Brien. 2019b. "Developing Regulatory Standards for the Concept of Security in Online Dispute Resolution Systems." *Computer Law & Security Review* 35 (5): 105328.

Abedi, Fahimeh, John Zeleznikow, and Emilia Bellucci. 2019c. "Universal Standards for the Concept of Trust in Online Dispute Resolution Systems in E-Commerce Disputes." *International Journal of Law and Information Technology* 27 (3): 209–37.

Boafo-Anang, Mark, Emmanuel Offei Akrofi, and Forster Junior Shitsi. 2021. "Land Disputes and Conflicting Land-Court Judgement Resolution in Ghana: Perspectives of Regulatory Stakeholders." *African Journal of Land Policy and Geospatial Sciences* 4 (3): 331–42.

British Columbia Civil Resolution. n.d. Civil Resolution Tribunal. Accessed May 3, 2025. https://civilresolutionbc.ca/crt-process/

Cashman, Peter, and Eliza Ginnivan. 2019. "Digital Justice: Online Resolution of Minor Civil Disputes and the Use of Digital Technology in Complex Litigation and Class Actions." *Macquarie Law Journal* 19: 1–39.

Figueroa, Robert Melchior. 2011. "Indigenous Peoples and Cultural Losses." In *The Oxford Handbook of Climate Change and Society*, edited by John S. Dryzek, Richard B. Norgaard, and David Schlosberg, 232–49. Oxford: Oxford University Press.

Kartika, Nana. 2021. "Analysis of Legal Aspects on Sale and Purchase of High Heritage Land Disputes." *International Journal of Economic, Technology and Social Sciences (Injects)* 2 (1): 75–80.

Munir, Misbakhul, Azis Budianto, and Rineke Sara. 2023. "Juridical Overview of Land Dispute Settlement." *Edunity Kajian Ilmu Sosial dan Pendidikan* 2 (1): 145–52.

Ojelabi, Lola Akin, Fahimeh Abedi, Louis De Koker, and John Zeleznikow. 2023b. "Effectiveness of Dispute Resolution Processes Conducted Online: Findings from Empirical Research with Implications for Small Business Providing Dispute Resolution Services." *Australian Business Law Review* 51: 345–59.

Ojelabi, Lola Akin, John Zeleznikow, Louis De Koker, and Fahimeh Abedi. 2023a. "Dispute Resolution Practitioners, Online Technologies and COVID-19: Findings from Empirical Research on Use of Technology." *Monash University Law Review* 49: 225.

Rubakula, Gelas, Zhanqi Wang, and Chao Wei. 2019. "Land Conflict Management through the Implementation of the National Land Policy in Tanzania: Evidence from Kigoma Region." *Sustainability* 11 (22): 6303–15.

The Legal Lo. 2024. Comprehensive Case Studies of Land Disputes and Their Resolutions. Accessed May 5, 2025. https://thelegallo.com/case-studies-of-land-disputes/

Thomas, Julian, Josephine Barraket, Chris Wilson, Kay Cook, Yee Man Louie, Indigo Holcombe-James, Scott Ewing, and Trent MacDonald. 2018. *Measuring Australia's Digital Divide: The Australian Digital Inclusion Index*, 1–52. Melbourne, Australia: RMIT University.

United Nations Commission on Legal Empowerment of the Poor. 2008. Making the Law Work for Everyone. Accessed 15 May 2025. https://ycsg.yale.edu/sites/default/files/files/making_the_law_work.pdf

Wehrmann, Babette. 2008. *Land Conflicts: A Practical Guide to Dealing with Land dispute.* Germany: Deutsche Gesellschaft für Technische Zusammenarbeit (GTZ).

Yusoff, Sakina Shaik Ahmad, Fahimeh Abedi, Rahmah Ismail, Azimon Abdul Aziz, and Suzanna Mohamed Isa. 2013. "Resolving Consumer Trade Disputes through Online Dispute Resolution: Issues to be Considered." In *Proceedings of the 2nd Applied International Business Conference, University Malaysia Sabah, Malaysia, 7–8 December, 2013.*

15

Land Intelligence for Affordable Housing: An Integrated Optimization, Machine Learning and Policy Framework

Peyman Jafary, Davood Shojaei, Abbas Rajabifard, and Tuan Ngo

15.1 Introduction

The need for data-informed strategies to address the current global housing crisis has become increasingly critical. Particularly in developed countries like Australia, Canada, Germany, the United Kingdom (UK) and the United States (US), there has been a significant surge in housing prices, fueled by a mix of intertwined driving factors, including the COVID-19 pandemic and increased migration flows (Hardin 2023; Lakševics et al. 2023; Pawson et al. 2022). In Australia, population growth, the influence of monetary measures like interest rates and limited availability of developable lands near employment centers and essential infrastructure have led to an intense and complex housing crisis (ANZ-CoreLogic 2023; PEXA 2023). The rental housing shortage and subsequent sharp rises in rents and low vacancy rates are major dimensions of the country's housing crisis (AHURI 2022).

The ongoing mismatch between housing supply and rental demand highlights the need for data-driven, practical solutions and better alignment of dwelling types with optimal locations (AHURI 2022). However, achieving this requires a holistic approach that accounts for different stakeholders' expectations and priorities, from developers and investors to renters and government bodies (Jafary et al. 2024a). Profitability and feasibility are among the most important priorities for developers and investors. In contrast, affordability, accessibility and social equity are the major concerns for renters and public authorities. These priorities are somehow competing, creating an inherently multi-objective problem space where no single solution can satisfy all interests simultaneously (Jafary et al. 2024a). Accordingly, addressing this crisis depends on coordinated strategies, such as smart land use regulations, targeted construction incentives, mortgage reform and data-driven planning tools (Lima 2024; Saiz 2023). Multi-objective optimization could be employed to identify balanced trade-offs that serve the

 DOI: 10.1201/9781003638308-17

broader public interest without disproportionately benefiting one group (Konak et al. 2006). Evolutionary algorithms like Genetic Algorithms (GA) effectively handle complex, conflicting objectives by uncovering Pareto-optimal solutions across stakeholders (Deb 2011), to find optimal locations for developing residential buildings.

In addition, land value plays a central role in shaping urban form in modern economies, dictating where cities grow and how. Lower land prices often lead to horizontal sprawl, while higher values drive vertical growth and density, so land price distribution strongly shapes urban form and infrastructure demand (Deng and Huang 2004; Hu et al. 2013; Silveira and Dentinho 2018). Traditional valuation models, often reliant on hedonic modeling, struggle to reflect the land markets' complex and dynamic nature. Today, advanced Machine Learning (ML) models offer great potential in mass land valuation through handling spatial dependency and capturing non-linear relationships between land value drivers (Jafary et al. 2024b). Accurate land valuation is pivotal in enabling equitable and efficient housing policies. As urban areas expand and infrastructure investments enhance land desirability, the resulting uplift in land values is often unequally distributed, benefiting landowners and developers while placing affordability further out of reach for others (Vejchodská et al. 2022). Land Value Capture (LVC) policies employ tools such as betterment levies, developer contributions and inclusionary zoning to reintegrate part of this unearned value into the public system (Canelas and Noring 2022; Hengstermann and Götze 2023).

Accordingly, this chapter combines interrelated components of stakeholder-responsive localization of housing development, land valuation and policy-oriented land use strategy into a unified, data-driven framework for advancing housing affordability. This integration forms the foundation of land intelligence—an approach that combines holistic planning with advanced analytics to understand and actively shape urban dynamics. The framework is designed to support equitable urban growth through more targeted investments, smarter regulations and adaptive policies such as LVC. The following sections describe the components of this framework in detail, beginning with the prioritization of potential sub-market areas, followed by land valuation modeling, and globally applicable land use recommendation scenarios.

15.2 Prioritization of Potential Sub-Market Areas

According to P. Jafary et al. (2024a), prioritizing sub-urban areas for affordable housing development through data-driven approaches is a key step in addressing the rental crisis. Nevertheless, this process requires considering different stakeholders' diverse and often conflicting interests and expectations, including builders/investors, renters and public authorities. Buyers/investors are primarily concerned with financial feasibility when looking for optimal locations to develop profitable residential construction projects. On the other hand, renters seek affordability and critical factors like accessibility and proximity to jobs and services. Consequently, planning for affordable housing emerges as an inherently multi-objective optimization problem that cannot be resolved through singular or linear policy tools. The authors proposed a generalizable, data-driven framework for prioritizing sub-urban areas based on a hybrid optimization approach, integrating GA with ML models of Random Forest (RF) and XGBoost through surrogate models.

15.2.1 Optimization Parameters: Balancing Competing Stakeholder Needs

The process accounts for the dual objective of enhancing residential developments' financial viability while improving renters' affordability and livability outcomes. A set of spatially informed indicators are selected and optimized to navigate this inherently multi-objective landscape using advanced modeling approaches. These indicators often reflect competing and occasionally conflicting objectives. Table 15.1 summarizes the core parameters used by P. Jafary et al. (2024a) in the prioritization framework, categorized according to stakeholder focus.

TABLE 15.1

Key Indicators for Multi-Objective Optimization in Affordable Housing Prioritization

Stakeholder	Parameter	Description
Builders and investors	Property value change	Measures market growth potential based on recent price trends.
	Rental yield	The ratio of rental income to property value, indicating investment return potential.
	Vacancy rate	The proportion of unoccupied housing units, indicating market saturation.
	Construction cost per square meter	Assesses feasibility by considering regional variations in building costs.
Renters	Median rent	Direct measure of rental affordability.
	Distance to Central Business District (CBD)	Spatial proximity to employment and activity hubs.
	Access to public transportation	Density or proximity of transportation infrastructure.
	Commuting distance to work	Average distance traveled between home and workplace.
	Access to essential services	Proximity to schools, hospitals, shopping centers etc.

15.2.2 Optimization Model Development and Evaluation

This section outlines developing the framework to balance profitability, affordability, accessibility and policy alignment through a hybrid process. This follows the model proposed by Jafary et al. (2024a):

- **Initial optimization using GA**: A multi-objective Genetic Algorithm ranks suburbs by maximizing builders/investors-related and minimizing renter-related indicators, using a linear weighted function to balance competing goals.
- **ML-based objective functions**: The initial GA-generated rankings are then used to train surrogate ML models. The GA rankings train ML models (e.g., RF, XGBoost, CatBoost), which learn non-linear relationships between indicators and prioritization.
- **Re-optimization**: GA is rerun using ML-derived objective functions, refining suburb rankings based on learned patterns.
- **Evaluation**: Spearman's rank correlation compares original and ML-augmented rankings, assessing consistency and revealing deeper indicator relationships.

This multi-objective optimization approach ensures that the final recommendations are robust and adaptable, thereby assisting urban planners and policymakers with a nuanced understanding of where affordable housing interventions could be most effectively targeted.

15.3 Land Valuation Using ML

The next stage is developing an ML-based Automated Valuation Model (AVM) to estimate land values at an urban scale. ML techniques allow for capturing complex, non-linear relationships among multiple spatial, socio-economic and regulatory factors influencing value. The valuation model developed by Jafary et al. (2024b) in Melbourne, Australia, provides a robust methodology that can be generalized to other contexts with appropriate local data inputs. This methodology is presented in Figure 15.1 and will be elaborated on in the following.

15.3.1 Identification of Valuation Factors

Based on an extensive literature review, Jafary et al. (2024b) proposed a comprehensive set of 37 valuation factors spanning five domains. Table 15.2 summarizes these factors and their categorization.

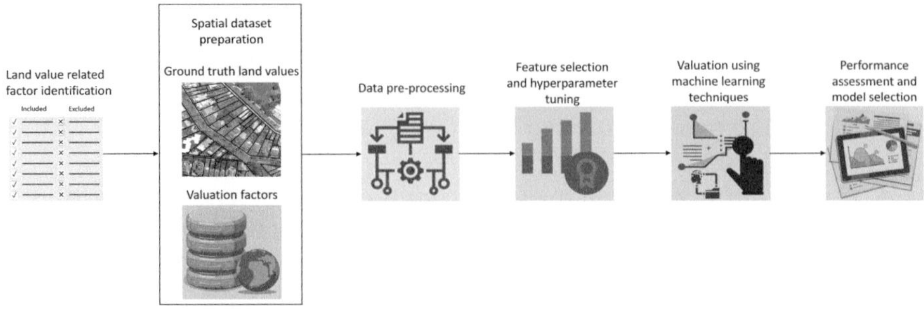

FIGURE 15.1
ML-based land valuation methodology. (Adapted from Jafary et al. 2024b).

15.3.2 Spatial Dataset Preparation

Two core spatial datasets are required for ML-based land valuation model development, enabling supervised learning. First, ground truth valuation data, representing the actual unimproved value of sample land parcels spatially distributed in each study area. Second, attribute data on valuation factors, spatially joined to each land parcel polygon. These could be obtained from various sources, including government databases, open data platforms (e.g., OpenStreetMap) and remote sensing products.

15.3.3 Data Preprocessing

Before training, the data undergo cleaning and transformation, including imputation, outlier removal, encoding, scaling and sampling to ensure model accuracy and robustness.

TABLE 15.2
Mass Land Valuation Factors Suggested by Jafary et al. (2024b)

Category	Factor	Number
Physical	Area, shape (convexity), elevation and slope	4
Geographical	Latitude, longitude, LGA, locality, road type, distance to CBD, distances to the different amenities (sustenance, shopping centers, higher-level education centers, health centers, childcares and kindergartens, public transportation, natural landscapes and water bodies) and walkability	15
Socioeconomic	Population density, employment rate, Gross Regional Product (GRP), mortgage rate, ethnicity, job opportunities, income, crash, criminal activities and offensive incidents	10
Environmental	Temperature, rainfall, green vegetation and air quality	4

15.3.4 Feature Selection and Hyperparameter Tuning

Feature selection is essential in ML to identify the most important features contributing to a predictive model's accuracy (Salcedo-Sanz et al. 2018) using techniques like Recursive Feature Elimination (RFE), RF and XGBoost (Dhal and Azad 2022). Hyperparameter tuning is another key step in optimizing models, as it identifies the most effective combination of hyperparameter values that enhances model performance on unseen data (Boehmke and Greenwell 2019).

15.3.5 Model Development

Following feature selection and hyperparameter tuning, various ML models are developed to estimate the land value for individual parcels. These models are trained using the preprocessed and spatially enriched dataset, where the known land values in the ground truth data serve as the target variable. Each model is trained on the same feature space derived from the previously selected important factors and evaluated against a common test set.

15.3.6 Evaluation and Model Selection

To ensure accuracy and reliability, the performance of each developed model is assessed using standard metrics like Coefficient of Determination (R^2), Mean Absolute Percentage Error (MAPE) and Root Mean Square Error (RMSE). Accordingly, the most suitable model for mass land valuation is selected.

15.4 Land Use Suggestion Scenarios

Once the most suitable sub-market areas for developing residential buildings have been prioritized and parcel-level land values have been accurately estimated, the next critical step is determining appropriate and actionable land use strategies for housing affordability. This step could be considered a shift from spatial analysis to spatial planning. Indeed, this step mainly focuses on recommending solutions for planning, repurposing or redeveloping parcels based on their distinct characteristics regarding current land use, land value and regulatory context.

This complex planning procedure can be addressed by categorizing parcels based on their land value and proposing tailored interventions for each category. These interventions should consider factors such as current land use, zoning constraints, ownership status and surrounding urban functions. Likewise, parcels with low current value may represent latent opportunities for housing or infrastructure. In contrast, high-value parcels may demand regulatory

TABLE 15.3

Parcel Action Scenarios for Affordable Housing Development

Land Use Type	Zoning/Planning Designation	Land Value	Suggested Action
Residential	Low-density	High	Upzone to allow higher density
	High-density	Low	Infill development or subsidize affordable housing development
Commercial	Commercial or mixed-use	High	Promoting vertical mixed-use
	Residential	Low	Considering adaptive reuse
Industrial	Industrial	High	Potential for rezoning or brownfield redevelopment
Vacant	Mixed-use or high-density	Low to moderate	Acquire for public housing, services or green space
	Any	High	Applying LVC tools

oversight or the application of LVC tools to ensure equity and sustainability. Translating parcel-level intelligence into strategic land use actions is essential for effective housing policy and urban development planning. To guide this process, Table 15.3 provides a generalized framework of land use suggestion scenarios that can be applied across various global urban contexts. It categorizes land use types, planning designations and relative land value classes (low, moderate and high) and matches them with recommended planning actions.

The actions suggested in Table 15.3 for each category of land parcels try to suggest efficient, equitable and sustainable land use planning. Regarding residential parcels, a distinction is made between low-density and high-density zoning contexts. In the case of the presence of high-value parcels in low-density areas, an upzoning strategy is recommended to activate underutilized development capacity and enhance the market's ability to meet growing housing needs. In such cases, policymakers should grant approvals and actively support multi-story developments, enabling builders/investors to accommodate population growth within existing urban boundaries. This approach enhances developers' profitability and improves renters' access to job centers and essential services, improving the impactful housing supply. In contrast, low-value parcels in high-density zones represent unrealized development capacity. This could arise from market failure, regulatory issues, frictions and/or barriers to residential development. In such situations, decision-makers are expected to adopt relevant strategies, such as affordable housing subsidies, public–private partnerships or infill development incentives to promote effective residential development. These measures could help ensure that low-value lands in high-density zones are effectively transformed into affordable housing solutions, especially for low-income renters who are being pushed out of the real estate market in central urban areas near job centers and services due to rising rents and intense competition.

Additionally, for land parcels experiencing high land values in commercial zones, vertical mixed-use development enables a balanced approach to merging commercial functions and housing supply. Permitting residential units to be built above retail or office spaces in multiplex buildings maximizes land use efficiency, creating mutual benefits by enhancing commercial vitality and easing rental demand. Meanwhile, commercial parcels with low values located in residential areas are suited to be supported by adaptive reuse or rezoning strategies. Transitioning these sites to housing or community infrastructure aligns land use with emerging needs and promotes more coherent planning.

For industrial lands, high-value parcels may be candidates for redevelopment, particularly if the industrial activity has ceased or become incompatible with adjacent residential uses. Such cases may no longer be productive and could support higher and better uses. Redevelopment into mixed-use or residential uses can increase returns while improving urban form.

Finally, and more importantly, vacant land parcels offer substantial potential for public intervention and informed decision-making. Unlike other land use types, vacant parcels provide a great opportunity and flexibility for planning authorities to shape future development outcomes effectively. Particularly, vacant parcels with low to moderate value located in mixed-use or high-density zones are ideal targets for public acquisition to support affordable housing. On the other hand, the elevated value of any high-value vacant parcels stems from surrounding public actions, such as infrastructure upgrades, land use changes or proximity to civic amenities. This form of value uplift, commonly referred to as an unearned increment, presents an opportunity for LVC. By applying instruments such as betterment levies, development charges or inclusionary zoning, planning authorities can ensure that private landowners do not disproportionately capture the financial benefits of public investment. For instance, developers may be required to contribute to affordable housing funds, allocate a portion of units for lower-income households or pay a levy based on the land's uplifted value. These captured funds can be reinvested into essential infrastructure, public services or affordable housing developments.

15.5 Conclusions

This chapter presents a data-driven framework designed to address the pressing issue of rental affordability by integrating location recommendation using optimization techniques, land valuation using ML and policy-oriented land use planning. Grounded in the principles of spatial intelligence, cutting-edge techniques and stakeholder-inclusive land use planning, the framework facilitates more equitable, efficient and scalable decision-making. Overall, the three main components of this chapter contribute to a new paradigm of land intelligence, where location selection, valuation and planning are not isolated tasks but interdependent processes within a unified system. The framework is adaptable across urban contexts and scalable with data availability, making it relevant for governments, housing agencies, developers and researchers seeking to tackle rental affordability through data-informed and evidence-based urban planning management.

References

AHURI. 2022. *Why Does Australia Have a Rental Crisis, and What Can Be Done about It?* Australian Housing and Urban Research Institute (AHURI). https://www.ahuri.edu.au/research/brief/why-does-australia-have-rental-crisis-and-what-can-be-done-about-it.

ANZ-CoreLogic. 2023. *Housing Affordability Report*. Australia and New Zealand Banking Group Limited (ANZ). https://news.anz.com/posts/2023/05/anz-news-corelogic-housing-affordability-report-2023.

Boehmke, B., and B. M. Greenwell. 2019. *Hands-On Machine Learning with R*. CRC Press.

Canelas, Patricia, and Luise Noring. 2022. "Governmentalities of Land Value Capture in Urban Redevelopment." *Land Use Policy* 122: 106396. https://doi.org/10.1016/j.landusepol.2022.106396.

Deb, Kalyanmoy. 2011. "Multi-Objective Optimisation Using Evolutionary Algorithms: An Introduction." In *Multi-Objective Evolutionary Optimisation for Product Design and Manufacturing*, edited by Lihui Wang, Amos H. C. Ng and Kalyanmoy Deb, 3–34. Springer London.

Deng, Feng, and Youqin Huang. 2004. "Uneven Land Reform and Urban Sprawl in China: The Case of Beijing." *Progress in Planning* 61: 211–36. https://doi.org/10.1016/j.progress.2003.10.004.

Dhal, Pradip, and Chandrashekhar Azad. 2022. "A Comprehensive Survey on Feature Selection in the Various Fields of Machine Learning." *Applied Intelligence* 52 (4): 4543–81. https://doi.org/10.1007/s10489-021-02550-9.

Hardin, Herschel. 2023. "Breaking the Immigration Taboo: Canada Needs to Reduce Immigration Dramatically Until the Housing Crisis Is Resolved, Especially in Vancouver." *Inroads: A Journal of Opinion* (53): 159–78. https://inroadsjournal.ca/breaking-the-immigration-taboo.

Hengstermann, Andreas, and Vera Götze. 2023. "Planning-Related Land Value Changes for Explaining Instruments of Compensation and Value Capture in Switzerland." *Land Use Policy* 132: 106826. https://doi.org/10.1016/j.landusepol.2023.106826.

Hu, Shougeng, Qiuming Cheng, Le Wang, and Deyi Xu. 2013. "Modeling Land Price Distribution Using multifractal IDW Interpolation and Fractal Filtering Method." *Landscape and Urban Planning* 110: 25–35. https://doi.org/10.1016/j.landurbplan.2012.09.008.

Jafary, Peyman, Davood Shojaei, Sarah Pishgahi, Abbas Rajabifard, and Tuan Ngo. 2024a. "Data-Driven Strategies for Affordable Housing: A Hybrid Genetic Algorithm-Machine Learning Optimization Model in the Melbourne Metropolitan Area." *ISPRS Annals of the Photogrammetry, Remote Sensing and*

Spatial Information Sciences X-4-2024: 175–82. https://doi.org/10.5194/isprs-annals-X-4-2024-175-2024.

Jafary, Peyman, Davood Shojaei, Abbas Rajabifard, and Tuan Ngo. 2024b. "Automated Land Valuation Models: A Comparative Study of Four Machine Learning and Deep Learning Methods Based on a Comprehensive Range of Influential Factors." *Cities* 151: 105115. https://doi.org/10.1016/j.cities.2024.105115.

Konak, Abdullah, David W. Coit, and Alice E. Smith. 2006. "Multi-Objective Optimization Using Genetic Algorithms: A Tutorial." *Reliability Engineering & System Safety* 91 (9): 992–1007. https://doi.org/10.1016/j.ress.2005.11.018.

Lakševics, Kārlis, Yvonne Franz, Annegret Haase, Bahanur Nasya, Daniela Patti, Ursula Reeger, Ieva Raubiško, Anika Schmidt, and Andris Šuvajevs. 2023. "The Permanent Regime of Temporary Solutions: Housing of Forced Migrants in Europe as a Policy Challenge." *European Urban and Regional Studies* 31 (1): 81–87. https://doi.org/10.1177/09697764231197963.

Lima, Valesca. 2024. "Pandemic Housing Policies: Mitigation Strategies and Protection of Rights." *Housing and Society* 51 (1): 34–49. https://doi.org/10.1080/08882746.2023.2266307.

Pawson, Hal, Chris Martin, Fatemeh Aminpour, Kenneth Gibb, and Chris Foye. 2022. *COVID-19: Housing Market Impacts and Housing Policy Responses-An International Review.* Australian Council of Social Service.

PEXA. 2023. "Australia's housing crisis – Uncovering the real causes." Property Exchange Australia (PEXA). https://www.pexa.com.au/content-hub/real-causes-housing-crisis/.

Saiz, Albert. 2023. "The Global Housing Affordability Crisis: Policy Options and Strategies." *MIT Center for Real Estate Research Paper* (23/01). https://dx.doi.org/10.2139/ssrn.4402329.

Salcedo-Sanz, S., L. Cornejo-Bueno, L. Prieto, D. Paredes, and R. García-Herrera. 2018. "Feature Selection in Machine Learning Prediction Systems for Renewable Energy Applications." *Renewable and Sustainable Energy Reviews* 90: 728–41. https://doi.org/10.1016/j.rser.2018.04.008.

Silveira, Paulo, and Tomaz Ponce Dentinho. 2018. "A Spatial Interaction Model with Land Use and Land Value." *Cities* 78: 60–66. https://doi.org/10.1016/j.cities.2017.08.023.

Vejchodská, Eliška, Ana Paula Barreira, Armands Auziņš, Evelin Jürgenson, Steven Fowles, and Vida Maliene. 2022. "Bridging Land Value Capture with Land Rent Narratives." *Land Use Policy* 114: 105956. https://doi.org/10.1016/j.landusepol.2021.105956.

16

Ensuring 3D Spatial Data Consistency in Digital Cadasters

Ali Asghari

16.1 Introduction

Rapid urbanization, climate change, economic pressures, and population growth continue to challenge cities worldwide. These forces drive the need for more efficient use of urban spaces, leading to the proliferation of multi-story buildings with complex ownership arrangements. Managing these complexities is essential for resilient and sustainable urban development. Urban land administration systems play a critical role in documenting and managing property rights, restrictions, and responsibilities (RRRs). Traditionally, these systems have been built on two-dimensional (2D) frameworks that struggle to handle the three-dimensional (3D) spatial complexities of modern urban environments. The transition to 3D digital cadasters is not just a technological upgrade; it is a fundamental shift required to meet current and future urban challenges. This chapter summarizes the principles, challenges, and solutions for validating 3D spatial data in urban land administration, based on research conducted in Victoria, Australia.

16.2 Challenges of 2D-Based Land Administration

Current land administration practices rely heavily on 2D cadastral plans to represent ownership spaces within multi-story buildings. These plans typically use multiple 2D floor plans and cross-sections to define complex 3D ownership settings. This approach introduces several limitations:

- Inability to accurately represent volumetric ownership spaces.
- Ambiguity in defining legal boundaries in vertical arrangements.
- Manual, time-consuming examination processes are prone to human error.

 DOI: 10.1201/9781003638308-18

- Difficulty in querying, visualizing, and communicating complex ownership settings.
- Difficulty in examining the subdivision plans and validating the ownership settings in complex multi-story buildings.

As urban environments grow in complexity, these limitations hinder effective land administration, leading to inefficiencies and potential disputes over property rights.

16.3 The Need for 3D Digital Cadasters

Advances in 3D geospatial technologies, including Building Information Modelling (BIM) and Geographic Information Systems (GIS), offer solutions to these challenges. A 3D digital cadaster provides:

- Accurate volumetric representation of ownership spaces.
- Clear delineation of private, communal, and public property boundaries.
- Enhanced data communication and visualization capabilities.
- Opportunities for automation in data examination and validation.

However, the successful implementation of 3D cadasters requires not only technical tools but also robust principles, validation rules, and examination processes designed specifically for 3D cadastral data (Asghari et al., 2019).

16.4 Principles and Validation Rules for 3D Cadasters

To ensure the spatial and legal integrity of 3D cadastral data, a comprehensive set of validation principles should be developed (Asghari, 2022). These principles should address two main aspects:

- Semantical information
- Spatial information

The spatial information should mainly focus on two aspects: internal and external spatial consistency of cadastral data.

16.4.1 Internal Spatial Consistency

The internal spatial consistency principles assure 3D geospatial data is correctly and accurately defined based on the standards in each application (Asghari et al., 2020). The spatial extent of legal space, to which rights, restrictions, and responsibilities relate in a 3D digital cadaster needs to be accurately defined and geometrically closed, watertight. Several techniques are engaged to formulate watertight concept for 3D digital cadaster. According to this concept, each 3D parcel is geometrically valid if it is closed, not self-intersected, and correctly oriented.

A real cadastral dataset is used as our case study. This dataset includes 50 legal spaces representing both private ownership and common property. To testify the algorithms, different scenarios were designed to showcase how the geometry closure of 3D digital data associated with various complex legal spaces of a real multi-story building can be checked against the validation rules mentioned above (Figure 16.1).

In all cases, the algorithms developed using C++ language and CGAL libraries could correctly and independently detect the errors associated to each parcel.

16.4.2 External Spatial Consistency

3D geospatial data is progressively adopted in urban land administration to represent 3D ownership rights in multi-story buildings. However, the urban land administration's integrity highly depends on the validity and quality of cadastral data. In a 3D digital representation of cadaster, the 3D space is subdivided into 3D volumetric ownership spaces partitioning the 3D space without overlaps or gaps (Stoter and Salzmann, 2003). This concept is also called title boundary consistency in surveying and mapping. Checking double-defined ownership spaces (overlaps) and undefined ownership spaces (gaps) are essential as two critical external spatial consistency principles for checking the correctness of 3D subdivision (Figure 16.2). We integrated the principles into practice and developed methods to check the validity of 3D parcels and their relationships in complex ownership settings.

A real cadastral dataset is used as our case study to check the external spatial consistency (Figure 16.3). This dataset includes 248 lots and 3 types of common property in an 18-floor mixed-use apartment complex. Each lot's ownership space was represented as an aggregation of dwelling, balcony, car park, and storage spaces. Like lots, an aggregation of the spaces described common property areas. An outer geometry for each level, which represents the parcel/title boundary delineating the maximum legal right limit, was also developed in the BIM model. Developing each level's outer geometry in which all the legal spaces are located is a critical step in the developed

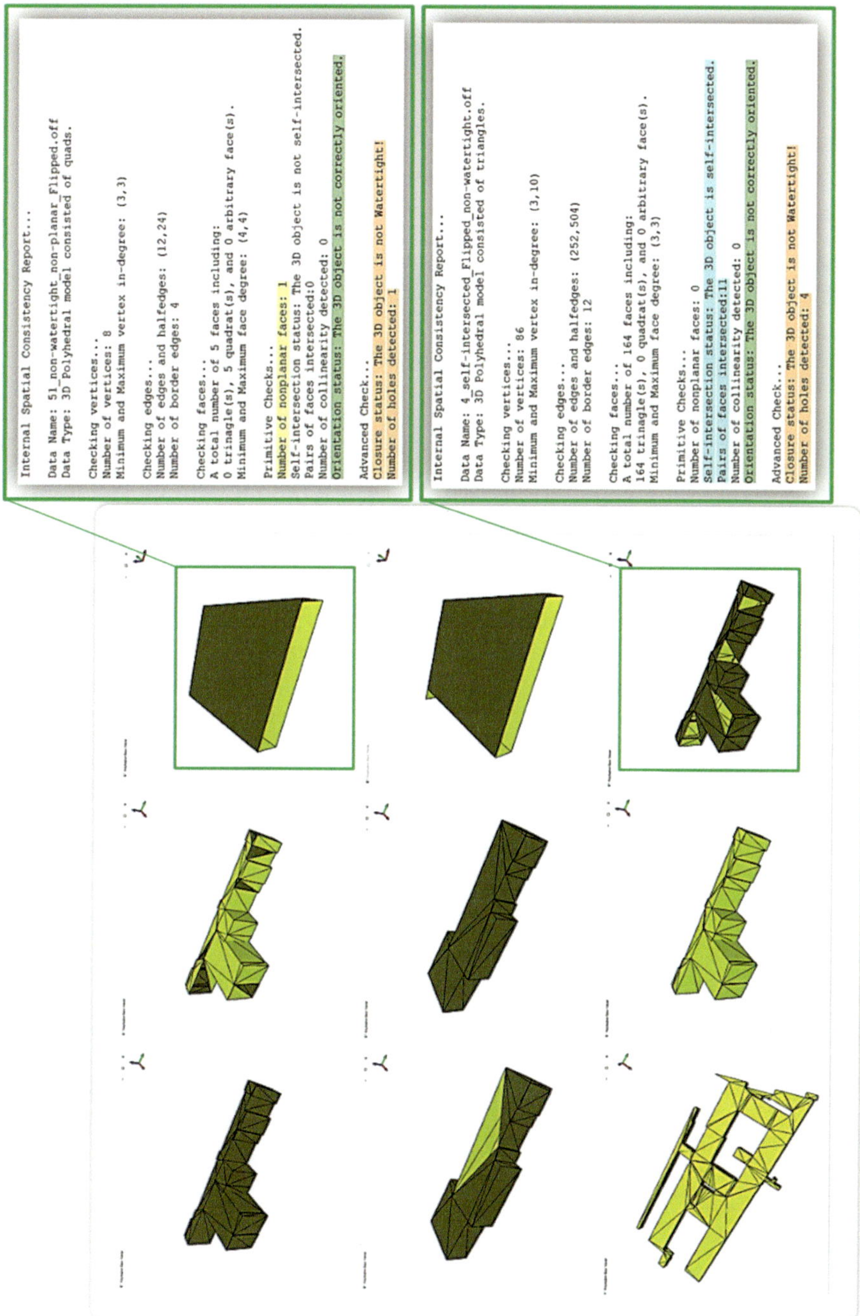

FIGURE 16.1
Testing different scenarios using the real cadastral parcels.

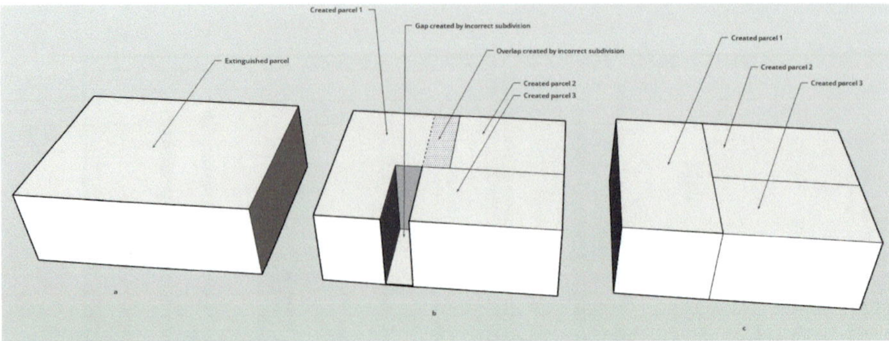

FIGURE 16.2
Title boundary consistency in 3D; (a) extinguished parcel, (b) incorrect subdivision, (c) correct subdivision.

method for identifying gaps. A polyhedral surface model with Object (OBJ) data structure was used to represent the ownership spaces.

Mathematical algorithms were developed to automate these validations. A novel method is developed for checking the gaps and overlaps among 3D parcels using a combination of 3D Boolean operations such as 3D Union, 3D Difference, and 3D Intersection. This method would help us to identify the undefined ownership spaces (gap), double-defined ownership spaces (overlap) and give us information about surrounding objects using some other functions. The methods developed in this chapter can be employed to check the validity of cadastral data and will help modelers and surveyors to check their native models for possible errors. The methods could also be exploited in an automatic examination process and could profit examiners in land registry.

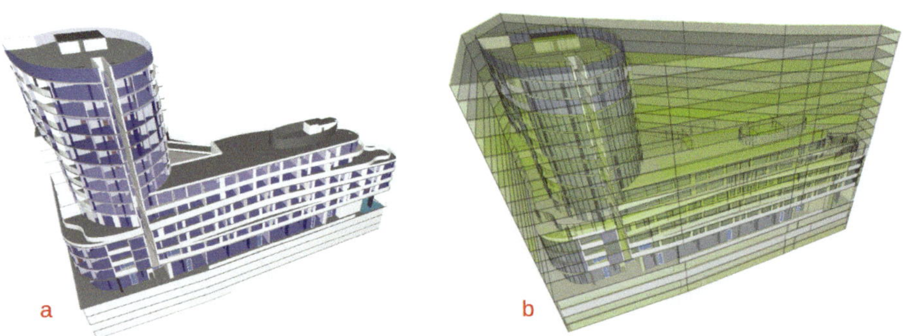

FIGURE 16.3
Case study; (a) physical model and (b) legal and physical model.

16.5 Integrating BIM into 3D Cadastral Validation

BIM offers a powerful platform for generating detailed 3D representations of built environments. However, aligning BIM data with cadastral requirements presents some challenges:

- BIM focuses on physical components, while cadasters require legal property boundaries.
- Data conversion from Industry Foundation Classes (IFC) to cadastral formats may introduce errors.
- Semantic enrichment is necessary to bridge the gap between physical and legal representations.

We developed validation workflows and guidelines to ensure BIM-derived cadastral data maintains spatial and legal integrity throughout its lifecycle (Asghari et al., 2022). In the proposed workflows, the validity of BIM-derived cadastral data could be assessed in two different approaches, during data creation or post-creation, respectively named formative and summative assessment. Summative assessment is the method that assesses the validity of the final cadastral output. However, a formative approach can be used during the process of data generation. Both formative and summative assessments play vital roles in the process of enriching BIM-based cadastral data models and improving the quality of the final output (Asghari et al., 2021).

16.6 Implications for Resilient and Sustainable Land Administration

The developed principles and methods were applied to complex cadastral datasets from Victoria. Key outcomes included:

- Successful detection of spatial inconsistencies (e.g., overlaps, gaps, non-watertight geometries).
- Automation of examination processes, significantly reducing manual effort.
- Improved accuracy and clarity in representing complex ownership arrangements.

These case studies validated the practicality of the proposed framework and demonstrated its potential to modernize cadastral practices in Victoria and beyond.

Transitioning to 3D cadasters supports broader objectives of resilient and sustainable urban development. Accurate and reliable cadastral data:

- Enhances transparency in property transactions.
- Supports infrastructure resilience by providing detailed spatial information.
- Facilitates urban planning and disaster management through accurate 3D models.

By addressing the limitations of 2D-based systems, 3D cadasters contribute to smarter, more sustainable cities capable of meeting the challenges of the 21st century.

16.7 Conclusion

The shift toward 3D digital cadasters is essential for modern urban land administration. This chapter has outlined a structured approach to validating 3D cadastral data, ensuring the spatial and legal integrity of property boundaries in complex urban environments.

Through the development of principles, validation rules, and computational methods, this chapter provides a practical foundation for modernizing land administration systems. Future work should focus on integrating these validation frameworks into operational systems, developing jurisdiction-specific standards, and promoting the adoption of 3D cadasters as a standard practice in urban land administration.

References

Asghari, A., Kalantari, M., & Rajabifard, A. (2019). A structured framework for 3D cadastral data validation – A case study for Victoria, Australia. Land Use Policy, 98, 104359. DOI: https://doi.org/10.1016/j.landusepol.2019.104359

Asghari, A., Kalantari, M., & Rajabifard, A. (2020). Advances in techniques to formulate the watertight concept for cadastre. Transactions in GIS, 25(1), 213–237. DOI: https://doi.org/10.1111/tgis.12695

Asghari, A., Kalantari, M., & Rajabifard, A. (2021). Formative and summative validation of building information model-based cadastral data. Land, 10(8), 822. DOI: https://doi.org/10.3390/land10080822

Asghari, A., Kalantari, M., Rajabifard, A., & Shin, J. (2022). Developing an integrated approach to validate 3D ownership spaces in complex multistorey buildings. International Journal of Geographical Information Science, 37(1), 215–242. DOI: https://doi.org/10.1080/13658816.2022.2109159

Asghari, A. (2022). 3D Spatial Data Validation in Urban Land Administration. PhD Thesis, The University of Melbourne.

Stoter, J. & Salzmann, M. (2003). Towards a 3D cadastre: Where do cadastral needs and technical possibilities meet? Computers, Environment and Urban Systems, 27, 395–410.

17

An Integrated 3D Data Model for Underground Land Administration

Bahram Saeidian

17.1 Introduction

Cities are rapidly expanding both horizontally and vertically due to increasing population pressures, limited land availability, and growing infrastructure needs. This expansion has driven cities to explore and utilise underground spaces for a wide range of infrastructure, including subway systems, utility networks, pedestrian tunnels, car parks, and commercial complexes (Kim and Heo 2019). In Australia alone, underground infrastructure spans approximately 740,000 km and is valued at over $340 billion (AADBYDS 2019). As urban environments become increasingly dependent on these subterranean systems, managing their spatial and legal complexities becomes imperative.

Despite their growing significance, underground spaces are currently managed using fragmented 2D approaches that lack integration between physical reality, ownership rights, and cadastral surveying elements. These 2D systems rely on disconnected floor plans, cross-sections, and textual notations, leading to cognitive challenges in interpreting underground information and planning infrastructure developments. There is a growing recognition that a shift towards integrated 3D digital models is essential for addressing the shortcomings of existing systems (Bieda et al. 2020; Guler 2024; Janečka and Bobíková 2018; Kim and Heo 2017; Radulović et al. 2019; Ramlakhan et al. 2023; Saeidian et al. 2022a; Yan et al. 2019; Yan, Van Son, and Soon 2021).

This chapter presents a solution to these challenges: the VicULA Application Domain Extension (ADE), an integrated 3D data model developed using CityGML 3.0. This model is designed to represent, manage, and communicate underground physical assets, legal spaces, and survey elements in a unified digital environment. By bridging the gaps between fragmented data sources, the model supports sustainable urban development, improves data communication, and facilitates effective underground land administration (ULA).

DOI: 10.1201/9781003638308-19

17.2 Underground Land Administration: Current Gaps

ULA refers to the management of legal boundaries and the rights, restrictions, and responsibilities (RRRs) associated with underground assets (Saeidian et al. 2021). ULA encompasses three key categories of data: physical, legal, and survey data. Physical data describe the spatial and semantic characteristics of underground utilities and structures, legal data represent ownership rights and boundary definitions, and survey data provide the geodetic measurements that support boundary delineation.

Current ULA practices in many jurisdictions, including Victoria, Australia, are constrained by 2D representations that separately manage these three categories of data (Figure 17.1). Legal data are stored and communicated at the parcel level using documents such as 2D survey (subdivision) plans with cross-sectional diagrams. Physical assets are recorded using 2D utility maps and engineering drawings, while survey data are documented in standalone field reports such as Abstracts of Field Records (AFRs).

These disjointed datasets must often be manually consolidated by professionals to derive a comprehensive understanding of an underground environment. This process is time-consuming, error-prone, and inaccessible to non-experts (Atazadeh et al. 2022; Saeidian et al. 2022b). The fragmented nature of current practices leads to several significant challenges (Geovation 2016; Phillips, Arancibia, and Janes 2018; Saeidian et al. 2023a):

- Ambiguities in interpreting the spatial extent of legal spaces.
- Difficulties in resolving overlapping ownership claims.
- Delays and disputes in infrastructure planning and construction.
- Lack of a scalable approach to represent underground information at a city level.

There is a clear need for a holistic data model that integrates these disparate data types into a single framework, supports 3D spatial representations, and aligns with existing standards.

17.3 Approach

This chapter adopts a design science methodology to develop a 3D integrated data model for ULA. The design science methodology is appropriate for problem-solving in information systems and has been successfully

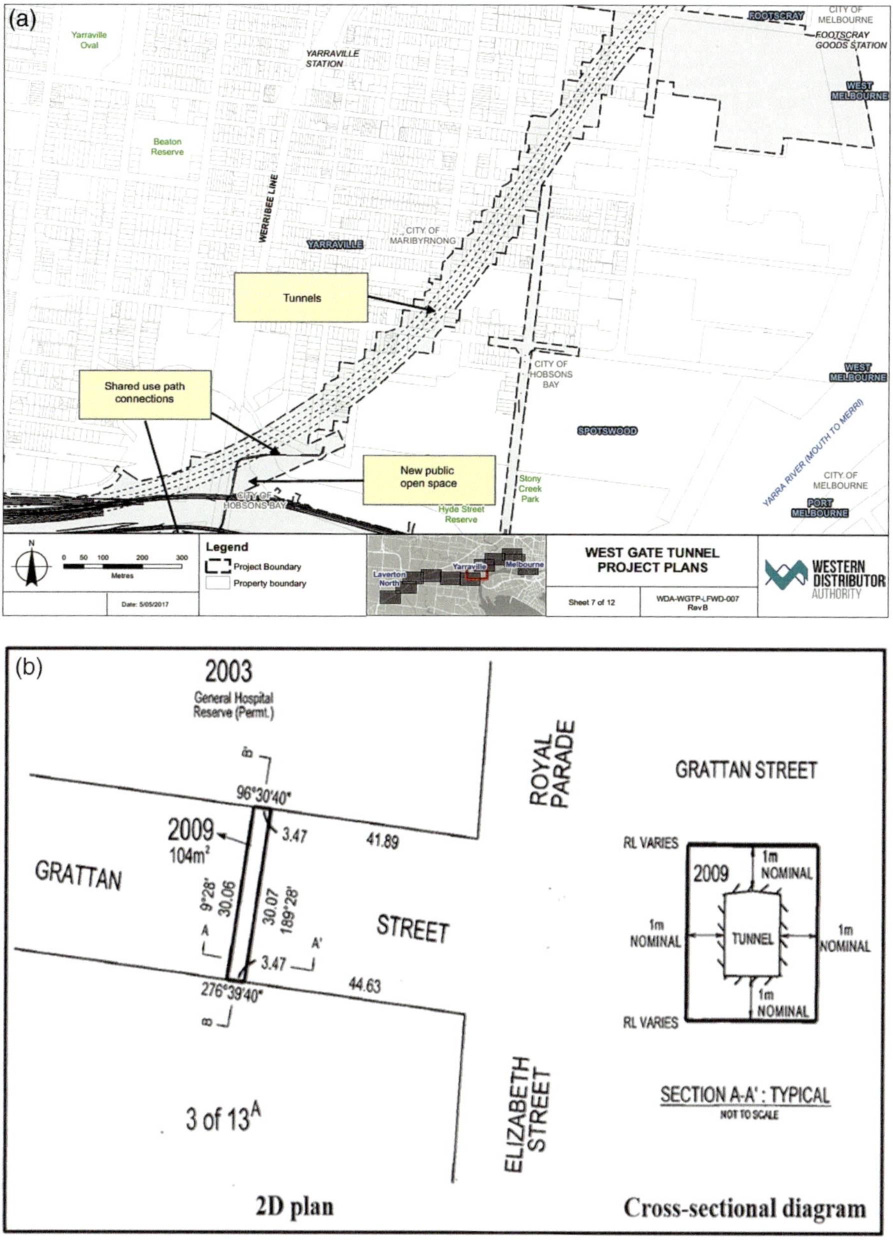

FIGURE 17.1
Examples of current practices: (a) The 2D spatial extent of a tunnel. (b) The legal space and boundaries of a tunnel.; (c) Survey report (AFR) of a tunnel. (Sources: (a) Saeidian et al. 2024, (b) Saeidian et al. 2024, and (c) Saeidian et al. 2023b.) *(Continued)*

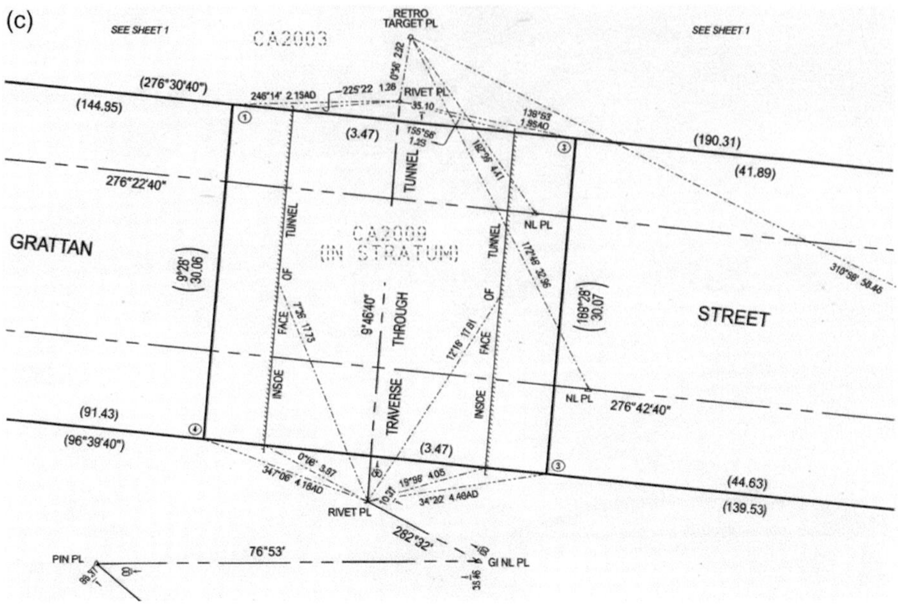

FIGURE 17.1 (*Continued*)

applied in land administration research (Atazadeh 2017). The process involved several key stages:

1. Problem explication: Based on a comprehensive review of literature and current practices, the limitations of 2D ULA methods were identified.
2. Requirement analysis: Physical, legal, and survey data requirements were elicited through analysis of use cases in Victoria, Australia.
3. Artefact design and development: A CityGML 3.0 ADE, termed VicULA ADE, was designed to model the required data elements.
4. Prototyping and evaluation: The VicULA ADE was implemented in selected case study areas using real-world data. Its feasibility and benefits were evaluated.

The methodology involved both top–down and bottom–up data modelling strategies (Silverston, Inmon, and Graziano 1997). The top–down approach ensured conceptual clarity and generalisability, while the bottom–up strategy allowed adaptation of existing models and standards. The model was developed at conceptual level as Unified Modelling Language (UML) diagrams and physical level as Extensible Markup Language (XML) schemas.

17.4 The VicULA ADE: A 3D Integrated Data Model

The VicULA ADE builds upon the modular architecture of CityGML 3.0 (Kolbe et al. 2021), extending it to model underground assets at a city scale. It introduces three core modules along with CityGML 3.0 modules for modelling underground physical, legal, and survey data (Figure 17.2):

- Physical assets: Represents underground utilities, tunnels, and structures with their geometric and semantic properties. CityGML 3.0 provides several modules such as *Tunnel* and *Building* to semantically model these assets in 3D. It also supports Level of Detail (LoD) differentiation to accommodate various planning and analysis needs (Kolbe et al. 2021).
- Legal spaces: The VicULA ADE introduces the *UndergroundParcel* module to encode different types of underground legal spaces, including primary underground parcels (e.g., lot, crown, common property, and reserve) and secondary underground interests (e.g., easement, restriction, and depth limitation) (Saeidian et al. 2023c).
- Legal boundaries: The VicULA ADE presents the *UndergroundBoundary* module to encode different types of underground legal boundaries in 3D. These boundaries are classified as surveyed, projected, or building boundaries, based on how they are delineated in real-world plans. The

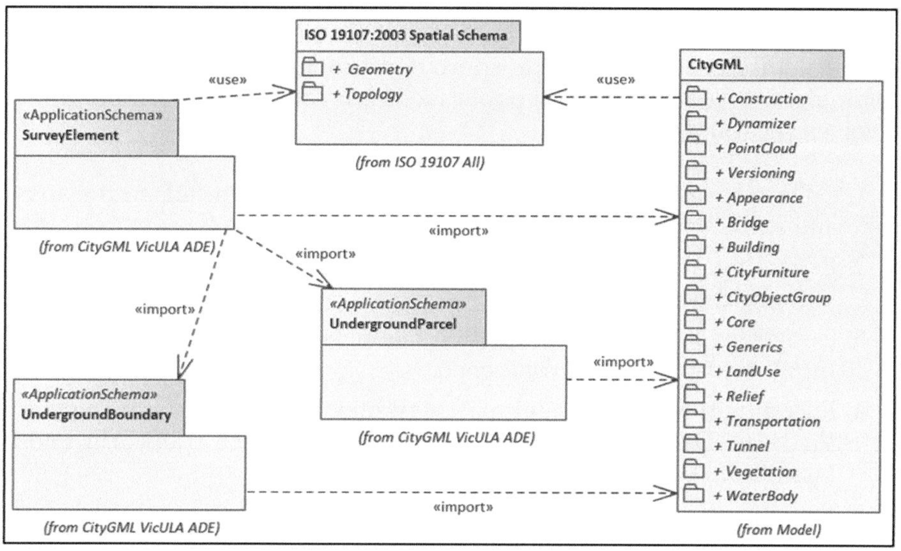

FIGURE 17.2
The usage of the VicULA ADE and CityGML 3.0 modules in the 3D integrated underground data model.

model captures boundary geometries and their references to physical (building) or survey elements (Saeidian et al. 2024).

- Survey elements: The *SurveyElement* module of the VicULA ADE integrates cadastral survey data, including survey observations (e.g., traverse, boundary, connection, and radiation), elevation information, and survey points (e.g., control, boundary, traverse, and reference points). These elements are critical for defining the spatial integrity of legal boundaries. The module covers the elements defined in AFRs, ensuring legal traceability (Saeidian et al. 2023b).

As shown in Figure 17.2, the VicULA ADE introduces semantic relationships across modules. For example, legal spaces are linked to their defining boundaries, which are in turn associated with physical or survey features. These relationships allow for complex spatial and legal queries within an integrated 3D framework. The conceptual data models (UML diagrams) and XML encodings (XSD schemas) of all modules are available at https://github.com/Bahram-Saeidian.

17.5 Case Study Implementation

To evaluate the feasibility of the VicULA ADE, implementation was carried out in several case study areas in Victoria. These areas included a mix of infrastructure types such as transportation tunnels, underground utilities, commercial developments, and public facilities. The implementation process involved the following steps:

1. Collecting and digitising legal and survey data from cadastral plans and AFRs.
2. Extracting physical asset data from engineering drawings, utility maps, and open data portals.
3. Designing 3D models using different tools such as ArcGIS, Autodesk InfraWorks, and FME Workbench.
4. Encoding the data in Geography Markup Language (GML) using the VicULA ADE and CityGML 3.0 schemas and Extract, Transform, and Load (ETL) pipelines.
5. Visualising and validating the model using 3D viewers such as FZKViewer.

Figure 17.3 shows 3D integrated models of some underground areas. The implemented prototypes with prepared datasets, the conversion pipelines,

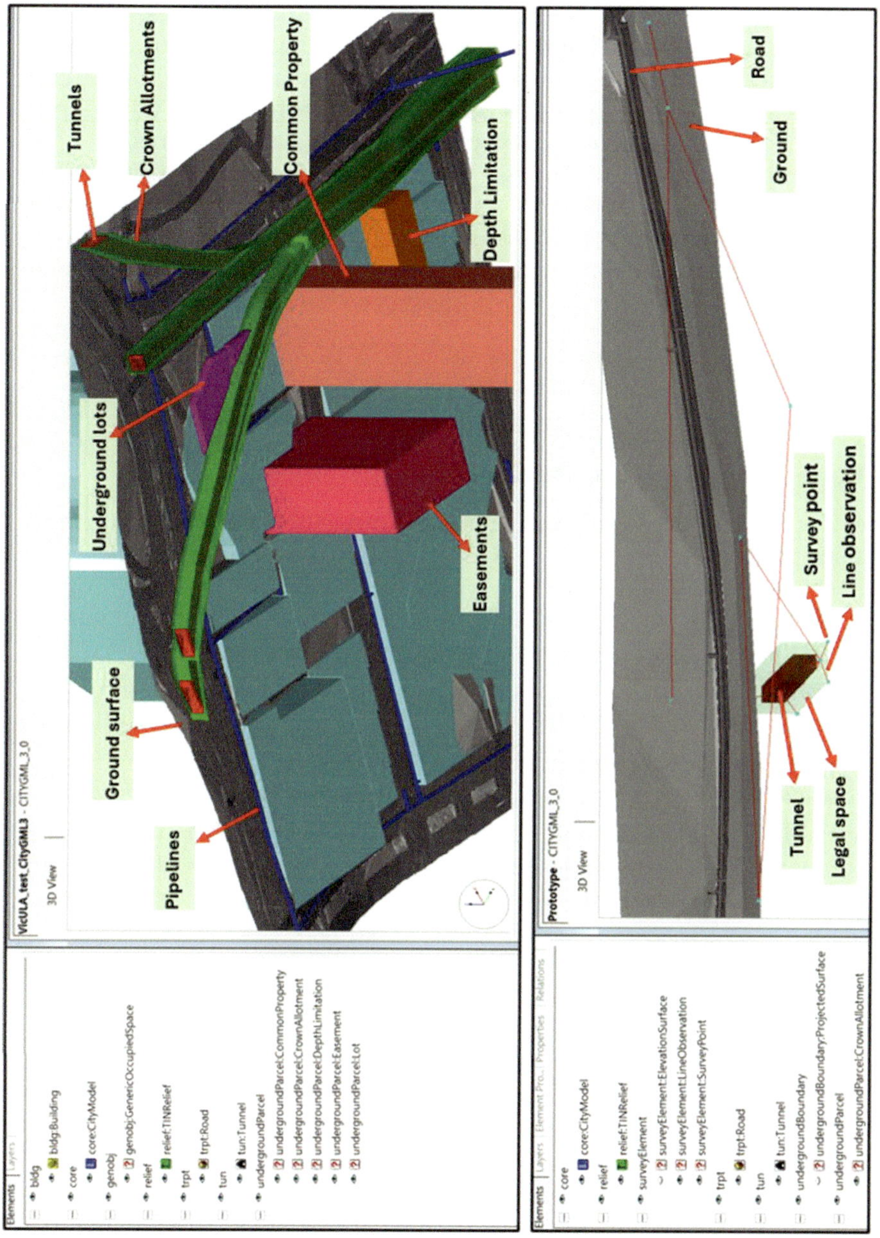

FIGURE 17.3
3D integrated models of some underground areas. (Sources: Saeidian et al. 2023b, 2023c.)

and the final GML files for all case studies are available at https://github.com/Bahram-Saeidian.

The implemented prototypes demonstrated that the VicULA ADE enables the accurate representation of legal ownership, boundary types, physical asset geometries, and survey references in a single 3D model. For instance, a tunnel defined by projected boundaries from a road centreline was linked to both legal space definitions and survey measurements or lots with depth limitations were modelled in conjunction with physical structures.

These implementations showcased the model's capacity to support spatial analysis, planning, and legal interpretation in complex subsurface environments. The models also enhanced communication between surveyors, planners, and utility managers by providing a common visual language.

17.6 Benefits and Future Outlook

The development of the VicULA ADE offers a range of benefits to urban land administration:

- Integrated data management: Consolidates physical, legal, and survey data, reducing redundancy and improving data quality.
- Enhanced decision-making: Supports complex spatial queries and 3D visualisations, aiding planning, construction, and conflict resolution.
- Legal certainty: Makes underground ownership information more accessible and verifiable, helping avoid disputes and project delays.
- Scalability and interoperability: Built on CityGML 3.0, the model can be adapted to other jurisdictions and linked with other standards such as LADM and BIM.

The study opens up several possibilities for future research:

- 4D modelling: Incorporating temporal dimensions to track changes in underground boundaries, ownerships, and infrastructure over time.
- Integration with BIM: Extend the ADE to enable interoperability with Building Information Models such as Industry Foundation Classes (IFC) for seamless integration with architectural workflows.
- Administrative data modelling: Future studies can extend the ADE to incorporate administrative elements such as actors (e.g., owners, surveyors, authorities) and documents (e.g., titles, engineering plans, certifications). This may be achieved either by integrating CityGML with LADM or by designing new ADE modules.

- Enrichment of physical asset representation: Further improvements are needed in modelling a broader range of underground utilities, incorporating and adapting existing ADEs such as the Utility Network ADE (Kutzner, Hijazi, and Kolbe 2018) to CityGML 3.0.
- Metadata modelling: Develop components to represent the accuracy and instrumentation methods used for data acquisition to increase trust in underground spatial data.
- Interoperability with other models: Research will explore harmonising the VicULA ADE with standards like MUDDI, LandInfra, and LADM to support diverse underground use cases across jurisdictions.

17.7 Conclusion

This chapter presented the development and implementation of the VicULA ADE, an integrated 3D data model for ULA. By extending CityGML 3.0, the model offers a scalable, semantically rich solution for managing physical, legal, and survey data in subterranean environments. Its application in Victorian case studies demonstrates how a unified 3D representation improves data communication, supports complex spatial queries, and enhances legal clarity.

The shift from fragmented 2D practices to integrated 3D models marks a significant step forward in managing underground space. As cities continue to build downward, models like the VicULA ADE will become essential tools for ensuring the efficient, equitable, and sustainable use of underground areas.

References

AADBYDS. 2019. "Association of Australian Dial before You Dig Services 20 Year History." https://www.1100.com.au/association-of-australian-dial-before-you-dig-services-20-year-history/.

Atazadeh, Behnam. 2017. "Building information modelling for urban land administration." Doctoral dissertation, Department of Infrastructure Engineering, The University of Melbourne.

Atazadeh, Behnam, Hamed Olfat, Abbas Rajabifard, and Bahram Saeidian. 2022. "Evaluation of the International 3D Geospatial Data Models and IFC Standard for Implementing an LADM-based 3D Digital Cadastre." 10th International FIG workshop on the Land Administration Domain Model, Dubrovnik, Croatia.

Bieda, Agnieszka, Jarosław Bydłosz, Artur Warchoł, and Monika Balawejder. 2020. "Historical Underground Structures as 3D Cadastral Objects." Remote Sensing 12 (10): 1547.

Geovation. 2016. "Deep Dig, Underground Assets Challenge." https://geovation.uk/wp-content/uploads/2017/01/A00031-Deep-Dig-Booklet_UPDATE.pdf.

Guler, Dogus. 2024. "3D Modelling of Subsurface Legal Spaces and Boundaries for 3D Land Administration." Tunnelling and Underground Space Technology 152: 105956.

Janečka, Karel, and Diana Bobíková. 2018. "Registering the Underground Objects in the 3D Cadastre: A Case Study of Wine Cellar Located in the Vineyard Area Tokaj." Acta Montanistica Slovaca 23 (3): 260–270.

Kim, Sangmin, and Joon Heo. 2017. "Development of 3D Underground Cadastral Data Model in Korea: Based on Land Administration Domain Model." Land Use Policy 60: 123–138.

Kim, Sangmin, and Joon Heo. 2019. "Registration of 3D Underground Parcel in Korean Cadastral System." Cities 89: 105–119.

Kolbe, Thomas H., Tatjana Kutzner, Carl Steven Smyth, Claus Nagel, Carsten Roensdorf, and Charles Heazel. 2021. OGC city geography markup language (CityGML) part 1: Conceptual model standard. Open Geospatial Consortium.

Kutzner, Tatjana, Ihab Hijazi, and Thomas H. Kolbe. 2018. "Semantic Modelling of 3D Multi-Utility Networks for Urban Analyses and Simulations: The CityGML Utility Network ADE." International Journal of 3-D Information Modeling (IJ3DIM) 7 (2): 1–34.

Phillips, Wilson, Gabriel Arancibia, and Hal Janes. 2018. "The 3D Cadastre for Underground Infrastructure: An Integrated, Reliable and Safety Proposal from Land Surveyors in Canada." FIG Congress 2018, Istanbul, Turkey.

Radulović, Aleksandra, Dubravka Sladić, Miro Govedarica, Aleksandar Ristić, and Dušan Jovanović. 2019. "LADM Based Utility Network Cadastre in Serbia." ISPRS International Journal of Geo-Information 8 (5): 206.

Ramlakhan, Rohit, Eftychia Kalogianni, Peter van Oosterom, and Behnam Atazadeh. 2023. "Modelling the Legal Spaces of 3D Underground Objects in 3D Land Administration Systems." Land Use Policy 127: 106537.

Saeidian, Bahram, Abbas Rajabifard, Behnam Atazadeh, and Mohsen Kalantari. 2021. "Underground Land Administration from 2D to 3D: Critical Challenges and Future Research Directions." Land 10 (10): 1101.

Saeidian, Bahram, Abbas Rajabifard, Behnam Atazadeh, and Mohsen Kalantari. 2022a. "Development of an LADM-Based Conceptual Data Model for 3D Underground Land Administration in Victoria." 10th International FIG workshop on the Land Administration Domain Model, Dubrovnik, Croatia.

Saeidian, Bahram, Abbas Rajabifard, Behnam Atazadeh, and Mohsen Kalantari. 2022b. "Extending CityGML 3.0 to Support 3D Underground Land Administration." International Archives of the Photogrammetry, Remote Sensing and Spatial Information Sciences XLVIII-4/W4-2022: 125–132.

Saeidian, Bahram, Abbas Rajabifard, Behnam Atazadeh, and Mohsen Kalantari. 2023a. "Data Lifecycle of Underground Land Administration: A Systematic Literature Review." Survey Review 55 (392): 396–415.

Saeidian, Bahram, Abbas Rajabifard, Behnam Atazadeh, and Mohsen Kalantari. 2023b. "Modelling Underground Cadastral Survey Data in CityGML." Transactions in GIS 27 (6): 1724–1747.

Saeidian, Bahram, Abbas Rajabifard, Behnam Atazadeh, and Mohsen Kalantari. 2023c. "A Semantic 3D City Model for Underground Land Administration: Development and Implementation of an ADE for CityGML 3.0." Tunnelling and Underground Space Technology 140: 105267.
Saeidian, Bahram, Abbas Rajabifard, Behnam Atazadeh, and Mohsen Kalantari. 2024. "Managing Underground Legal Boundaries in 3D – Extending the CityGML Standard." Underground Space 14: 239–262.
Silverston, Len, William H. Inmon, and Kent Graziano. 1997. The data model resource book: A library of logical data models and data warehouse designs. John Wiley & Sons, Inc.
Yan, Jingya, Rob Van Son, and Kean Huat Soon. 2021. "From Underground Utility Survey to Land Administration an Underground Utility 3D Data Model." Land Use Policy 102 (10526).
Yan, Jingya, Siow Wei Jaw, Kean Huat Soon, Andreas Wieser, and Gerhard Schrotter. 2019. "Towards an Underground Utilities 3D Data Model for Land Administration." Remote Sensing 11 (17): 1957.

18

Evolution of AI in Urban Design

Yiqun Chen, Abbas Rajabifard, Mazdak Ghasemi Tootkaboni, and Victoria Oberin

18.1 Introduction

Urban design, rooted in city planning and architecture, emerged as a distinct discipline following the influential 1956 Harvard conference, which formalized its identity by integrating planning and architectural practices (Cuthbert, 2007). Traditionally interdisciplinary, it draws from architecture, landscape design, and planning to shape functional, appealing urban spaces (Foroughmand Araabi, 2016). Early methods relied on manual tools and planners' judgement, using models and drawings. As urban challenges became more complex, spanning land use, transport, and sustainability, design methods evolved to become more systematic and iterative, involving multiple stakeholders (Mehaffy, 2008; Moura & de Santana, 2014). The integration of artificial intelligence (AI) into urban design represents a significant transformation in city design, management methods, and philosophies. With its capacity to process vast datasets and detect complex patterns, AI introduces new ways for designers to engage with urban environments. This shift stems from increasing digitalization and the growing complexity of urban challenges, including rapid urbanization, sustainability, and infrastructure adaptability (Salvini, 2018). AI enhances urban design through predictive modelling and data-driven insights, enabling more precise and innovative solutions (Gómez-Villarino et al., 2021).

This chapter scrutinizes the integration of AI with traditional urban design approaches, aiming to blend AI-driven techniques and conventional practices effectively, focusing on the expansive role of AI in urban design. The databases Scopus and Web of Science are selected to capture the most relevant and up-to-date literature on AI applications in this domain. The selection of keywords included urban design, AI, machine learning, artificial neural networks (ANNs), generative adversarial networks (GANs), large language models (LLMs), ChatGPT, GeoGPT, DALL-E, and Midjourney. The period for the literature search is restricted to those published in English between

 DOI: 10.1201/9781003638308-20

2010 and 2024 for their relevance to the current state of AI development and involvement in the urban design process.

The following sections provide a critical analysis of AI's transformative impact on urban design, aligning with the core aim of this review. It traces the recent evolution of AI in the field, highlighting key applications and their role in reshaping design practices. The discussion covers AI's strengths in processing complex urban data, guiding design decisions, addressing ethical concerns, and identifying current limitations and future challenges. It also examines the efficiencies gained through automation and how AI is redefining the strategic direction and outcomes of urban design.

18.2 Pre-AI Computational Models

Before the advent of AI, urban design practice primarily relied on heuristic-based methods. These traditional approaches were grounded in experiential knowledge and established practices, typically involving rules of thumb, subjective assessments, and generalized principles that stem from historical examples and the expertise of professionals. Heuristic methods often lack the granularity and specificity required to address diverse urban needs and may not effectively incorporate emerging trends and data (Gordon, 2011).

The advent of computer-aided design (CAD) systems and mathematical modelling marked a shift from traditional methodologies, characterized by hands-on processes such as physical models and detailed drawings. Huang et al. (2022) elaborate a detailed exposition of the conventional approach to urban design, emphasizing a hands-on, artisanal process where the expert judgement of urban planners and architects plays a significant role. Despite its human-centric design richness, this methodology encounters limitations in uniformly tackling diverse urban challenges across varied contexts and scales. In addressing the limitations, Kunze et al. (2011) emphasize a conceptual framework for participatory design. This framework integrates professional expertise in urban aesthetics, functionality, and spatial organization with stakeholder and community insights essential for sustainable urban development.

The work of Beirão et al. (2010) discusses the initial outcome of a generative urban design tool and the implementation of its prototype, which transforms into design patterns supported by grammar formalisms. Using rule-based design patterns for the urban design process is a significant aspect of research in this field. The adoption of rule-based systems and elementary machine learning applications characterizes the initial phase of AI applications in urban design. The generated method considers realistic design constraints and the input of human urban experts and generates a model by considering all aspects (Jiang et al., 2023). These early adaptations are pivotal in overcoming some inherent limitations of conventional methods, offering more consistent and scalable solutions. Their study highlights the early adaptability of AI systems in understanding and modelling urban dynamics, paving the way for the future sophistication of these technologies.

18.3 Emergence of AI in Urban Design

The integration of AI into urban design marks a key transition from traditional, heuristic-based, and rule-based methods to a data-driven, analytical approach. This evolution has been marked by the adoption of ANNs, showcasing their growing sophistication and expanding applications in this domain. ANNs are instrumental in automating and enhancing the accuracy of urban layout generation, site selection, and infrastructure planning. As highlighted by Kitchin (2014), early applications demonstrate the potential of ANNs to transform urban design through enhanced data analysis and decision-making capabilities.

Expanding on early ANN applications, GANs have significantly advanced urban design by generating adaptable layouts that respond to city morphology without predefined parameters. Fedorova (2021) developed a flexible model that learns urban block patterns directly from context using data from sources like OpenStreetMap, enabling both quantitative and qualitative evaluations. Ye et al. (2022) introduced MasterplanGAN, which automates the creation of urban master plans, traditionally reliant on expert judgement, offering a more standardized and efficient approach. As illustrated in Figure 18.1, the trained model generates AutoCAD-compatible master plan renderings from uncoloured input files.

Boim et al. (2022) applied GANs to non-planned settlements, training on diverse urban datasets to replicate complex patterns resembling existing urban fabrics. Zhang et al. (2022) introduced MetroGAN, a framework that integrates GANs with geographic knowledge, using a progressive structure and geographical loss function to simulate urban morphology. Quan (2022) developed Urban-GAN, combining deep convolutional neural networks and GANs to support public participation in urban design. Miletic (2023) discusses the crucial nature of human decision-making in generative design. While iterations can output an optimized volume of design aspects, there remains a critical role for humans, who must manage the iterations. This enhances the decision-making process, while at the same time, it also requires urban designers to adjust their responsibilities. The evolution of AI technologies continues to offer promising solutions to these challenges, driving the field towards more sustainable, efficient, and inclusive urban environments.

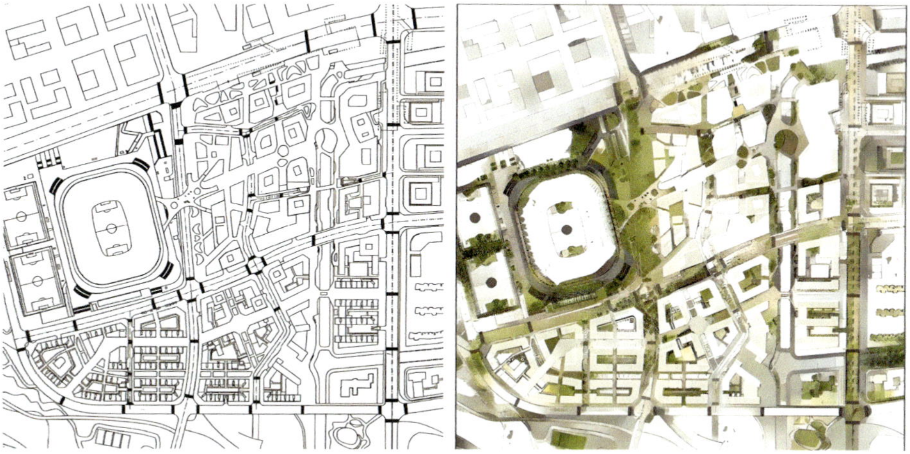

FIGURE 18.1
Design input file from AutoCAD (left) and generated output based on MasterplanGAN (right) (Ye et al., 2022).

18.4 Advancements of AI Applications in Urban Design

The landscape of urban design has been profoundly transformed by recent advancements in AI, leading to more sophisticated applications in predictive analysis and modelling. Among these advancements, the emergence of LLMs (such as ChatGPT, Gemini, Qwen, and Llama) and other AI advancements, including text-to-image models (e.g., DALL-E, Midjourney, and Stable Diffusion) and text-to-video models (e.g., Sora, Runway, and Pika), marks a significant milestone, broadening the horizon of AI applicability in the scope of urban design.

Numerous studies highlight the use of LLMs in urban design, particularly for stakeholder engagement and interpreting regulations through natural language processing. Lee et al. (2022) explored ChatGPT-3's potential in supporting collaborative and creative urban design tasks. LLMs offer intuitive interfaces that aid conceptualization and dialogue with stakeholders. However, their understanding is limited by training data, and outputs may overlook contextual nuances. As such, human oversight remains essential to ensure relevance and accuracy.

Verma et al. (2023) demonstrate the potential of LLMs as instruments for collecting and analysing perceptions of urban environments. They emphasize the novel use of LLMs in gathering insights about urban areas and assisting in the conceptual design phase. Similarly, the exploration of ChatGPT4 by Ching and CHUA (2023) for urban design tasks demonstrates the AI's ability to generate spatial layouts and perform calculations, supporting urban analysis and planning. The capacities of ChatGPT4 are assessed through a series of tasks ranging from transportation demand modelling to spatial plan layout generation. Another notable study by Wang et al. (2023) on optimizing and fine-tuning LLMs for urban renewal tasks introduces a sophisticated approach that integrates Prefix and LoRA (Low-Rank Adaptation) techniques.

A key development is GeoGPT, introduced by Ouchra et al. (2023), which extends LLMs into the spatial domain. GeoGPT automates the analysis of geospatial data, such as satellite imagery and infrastructure patterns, to support predictive urban modelling. Mortaheb and Jankowski (2023) note its efficiency in processing demographic and environmental data, enabling planners to identify land use trends and assess impacts with greater precision. While promising, the adoption of GeoGPT also raises concerns around data privacy, result reliability, and the continued need for human oversight.

The discussion then highlights the dynamic visualization capabilities of text-to-image models like DALL-E and Stable Diffusion, which have notably impacted urban design representation, as shown in Figure 18.2. Two prompts were provided to these models to generate urban layouts for both high- and low-density areas:

Prompt 1: *Design an urban layout for a low-density area in landscape mode. Feature single-family homes with large gardens, community parks, and local businesses like shops and cafes. Include winding roads, green spaces,*

FIGURE 18.2
Low-density (left) and high-density (right) urban layouts created by DALL-E 3 (top) and Stable Diffusion (bottom), created in March 2025.

pedestrian paths, and bike trails for a relaxed, eco-friendly feel. Blend modern and traditional architecture, showcasing areas for community gardens and markets. Focus on spaciousness and nature connectivity.

Prompt 2: *Design an urban layout for a high-density area in landscape mode. Include high-rise buildings, narrow pedestrian streets, and public transport like buses and trams. Feature green spaces, parks, and rooftop gardens. Blend modern and traditional architecture to show diverse styles. Add bustling marketplaces, outdoor cafes, and bike lanes to reflect a lively atmosphere. Focus on space efficiency and sustainability.*

While AI enhances efficiency and creativity in urban design, key limitations remain, especially regarding the accuracy of generated visuals and the essential role of architects in refining outputs. Dilaveroglu (2024) explores how text-to-image models can support architectural storytelling by converting written design narratives into visual representations. The study assesses these models' ability to capture stylistic and conceptual nuances, noting challenges in representing complex architectural ideas due to inherent biases and technical limitations. These findings highlight the potential of text-to-image AI to transform design visualization, while underscoring the need for critical, informed use.

18.5 Conclusion

Integrating AI into urban design marks a paradigm shift from manual, intuition-driven methods to data-driven, predictive approaches. This review has traced AI's evolution in the field, examining technologies such as ANNs, GANs, LLMs, and their roles in enhancing scalability, efficiency, and adaptability. The rise of text-to-image tools further supports concept visualization, highlighting a clear transition from early computational models to advanced AI-driven design methods. Case studies illustrate AI's broad application, from intuitive interfaces to collaborative design, while also revealing its limitations. Despite AI's potential to promote creativity, sustainability, and resilience, human oversight remains essential to ensure contextual relevance and ethical use. Challenges persist, including privacy concerns, technical constraints, and the need to balance innovation with human-centric principles. Future research should explore AI's role in urban resilience, ethical design, integration with traditional planning, and the development of more advanced, context-aware models. In summary, this chapter underscores AI's transformative potential in urban design and calls for responsible, forward-thinking integration to shape more sustainable and livable urban futures.

References

Beirão, J., Mendes, G., Duarte, J., & Stouffs, R. (2010). *Implementing a Generative Urban Design Model Grammar-based design patterns for urban design*. Paper presented at the 28th Conference on Education in Computer Aided Architectural Design in Europe, eCAADe 2010.

Boim, A., Dortheimer, J., & Sprecher, A. (2022). *A machine-learning approach to urban design interventions in non-planned settlements*. Paper presented at the Post Carbon-Proceedings of the 24th International Conference on Computer-Aided Architectural Design Research in Asia. Sydney, Australia: CAADRIA.

Ching, T.-Y., & CHUA, A. (2023). *Early Exploration of GPT-4's Abilities in the Context of Urban Planning and Design in Singapore*. Singapore, Urbex.

Cuthbert, A. R. (2007). Urban design: Requiem for an era–review and critique of the last 50 years. *Urban Design International*, 12, 177–223.

Dilaveroglu, B. (2024). The architecture of visual narrative: Can text-to-image algorithms enhance the power of stylistic narrative for architecture. *International Journal of Architectural Computing*, 14780771241234449.

Fedorova, S. (2021). *Generative adversarial networks for urban block design*. Paper presented at the SimAUD 2021: A Symposium on Simulation for Architecture and Urban Design.

Foroughmand Araabi, H. (2016). A typology of Urban Design theories and its application to the shared body of knowledge. *Urban Design International*, 21, 11–24.

Gómez-Villarino, M. T., Urquijo, J., Gomez Villarino, M., & García, A. (2021). Key insights of urban agriculture for sustainable urban development. *Agroecology and Sustainable Food Systems*, 45(10), 1441–1469.

Gordon, D. (2011). *Urban Design Reclaimed: Tools, Techniques and Strategies for Planners Grid/Street/Place: Elements of Sustainable Urban Districts*. Taylor & Francis.

Huang, X., Yuan, W., White, M., & Langenheim, N. (2022). A parametric framework to assess generative urban design proposals for transit-oriented development. *Buildings* 2022, 12, 1971.

Jiang, F., Ma, J., Webster, C. J., Chiaradia, A. J., Zhou, Y., Zhao, Z., & Zhang, X. (2023). Generative urban design: A systematic review on problem formulation, design generation, and decision-making. *Progress in Planning*, 180, 100795.

Kitchin, R. (2014). *The Data Revolution: Big Data, Open Data, Data Infrastructures and Their Consequences*. London: SAGE Publications Ltd.

Kunze, A., Halatsch, J., Vanegas, C., Jacobi, M. M., Turkienicz, B., & Schmitt, G. (2011). *A Conceptual Participatory Design Framework for Urban Planning*. Paper presented at the Proceedings of eCAADe.

Lee, M., Liang, P., & Yang, Q. (2022). *Coauthor: Designing a human-ai collaborative writing dataset for exploring language model capabilities*. Paper presented at the Proceedings of the 2022 CHI conference on human factors in computing systems.

Mehaffy, M. W. (2008). Generative methods in urban design: A progress assessment. *Journal of Urbanism*, 1(1), 57–75.

Miletic, B. (2023). Generative design in urban design, planning and sustainability. Retrieved from https://www.architectureanddesign.com.au/features/comment/generative-design-in-urban-design-planning

Mortaheb, R., & Jankowski, P. (2023). Smart city re-imagined: City planning and GeoAI in the age of big data. *Journal of Urban Management*, 12(1), 4–15.

Moura, A. C. M., & de Santana, S. A. (2014). From authorial drawings to the parametric modeling of territorial occupation: Representation and modeling influences in the process of designing the urban space. *Rev. Brasileira. Cartogr*, 66(7), 1451–1463.

Ouchra, H., Belangour, A., & Erraissi, A. (2023). *An overview of GeoSpatial Artificial Intelligence technologies for city planning and development*. Paper presented at the 2023 Fifth International Conference on Electrical, Computer and Communication Technologies (ICECCT).

Quan, S. J. (2022). Urban-GAN: An artificial intelligence-aided computation system for plural urban design. *Environment and Planning B: Urban Analytics and City Science*, 49(9), 2500–2515.

Salvini, P. (2018). Urban robotics: Towards responsible innovations for our cities. *Robotics and Autonomous Systems*, 100, 278–286.

Verma, D., Mumm, O., & Carlow, V. M. (2023). Generative agents in the streets: Exploring the use of Large Language Models (LLMs) in collecting urban perceptions. *arXiv preprint arXiv:2312.13126*.

Wang, X., Ling, X., Zhang, T., Li, X., Wang, S., Li, Z., …, & Gong, P. (2023). Optimising and Fine-tuning Large Language Model for Urban Renewal. *arXiv preprint arXiv:2311.15490*.

Ye, X., Du, J., & Ye, Y. (2022). MasterplanGAN: Facilitating the smart rendering of urban master plans via generative adversarial networks. *Environment and Planning B: Urban Analytics and City Science*, 49(3), 794–814.

Zhang, W., Ma, Y., Zhu, D., Dong, L., & Liu, Y. (2022). *Metrogan: Simulating urban morphology with generative adversarial network*. Paper presented at the Proceedings of the 28th ACM SIGKDD Conference on Knowledge Discovery and Data Mining.

19

BIM for 3D Digital Land Administration in Multi-Story Developments and Major Infrastructure Projects

Behnam Atazadeh, Lanxuan Shen, and Abbas Rajabifard

19.1 Introduction

Urban land administration is a critical function in the sustainable governance of rapidly growing cities. It provides technical, legal, and institutional frameworks to manage land ownership, use, value, and development. As urban environments become increasingly vertical with high-rise buildings, underground facilities, and complex public infrastructures, traditional 2D land administration systems are no longer sufficient in these environments. In urban areas, legal interests in land are often multilayered, spanning both horizontal and vertical dimensions, necessitating sophisticated approaches to delineate, manage, and communicate these interests. The legal intricacies in multi-story developments, such as apartments, office towers, and shopping complexes, include overlapping ownership boundaries, easements, rights of way, and common property zones (Figure 19.1).

In addition, major infrastructure assets (Figure 19.2), such as roads, railways, and tunnels, present growing spatial and legal complexity, driven by rapid network expansion and service efficiency demands. These linear assets frequently traverse multiple strata and intersect with other developments above and below the ground, which strains the capacities of traditional land administration systems (Shen et al. 2025).

Currently, most jurisdictions rely on 2D subdivision plans that flatten this spatial and legal complexity into abstract and static 2D representations. These plans are difficult to interpret, especially for non-expert stakeholders, leading to miscommunication, errors in legal transactions, and costly disputes. With the expansion of smart cities and digital governance, there is an urgent need for 3D digital representations of land and property ownership in urban environments.

On the other hand, building information modeling (BIM), as a transformative technology in the Architecture, Engineering, and Construction (AEC)

DOI: 10.1201/9781003638308-21

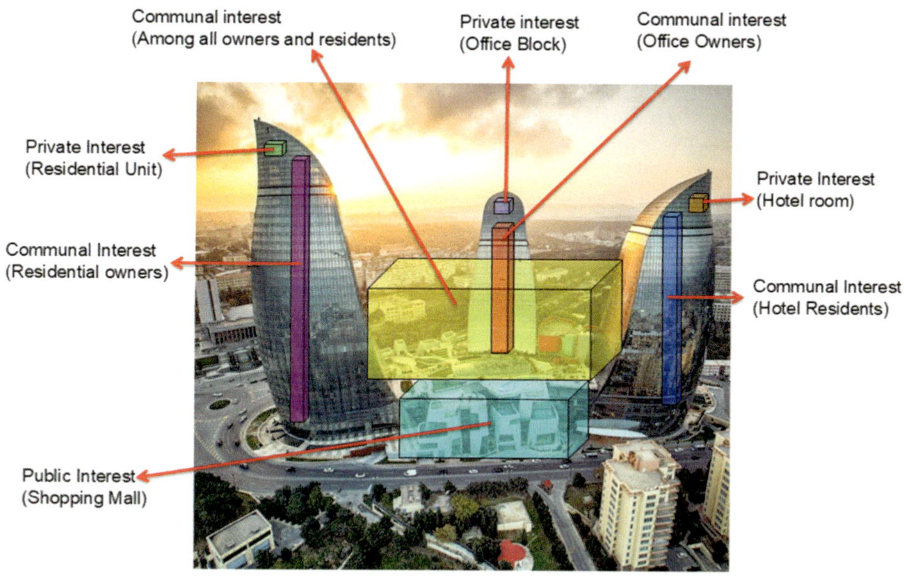

FIGURE 19.1
An example of a complex urban development comprising various private, communal, and public legal interests (Flame Towers in Azerbaijan).

FIGURE 19.2
An example of the junction of the Tunnel, Airport Link and Inner-City Bypass in Brisbane. (Adopted from ROADS AUSTRALIA (2015)).

sectors, offers a powerful platform to integrate spatial, legal, and administrative data throughout the entire lifecycle of buildings and infrastructure projects. This chapter investigates the potential of extending an open data model in BIM, known as the Industry Foundation Classes (IFC) standard, to support 3D digital management of land and property ownership in multistory buildings and infrastructure projects.

19.2 Significance of 3D Digital Data in Urban Land Administration

Traditional land administration systems were built to manage simple land parcels defined in two dimensions. These systems served their purpose during the industrial and post-industrial eras when urban developments were predominantly horizontal. However, modern cities are characterized by vertical densification, where ownership units are stacked above and below ground. Underground shopping centers, subway stations, pedestrian tunnels, and vertical mixed-use towers are common in metropolises such as Tokyo, New York, and Melbourne. The spatial complexity of these assets makes it challenging to manage ownership, maintenance responsibilities, and usage rights using 2D paradigms.

Urban land administration must evolve to address the following 3D-specific challenges:

- Ambiguity in ownership boundaries: Multiple units within a single development often share their ownership boundaries via physical elements (e.g., walls, ceilings). Determining whether these elements are part of one unit, another, or common property is often unclear in 2D.
- Complex strata subdivisions/consolidation: High-rise buildings involve numerous title units. Representing and validating their spatial arrangements in 2D results in error-prone and hard-to-understand legal records. Infrastructure (linear) projects must amalgamate multiple titles and strata into continuous 3D legal spaces across many land parcels located above and below the ground. Ensuring chainage/vertical continuity and resolving overlaps and encumbrances is error-prone in 2D.
- Subterranean assets and airspace rights: Developments may extend below ground (e.g., parking garages, storage rooms) or into the air (e.g., penthouse extensions, signage rights). Such vertical interests cannot be accurately represented without 3D modeling.

- Information flows: Major infrastructure projects progress through long, multi-phase lifecycles in which various physical, legal, planning, and surveying aspects frequently change. These are captured in disparate systems and formats, causing version conflicts, loss of provenance, and delays when reconciling records across planning, design, construction, and operations.
- Stakeholders: Major infrastructure involves many actors (e.g., transport and road authorities, rail operators, utilities, contractors, surveyors, local councils, land registries, and private owners), each with distinct mandates and approval requirements. Coordinating responsibilities, permissions, and audits across jurisdictions and project phases is difficult to express and verify without a shared, 3D digital model-based representation.

These challenges have given rise to the need for 3D digital land administration, defined as systems that enable storage, analysis, and visualization of legal rights, restrictions, and responsibilities in 3D digital environments. 3D digital land administration has gained significant attention globally but implementation remains fragmented due to a lack of standards, tools, and institutional capacity. BIM, with its inherently 3D nature and extensible schema, presents a promising solution (Atazadeh 2017).

19.3 BIM as a Solution: Framework and Data Requirements

BIM goes beyond simple 3D modeling by embedding semantic information, relationships, and metadata into digital representations of physical environments. In a typical BIM model, components such as walls, doors, windows, and slabs are not only geometrically defined but also semantically enriched with attributes and their connectivity relationships.

From a 3D land administration standpoint, BIM environment is ideal for embedding land and property associated with vertically placed developments. For example, a wall could serve as a legal boundary between two ownership units. The ability to reference legal titles, ownership metadata, usage restrictions, and responsible actors within the same environment allows for a holistic view of the built environment. To support this, the following data requirements were identified (Figure 19.3):

- Legal objects: Including lots (private properties), common property areas, and legal boundaries.
- Physical objects: These include building elements, distribution elements, geographic features, civil and site elements.

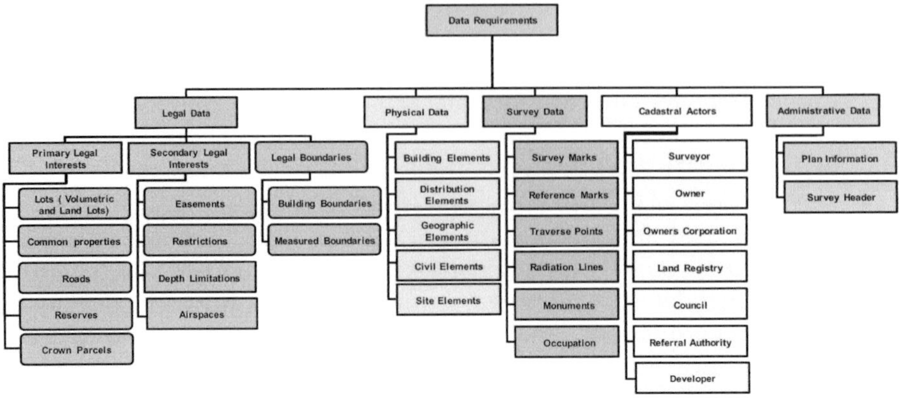

FIGURE 19.3
Data requirements in 3D land administration.

- Legal documents: References to title documents, strata plans, permits, and easement documents.
- Actors and roles: Such as owners, surveyors, property managers, and approving authorities.
- Administrative data: Plan registration details, strata scheme identifiers, zoning classifications, and survey metadata.

To model this information in BIM, the IFC standard was selected for extension. IFC, developed by buildingSMART, is the leading open standard for BIM interoperability (ISO16739 2013). It supports object-based modeling with a structured hierarchy, making it suitable for linking legal and physical components. Property sets (Psets) within IFC were used to define custom legal attributes, while new entities were introduced to model legal volumetric spaces. Not only for multi-story building developments, the design of current IFC standards preserved compatibility with existing BIM tools while enabling 3D cadastral functionality for major infrastructure projects. In planning, IFC-based models support visualization and simulation that clarify overall layout and design; open interoperability improves coordination quality and reduces errors. Recent versions of IFC have also been applied to encode key 3D attributes of infrastructure assets (Shen et al. 2025). In construction, coupling BIM/IFC with 4D/5D enables schedule and cost control, real-time tracking, and more efficient resource use (Bradley et al. 2016; Lee et al. 2017). Across the lifecycle, IFC supports managed change, such as updating legal spaces when parcel boundaries are subdivided or consolidated (Paasch and Paulsson 2021), by linking legal and physical data for 3D digital land administration.

19.4 Prototype Development: IFC-Based BIM Models

The prototype development phase aimed to demonstrate the feasibility of extending the IFC schema to support legal data elements in a BIM environment for both high-rise and major infrastructure development. This required both technical modeling and domain-specific knowledge of cadastral systems, particularly as practiced in Victoria, Australia.

19.4.1 High-Rise Buildings

The first prototype focused on a high-rise development with mixed-use strata lots and involved several key steps.

First, the physical BIM model of the building was authored in Autodesk Revit. This model contained the architectural and structural components, such as walls, slabs, columns, stairs, and lifts, organized by levels and spatial zones. The spatial hierarchy was structured using 'IfcProject', 'IfcSite', 'IfcBuilding', 'IfcBuildingStorey', and 'IfcSpace' entities. These were enriched with legal semantics by defining custom property sets.

Legal boundaries were encoded using geometric primitives and linked to physical structures (e.g., walls, doors, and slabs). Each 'IfcSpace' was tagged with metadata such as ownership ID, lot number, and title document reference. The enriched BIM model was then exported to IFC and visualized using Solibri viewer. This tool provides the capability to navigate the model, inspect ownership information, and query relationships. The prototype showed how BIM could act as both intelligent and visually interactive environment, enabling accurate, navigable, and interoperable 3D digital representations of legal objects (Figure 19.4).

19.4.2 Major Infrastructure Projects

The second BIM model was authored in Autodesk InfraWorks and included the tunnel lining, track/roadway, portals, station boxes, surface context (terrain, roads, buildings), and cadastral parcel volumes derived from open data (Figure 19.5).

Second, the legal layer was then constructed in Autodesk Revit by converting 2D records into volumetric legal units along the alignment. Ownership volumes, easements (permanent legal corridors), depth/height restrictions, and temporary occupation areas were represented as 3D legal spaces.

Finally, the physical and legal models were exported to IFC and integrated into Solibri for navigation, interrogation, and quality checks, which shows the capabilities of IFC to support 3D modeling and representation of 3D land and property data. A typical example of legal spaces, crown allotment in Figure 19.6, is presented in the selected area as legal spaces of tunnel in underground environment.

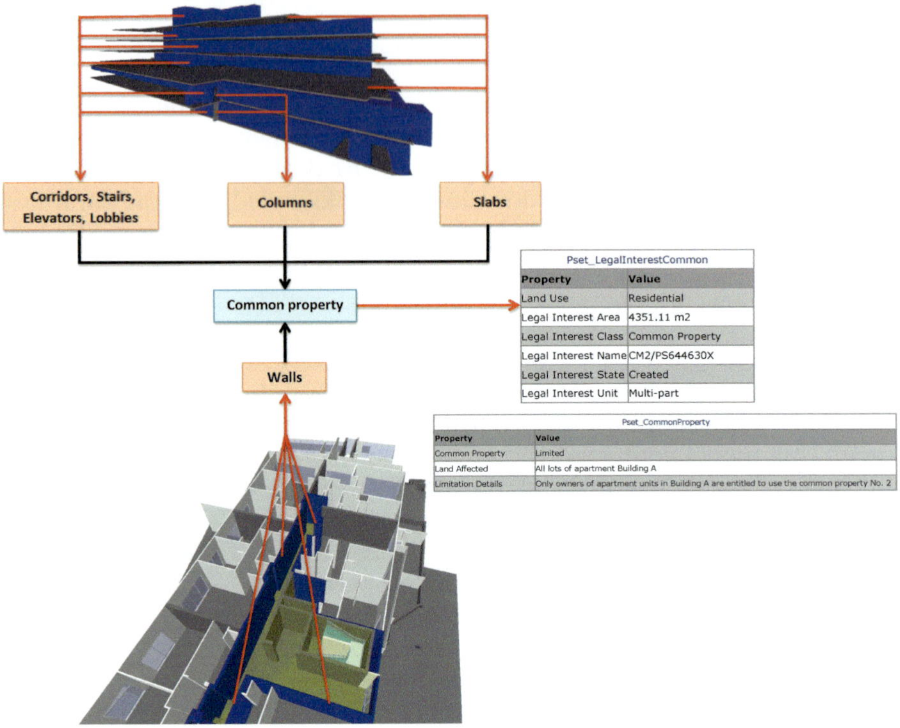

FIGURE 19.4
Example of legal objects (common property) modeled in BIM.

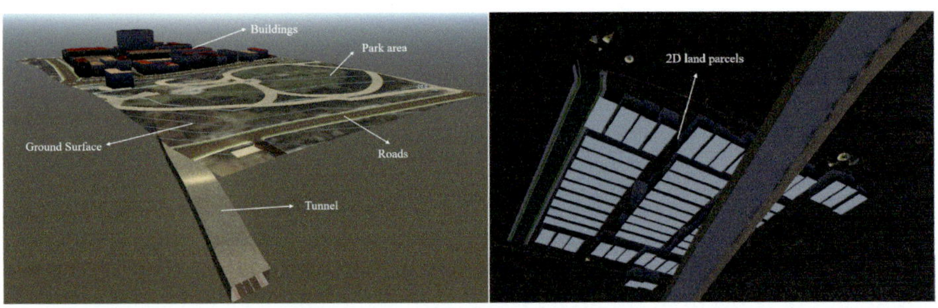

FIGURE 19.5
The 3D physical model of the selected area for major infrastructure project.

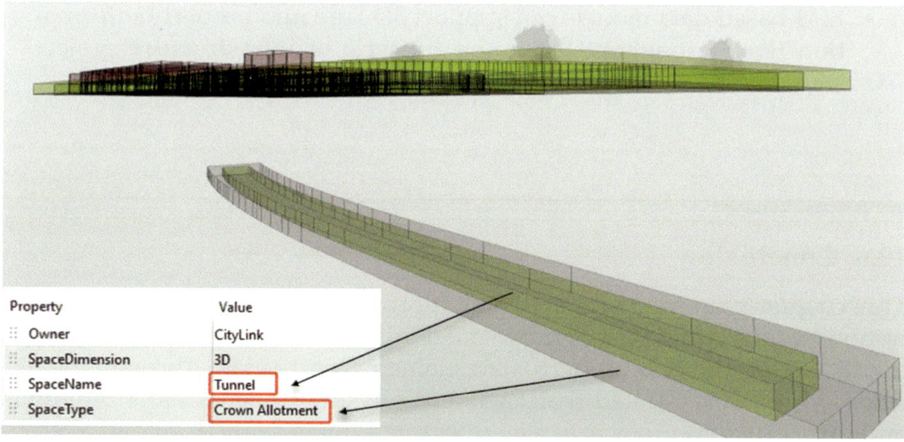

FIGURE 19.6
Tunnel and its underground external space of the selected site.

19.5 Evaluation

The BIM-based prototype models for multi-story buildings and major infrastructure projects were evaluated through both expert feedback and comparative analysis with traditional 2D plans (Atazadeh 2017).

In the expert review, professionals from surveying, urban planning, and land registration domains reviewed the prototype's capabilities (Atazadeh 2017). Feedback highlighted the clarity of legal boundary visualization, the ability to model complexity of common property areas, and the inclusion of legal documentation metadata. Experts acknowledged that the BIM model enabled an intuitive understanding of legal boundaries, especially for non-surveying professionals. They noted that visual queries, such as identifying all easements impacting a lot, were straightforward using the BIM viewer.

The comparison involved a side-by-side analysis of the BIM model versus the legally registered 2D subdivision plan (Atazadeh 2017). Several insights are drawn from this comparison:

- The 2D plan required multiple pages to depict ownership units and boundaries; however, the BIM models present it interactively in a single 3D digital environment.
- Boundary ambiguity was resolved more effectively in 3D. For example, projection boundaries and vertical airspace limits are visually intuitive in BIM.
- Spatial queries, such as counting lots per floor or identifying legal units within a volumetric boundary, can be automated in BIM.

- BIM-based data models can support 3D land and property information flow throughout the entire lifecycle of infrastructure project stages.

19.6 Conclusion

This chapter presented an intelligent approach to 3D digital land administration by leveraging BIM/IFC extended with legal data elements. By aligning BIM with land and property data elements, particularly through the IFC standard, it demonstrated that legal and physical dimensions of buildings and infrastructure projects can be integrated into a unified 3D digital environment. This integration supports better visual communication, enhances stakeholder understanding, and enables semi-automated validation processes. Compared to traditional 2D subdivision plans, BIM offers an interactive, scalable, and standards-based solution that reflects the complexity of contemporary urban land developments in particular multi-story buildings and major infrastructure projects.

However, adoption of BIM for 3D digital land administration requires significant transformation in law, policy, data governance, and professional practice. Regulatory frameworks must be updated to permit 3D digital representations as official land and property records. Further research should focus on creating common ontologies and workflows between BIM and GIS platforms. Looking ahead, BIM-enabled 3D digital land administration has the potential to become a critical enabler for digital governance and smart city infrastructure. This chapter contributes a foundational step toward that vision by offering a technical methodology and a case-proven BIM prototype for 3D land administration.

References

Atazadeh, Behnam. 2017. "Building Information Modelling for Urban Land Administration." *PhD thesis*, Department of Infrastructure Engineering, The University of Melbourne. http://hdl.handle.net/11343/150614.

Bradley, A., H. Li, R. Lark, and S. Dunn. 2016. "BIM for Infrastructure: An Overall Review and Constructor Perspective." *Automation in Construction* 71: 139–52.

ISO. 2013. *Industry Foundation Classes (IFC) for Data Sharing in the Construction and Facility Management Industries (ISO 16739)*. BuildingSMART.

Lee, S.-H., S. I. Park, T.-H. Kwon, and K.-W. Seo. 2017. "Civil Infrastructure Information Modeling Method Based on Extended IFC Entities Using BIM Authoring Software." *Journal of the Computational Structural Engineering Institute of Korea* 30 (1): 77–86.

Paasch, J. M., and J. Paulsson. 2021. "New Trends in 3D Cadastre Research—A Literature Survey." In *Proceedings of the 7th International FIG 3D Cadastre Workshop*, 129–42. International Federation of Surveyors. https://doi.org/10.4233/UUID:4017E461-8AD3-44EB-9CD4-B4CC4F3E4E77

Roads Australia. 2015. *Roads Australia*. Accessed November 28, 2024. https://roadsaustralia.weebly.com/

Shen, L., B. Atazadeh, S. Ho, and A. Rajabifard. 2025. "Utilizing the BIM-Based Standard for Supporting 3D Digital Representation of Legal Spaces in Major Infrastructure Projects." In Proceedings of FIG Working Week - *Collaboration, Innovation and Resilience: Championing a Digital Generation*, Brisbane, Australia, 6–10 April 2025.

20

Advancements in 3D Cadastral Data Visualization

Davood Shojaei

20.1 Introduction

Population growth combined with limited land availability commonly challenges urban environments, resulting in intensified property development extending vertically above and below the ground, such as high-rise buildings and underground infrastructure. These developments involve complex sets of rights, restrictions, and responsibilities (RRRs) to delineate ownership clearly. However, current two-dimensional (2D) cadastral systems struggle to manage and accurately register these intricate multi-level structures effectively (Shojaei et al. 2017). Typically, cadastral systems represent RRRs through 2D building plans, cross-sectional diagrams, isometric views, and textual descriptions, primarily recorded on paper or in PDF format (Shojaei 2014). This method has several drawbacks, including difficulty for non-experts to comprehend property boundaries, complexity in interpreting numerous floor plans and cross-sections, and limitations on conducting digital queries and analyses. Consequently, there is a strong need for more efficient representation methods to facilitate the registration and clear comprehension of RRRs, especially in complex vertical developments. Three-dimensional (3D) visualization offers an enhanced approach, significantly improving the understanding of property ownership in complex multi-story structures (Rajabifard et al. 2014).

This shift toward 3D cadastral data visualization is a critical enabler within the broader framework of land intelligence and the geospatial ecosystem. As urban systems become increasingly data-driven and interconnected, traditional methods of land management must evolve to support smarter governance, inclusive planning, and sustainable outcomes. 3D visualization technologies contribute to this evolution by offering more intuitive, interoperable, and accessible tools for visualizing, analyzing, and communicating spatial data. They also serve as foundational infrastructure for implementing sustainability-focused urban policies, helping cities better manage land use,

DOI: 10.1201/9781003638308-22

reduce spatial conflicts, and optimize the allocation of vertical and underground spaces.

Moreover, the integration of digital innovations and artificial intelligence (AI) further enhances the value of 3D cadastral systems. AI can be leveraged to automate the extraction, classification, and interpretation of spatial data, tailoring visualizations to user-specific needs, predicting urban growth patterns, or even optimizing planning scenarios for energy efficiency and environmental resilience. These innovations amplify the functionality and scalability of 3D cadastral tools, transforming them from passive records into intelligent decision-support systems within a geospatial ecosystem.

In this context, 3D cadastral visualization is not merely a technical advancement; it exemplifies the dynamic convergence of spatial data science, AI technologies, and sustainability imperatives. As such, it plays a pivotal role in the modernization of land administration systems and supports a more equitable, resilient, and intelligent geospatial future.

20.1.1 The Importance of 3D Visualization

3D visualization transforms cadastral data, presenting spatial information in ways that 2D representations cannot achieve. It allows stakeholders to understand complicated property arrangements, especially in densely populated areas, where multi-story buildings and underground infrastructures pose challenges for traditional mapping.

Accurate 3D models reduce disputes by offering visual clarity regarding ownership boundaries, responsibilities, and restrictions. Enhanced understanding through visualization also supports decision-making in urban planning and disaster management scenarios. Figure 20.1 represents the current methods of representing ownership spaces in cadastral plans in Victoria, Australia.

As shown in Figure 20.1, the overlapping ownership boundaries require significant knowledge and experience to interpret the location of these boundaries. This can be significantly difficult for the public to understand their RRRs. With having complex developments above and below the ground, the complexity will increase, and it would be extremely difficult to imagine the extent of ownership spaces. Figure 20.2 also represents a volumetric format plan for representing ownership spaces using isometric diagrams in Queensland, Australia. Although this represents ownership spaces in a more imaginable way, it is still difficult to bring a very large and complex development together in this method.

In addition, these methods of representation are not interactive. You cannot efficiently represent specific RRRs in a static PDF plan. Furthermore, queries and analyses are not easily possible. It means you cannot query a PDF document to measure distances or find the smallest lot in a subdivision plan. These are some of the limitations identified in existing (traditional) methods of representing ownership spaces.

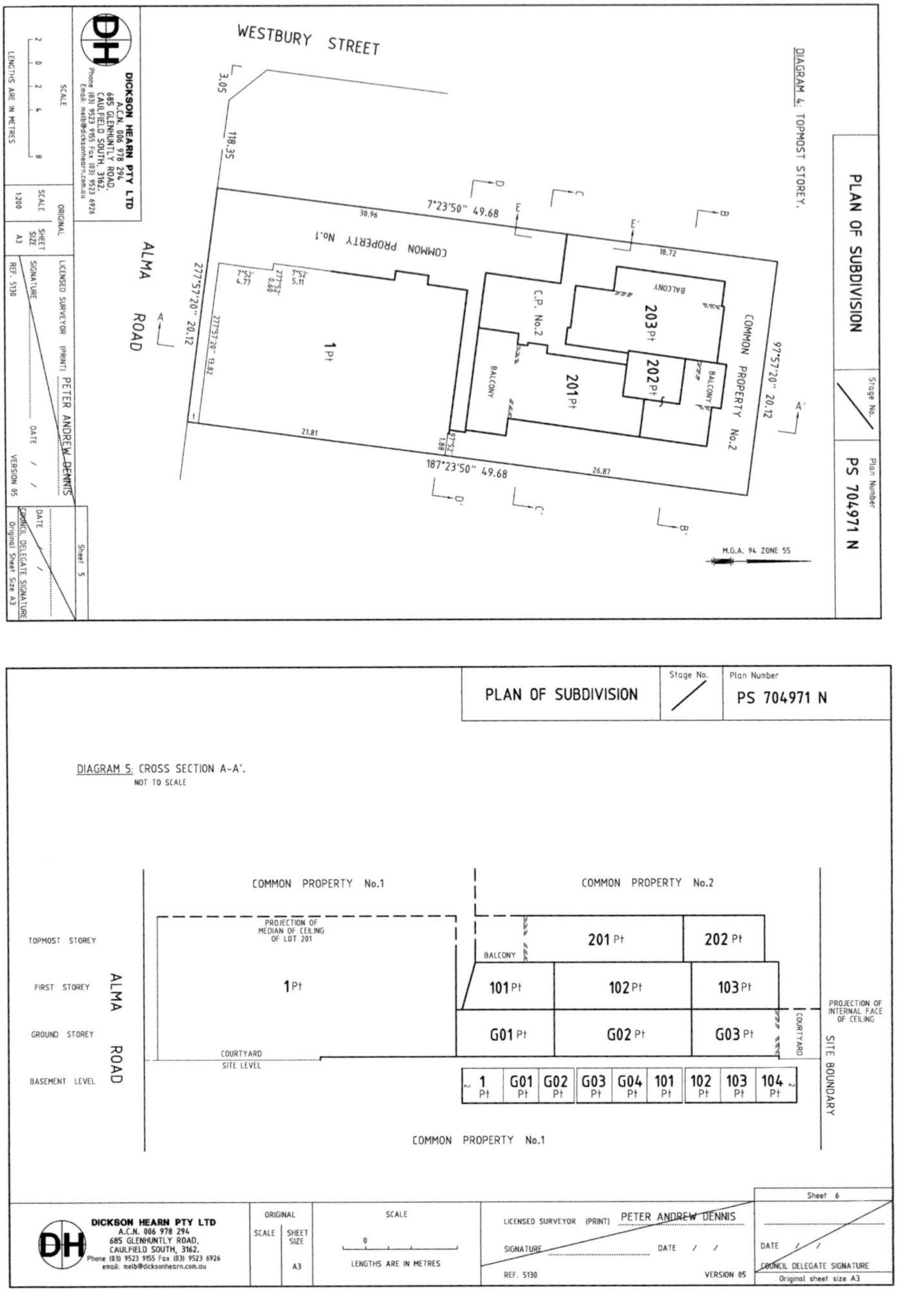

FIGURE 20.1
A sample of Plan of Subdivision in Victoria, Australia. The top figure represents a floor plan, and the bottom figure represents a cross-section plan (Department of Transport and Planning 2025).

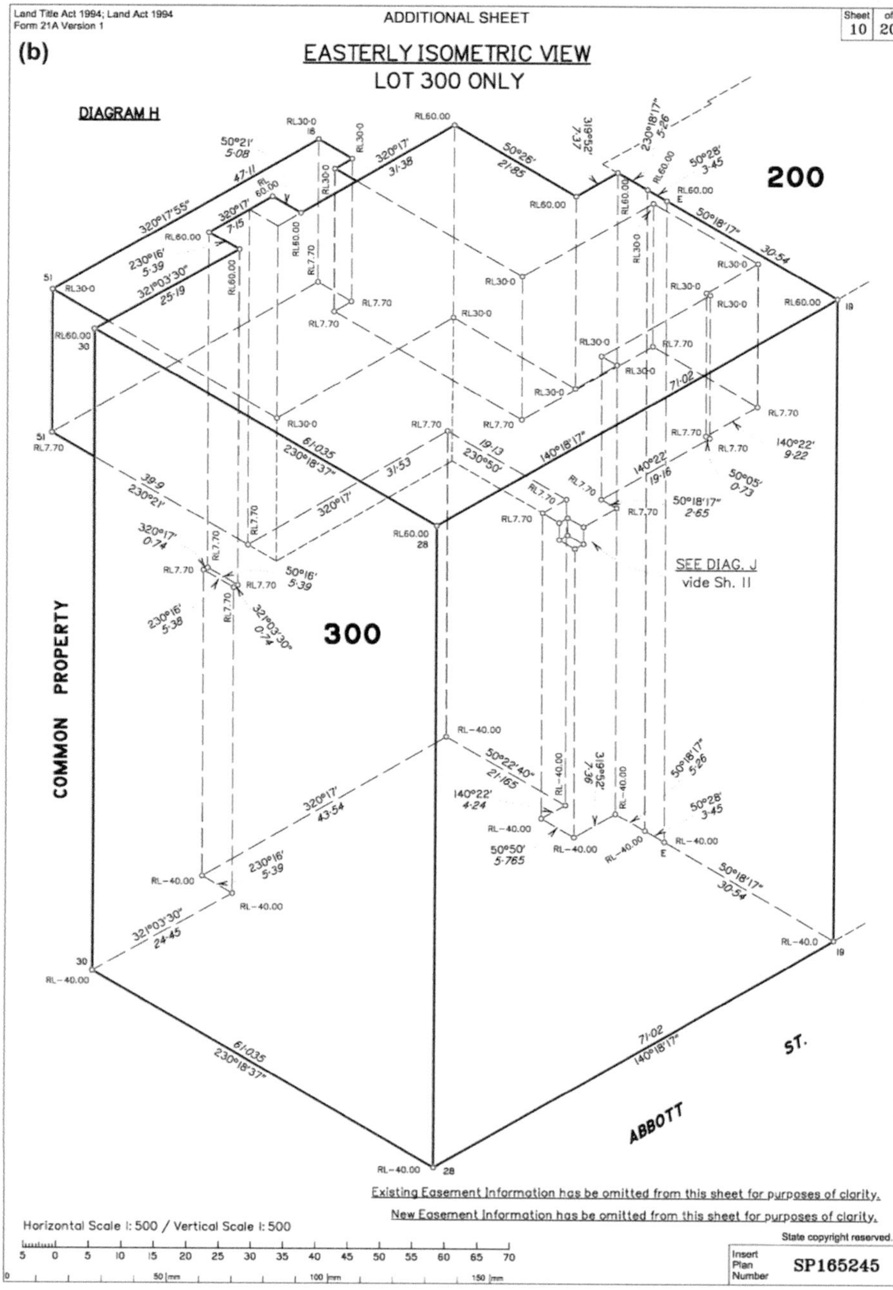

FIGURE 20.2
Volumetric Format Plan (Isometric Diagram), Queensland, Australia (Queensland Government 2025).

The shift toward 3D visualization offers significant advantages, particularly in enhancing the efficiency of visualizing complex 3D objects. This technology is increasingly adopted across various domains, including architecture, urban planning, building development, and disaster management. Professionals are actively seeking compatible 3D visualization tools to better manage ownership rights in 3D space (Shojaei et al. 2013). Furthermore, 3D visualization creates a compelling visual impact that captures and maintains audience attention (Pouliot et al. 2018). It also provides a robust infrastructure to support the examination and registration of spatial plans.

This chapter presents several innovative visualization methods aimed at effectively representing RRRs for end users within the cadastral domain. Section 20.2 reviews contemporary visualization technologies, while Section 20.3 concludes the chapter with key insights and directions for future research.

20.2 Recent 3D Visualization Technological Innovations

3D visualization technologies began their major developments primarily after the initial developments in computer graphics in the 1960s (Machover 1994). From this point onward, 3D visualization technologies continually evolved with increasing computing power, eventually encompassing photorealistic rendering, virtual reality (VR), augmented reality (AR), and real-time interactive 3D visualization widely used today. This section looks at some of the common 3D visualization technologies that can be used for interactive 3D visualization.

20.2.1 3D Web-Based Visualization

3D web-based visualization utilizes modern web technologies—most notably WebGL (Web Graphics Library)—to deliver interactive and dynamic 3D graphics directly within web browsers. This method significantly enhances accessibility, as it eliminates the need for specialized software installations and allows content to be distributed seamlessly through standard web channels. Platforms such as CesiumJS and three.js demonstrate the capabilities of WebGL by providing powerful tools for real-time rendering of complex spatial data, interactive maps, and property visualizations.

These tools play a crucial role in improving collaboration, data sharing, and accessibility, enabling complex 3D cadastral data to be effectively communicated to a wide range of stakeholders, including urban planners, property managers, and the general public. For instance, Shojaei et al. (2015) developed a WebGL-based platform to visualize cadastral RRRs. In addition, several jurisdictions have created WebGL-based prototypes to represent ownership rights in 3D environments.

With web-based applications now offering most of the core functionalities required in cadastral systems, they are increasingly becoming the primary platform for 3D ownership visualization. This trend is further driven by technological advancements in efficiently rendering large datasets, particularly through the development of 3D Tiles, a format standardized by the Open Geospatial Consortium (OGC) to support scalable, high-performance 3D content streaming.

Figure 20.3 illustrates a prototype developed in Victoria, Australia, designed to visualize ownership spaces within a web-based 3D environment.

20.2.2 3D Mobile-Based Visualization

3D mobile-based visualization involves delivering interactive 3D content through mobile devices, such as smartphones and tablets, enhancing the accessibility and flexibility of spatial data interaction. Rapid advancements in mobile processing power, graphics performance, and intuitive touch-screen interfaces have transformed mobile devices into powerful platforms capable of handling sophisticated and complex spatial visualizations.

Modern mobile devices offer high-resolution displays and efficient processing power, allowing users to engage with 3D spatial data intuitively. This technology is particularly beneficial in sectors like real estate, urban planning, surveying, and construction, where professionals require accurate and immediate visualization of property boundaries, spatial layouts, building structures, and ownership information.

The portability and connectivity of mobile devices provide unmatched convenience, enabling professionals and end users to explore, query, annotate,

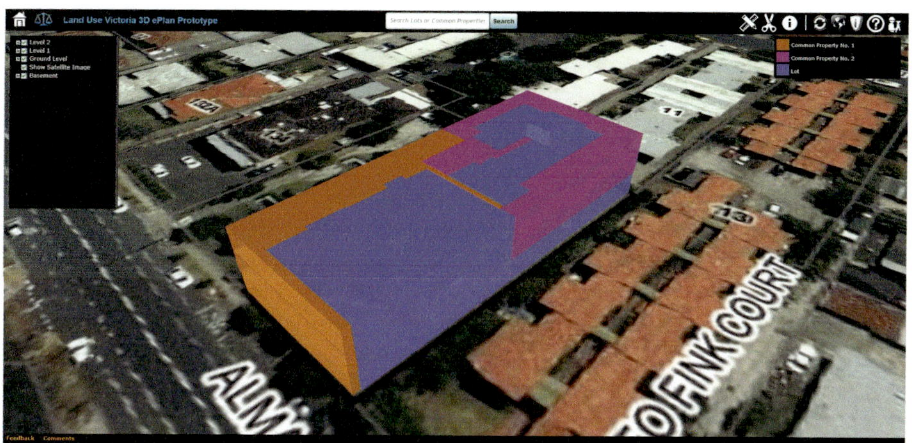

FIGURE 20.3
A web-based 3D visualization prototype in Victoria, Australia (Department of Transport and Planning 2025).

and interpret spatial data in real time, regardless of location. With seamless integration of GNSS and location-based services, users can also align digital 3D visualizations directly with physical environments, significantly enhancing situational awareness and decision-making accuracy.

Consequently, 3D mobile-based visualization facilitates operational efficiency by streamlining workflows, reducing reliance on stationary computing systems, and minimizing the necessity for physical site visits. By offering powerful visualization capabilities on-the-go, this technology significantly improves productivity, collaboration, and overall data-driven insights, driving innovation across multiple industries.

20.2.3 Virtual Reality (VR)

VR immerses users in a completely digital, simulated environment, typically experienced through specialized VR headsets. VR enables realistic, interactive exploration of property environments, allowing users to virtually navigate, manipulate, and investigate 3D spatial data and cadastral details. In cadastral applications, VR provides unparalleled visualization clarity for understanding complex multi-layered ownership structures, underground utilities, or high-rise building components. Moreover, VR aids in stakeholder engagement and public consultation by providing vivid, experiential insights into proposed developments or existing property configurations. Several studies looked at VR as an option for representing ownership boundaries. Among these, we can refer to Meliana et al. (2024) for representing ownership spaces using VR.

20.2.4 Augmented Reality (AR)

AR overlays digital visualizations onto real-world views, seamlessly integrating virtual cadastral information with physical surroundings. AR is typically experienced through smartphones, tablets, or specialized AR glasses, enabling users to perceive digital property boundaries, ownership data, and spatial relationships in the context of the actual built environment. Applications leveraging ARKit (iOS) and ARCore (Android) allow users to interact with property and cadastral data directly in their real-world context. This contextual visualization significantly enhances user comprehension and decision-making capabilities, particularly beneficial during property inspections, planning approvals, or boundary validations. The tangible interaction enabled by AR also improves stakeholder communication, as users directly visualize property information overlaying real-world features. Figure 20.4 demonstrates an AR application to present ownership spaces using an iPad.

20.2.5 Mixed Reality (MR)

Mixed reality (MR) blends elements of VR and AR, enabling digital and physical objects to interact dynamically within the same environment. Using

FIGURE 20.4
A screenshot of the AR application to present ownership spaces (Shojaei et al. 2019).

advanced MR headsets such as Microsoft HoloLens or Apple Vision Pro, users can visualize, interact with, and manipulate cadastral data directly within their real-world surroundings. MR provides an advanced interface for complex spatial analyses, enabling real-time visualization of property boundaries, building information, and underground infrastructure with an unprecedented level of interactivity and precision. It particularly benefits scenarios requiring dynamic adjustments, collaborative planning, or real-time decision-making, effectively bridging the gap between virtual models and physical reality.

20.2.6 Hologram Visualization

Hologram visualization involves projecting 3D images into real-world spaces, providing a natural and intuitive interface for viewing complex spatial data. This technology typically employs specialized holographic displays or projectors, enabling users to visualize and interact with digital cadastral information as if physically present in space. Holographic visualization significantly simplifies comprehension of complex vertical and horizontal relationships within urban environments, making it easier to understand intricate property boundaries and ownership rights. Industries including real estate, urban planning, and property management increasingly leverage holographic displays to improve decision-making, stakeholder engagement, and effective communication of spatial information.

20.2.7 3D Cave

A 3D cave visualization refers to an immersive, interactive environment where 3D spatial data is projected onto multiple surrounding screens, creating a realistic, virtual representation that completely envelops users. This technology utilizes stereoscopic projection systems, sophisticated graphics processing, and specialized eyewear, enabling participants to explore and interact with complex datasets in a lifelike manner. Widely employed in industries such as geology, mining, architecture, and engineering, 3D cave environments enhance spatial understanding and collaboration by offering real-time interaction with simulated data, including geological structures, mine layouts, architectural designs, and urban planning scenarios. By simulating real-world spatial conditions, this visualization method significantly improves decision-making accuracy, reduces operational risks, facilitates better project planning, and enhances the overall efficiency of complex spatial analyses.

20.3 Conclusion and Future Direction

Advancements in 3D cadastral data visualization significantly improve how we manage and interact with property information. The continuous integration of innovative technologies like VR, AR, and MR, along with sophisticated data acquisition methods, is paving the way for smarter, more efficient, and transparent cadastral systems. Addressing ongoing challenges related to data interoperability, accuracy, and privacy will ensure that 3D visualization remains robust and reliable, supporting increasingly complex property rights management and urban planning scenarios.

Future advancements in 3D cadastral visualization promise further enhancements in perception, accuracy, accessibility, and utility. The incorporation of AI for providing a suitable visualization based on user requirements will significantly streamline cadastral processes and make them easier to understand and interpret. Automation in the process of acquisition, visualization, and registration are the future advancements in cadaster and it will benefit significantly from advancements in AI.

References

Department of Transport and Planning 2025. "3D ePlan Prototype." https://www.spear.land.vic.gov.au/spear/pages/eplan/3d-digital-cadastre/3dprototype/prototype.htmlMachover, Carl 1994. "Four Decades of Computer Graphics." *IEEE Computer Graphics and Applications* 14 (6): 14–19. https://doi.org/10.1109/38.329088

Meliana, Imane, Rafika Hajji, and Davood Shojaei. 2024. "Exploring Spatial Interaction and Visualization Paradigms for 3D Cadastral Visualization." *ISPRS Annals of the Photogrammetry, Remote Sensing and Spatial Information Sciences* X-4/W5-2024: 237–246. https://doi.org/10.5194/isprs-annals-X-4-W5-2024-237-2024

Pouliot, Jacynthe, Claire Ellul, Frédéric Hubert, Chen Wang, Abbas Rajabifard, Mohsen Kalantari, Davood Shojaei, Behnam Atazadeh, Peter van Oosterom, and Marian De Vries. 2018. "Visualization and New Opportunities." *Best Practices 3D Cadastres: Extended Version*. Copenhagen: FIG Publication: 183–230.

Queensland Government. 2025. "Volumetric Format Plan (Isometric Diagram)." http://www.digitalamigo.com.au/WEB%20SITE/PDFs/Volumetric%20Example%20Sh10.pdf

Rajabifard, Abbas, Ian Williamson, Brian Marwick, Mohsen Kalantari, Serene Ho, Davood Shojaei, Behnam Atazadeh, Sam Amirebrahimi, and Alireza Jamshidi. 2014. "3D-Cadastre, A Multifaceted Challenge." FIG Congress.

Shojaei, D, H Olfat, A Aien, N Nimorakiotakis, and A Rajabifard. 2019. "Utilising Virtual and Augmented Reality Technologies to Improve the Visualisation of Digital Cadastre—The Victorian Cadastre Experience." Proceedings of the 2nd International IAG Workshop on BIM and GIS Integration, 3D GeoInfo Conference.

Shojaei, D., H. Olfat, M. Briffa, and A. Rajabifard. 2017. "3D Digital Cadastre Journey in Victoria, Australia." *ISPRS Annals of the Photogrammetry, Remote Sensing and Spatial Information Sciences* IV-4/W5: 117–123. https://doi.org/10.5194/isprs-annals-IV-4-W5-117-2017

Shojaei, Davood. 2014. "3D Cadastral Visualisation: Understanding Users' Requirements." PhD, Infrastructure Engineering, The University of Melbourne.

Shojaei, Davood, Mohsen Kalantari, Ian D. Bishop, Abbas Rajabifard, and Ali Aien. 2013. "Visualization Requirements for 3D Cadastral Systems." *Computers, Environment and Urban Systems* 41: 39–54. https://doi.org/10.1016/j.compenvurbsys.2013.04.003. https://www.sciencedirect.com/science/article/pii/S0198971513000422

Shojaei, Davood, Rajabifard Abbas, Kalantari Mohsen, Bishop Ian D., and Ali Aien. 2015. "Design and Development of a Web-Based 3D Cadastral Visualisation Prototype." *International Journal of Digital Earth* 8 (7): 538–557. https://doi.org/10.1080/17538947.2014.902512

21

3D Cadastre Database Development Lifecycle and Data Migration

Javad Shahidinejad

21.1 Introduction

A fundamental component of any land intelligence and geospatial ecosystem is ensuring access to clean, accurate, and up-to-date data. Developing a well-structured 3D cadastral database serves as a foundation for subsequent processes such as spatial data analysis, visualization, reporting, business intelligence, and AI-driven solutions. In distributed or multi-platform environments, such as Victoria's cadastral system, implementing a 3D cadastral Database Management System (DBMS) will ensure data consistency and integrity. In addition, this facilitates collaboration across a variety of organizations and stakeholders. Ultimately, reliable and well-managed spatial data supports sustainability by saving time, reducing resource consumption, and improving the quality of insights derived from geospatial applications.

Migrating from a 2D to a 3D cadastre is a complex undertaking that extends beyond mere data modelling and database design. The database development lifecycle (DDLC) represents a systematic process and offers a comprehensive framework for managing this significant transition. The lifecycle typically involves several stages, including initial study, database design, implementation and data loading, data migration, testing and evaluation, operation, and ongoing maintenance and evolution (Coronel and Morris 2017). File-based systems or spatial databases can be used to generate and manage 3D cadastral data. For instance, surveyors may submit 3D data in file formats, while land registration authorities typically maintain cadastral records in databases (Olfat et al. 2021). These file-based datasets must eventually be ingested into a spatially enabled database or data lake environment to be accessible for broader use and analysis.

Even though cadastral database schemas and exchange formats are being developed, a comprehensive solution addressing the needs for storing,

 DOI: 10.1201/9781003638308-23

editing, manipulating, querying, retrieving, managing, and analysing this data remains elusive. Specifically, the ability to fully support 3D functionalities, topologically and geometrically model 3D cadastral data within a DBMS, handle large datasets efficiently, and manage 2D and 3D data with optimal performance are significant knowledge gaps (Shahidinejad, Kalantari, and Rajabifard 2024a). Furthermore, current solutions and best practices often lack a defined strategy for migrating data from legacy systems to newly developed 3D cadastral systems.

This chapter contributes to advancing smart and sustainable land administration by embedding the 3D cadastre development lifecycle into broader digital transformation initiatives, enabling improved land governance. It suggests a framework that focuses on requirement analysis, schema design, data migration, the extract, transform, load (ETL) process from distributed sources, as well as data validation and database evaluation.

21.2 Cadastral Database Schema Design and Implementation

As illustrated in Figure 21.1, the cadastral database design process commences with an initial study of diverse documents, standards, expert knowledge, and user needs to identify requirements. These requirements are then translated into data models for database implementation, preparing it for eventual data insertion.

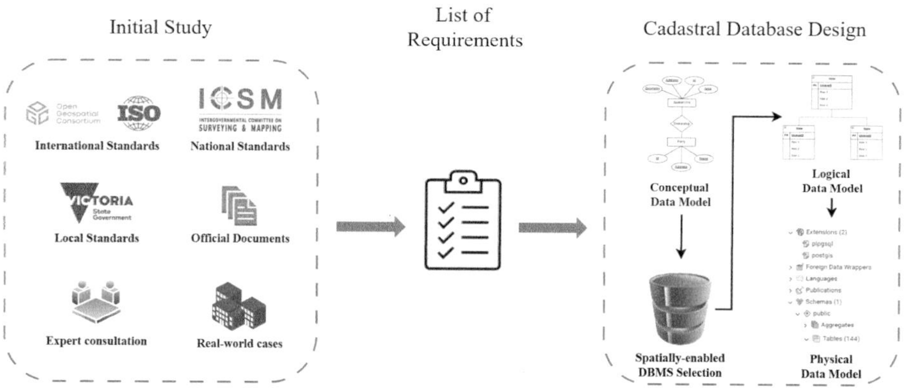

FIGURE 21.1
Cadastral database design process.

21.3 Requirement Specification

Initial study and requirement collection and analysis is a key step in 3D cadastral database implementation. This is particularly important due to the increasing use of underground and airspace, leading to complexity in defining associated rights, restrictions, and responsibilities (RRRs) in vertical dimensions (Shahidinejad, Kalantari, and Rajabifard 2025). A cadastral data model is required to integrate legal spaces with physical elements, as well as survey observations and measurements (Atazadeh et al. 2021). Furthermore, spatial relationships, including geometrical, topological, and directional, should be explicitly modelled and stored. As some real-world scenarios such as modern high-rise developments may exceed existing regulations, analysing case studies and anticipating spatial queries is essential to inform system requirements (Shahidinejad, Kalantari, and Rajabifard 2025).

This phase involves a comprehensive review of relevant regulations. In Australia/Victoria, for example, these include the *subdivision register requirements* (Land Victoria 2021), local standards like *ePlan* (SPEAR 2024) and the national standard *ICSM Conceptual Model for 3D Cadastral Survey Dataset Submissions* (ICSM 2022), alongside international ISO standards such as the *Land Administration Domain Model* (LADM) (ISO 2024). It also encompasses official documents such as the *abstract of field record* which is prepared by licensed surveyors to detail field conditions (DELWP 2021). In addition, consultation with domain experts through interviews, workshops, surveys, and questionnaires forms a vital part of this stage (Shahidinejad, Kalantari, and Rajabifard 2024b). These activities help identify functional, non-functional, and user requirements needed to accurately model both legal and physical aspects of land and property.

21.4 Data Modelling and Database Implementation

Following requirements specification, the next step involves translating those requirements into data solutions by identifying relevant entities, defining their relationships, and modelling spatial interactions within a 3D environment. The *conceptual data model* represents the highest level of abstraction, focusing on core entities, their relationships, and business rules aligned with the requirements. This phase provides a simplified, technology-neutral overview of the system, often using Entity-Relationship (ER) or Unified Modelling Language (UML) diagrams to capture essential concepts such as parties, spatial units, RRRs, physical elements, and surveying observations.

Despite ongoing efforts, a fully comprehensive solution for storing, querying, and managing 3D cadastral data remains under development

(Shahidinejad, Kalantari, and Rajabifard 2024a). Evolving standards aim to address this gap, including the internationally recognized LADM – an ISO standard first released in 2012 and expanded in the multi-part Edition II (ISO 2024) – some parts of which are still under development. At the national and state levels, Australia's Intergovernmental Committee on Surveying and Mapping is leading the development of a 3D Cadastral Survey Data Model (3DCSDM) and a unified exchange standard across Australia and New Zealand which 3D aspect has not yet been implemented (ICSM 2024).

Once the conceptual model is defined, the appropriate DBMS must be chosen based on user requirements, system architecture (centralized or distributed), and intended usage (e.g., operational or analytical) (Coronel and Morris 2017). One common characteristic of 3D cadastral systems is that they are distributed, with data coming from different sources and with different structures, including structured, semi-structured, and unstructured data which is illustrated in Table 21.1.

DBMSs can also be categorized by their data model as relational or non-relational (NoSQL). Relational databases like PostgreSQL/PostGIS and Oracle Spatial are suitable for managing structured spatial data, while NoSQL options such as MongoDB (document-based) and Neo4j (graph-based) offer greater flexibility for semi-structured and unstructured data. The selection of a database type primarily depends on the intended application and project requirements. Secondly, legal and regulatory considerations within each jurisdiction also play a crucial role. For example, some countries or organizations may not permit the use of cloud-based or commercial databases and may require the use of open-source or on-premises solutions instead.

Regardless of whether a relational or a non-relational system is chosen, the next step typically involves translating the conceptual model into a *logical data model* that reflects the structural and integrity constraints supported by

TABLE 21.1

Examples of Cadastral Data and Their Structures

Data Structure	Examples in Cadastre
Structured data	• Legacy data in current 2D cadastral databases • Shapefiles • Table of licensed surveyor information from the Surveyors Registration Board of Victoria
Semi-structured data	• ePlan files stored in XML format based on the LandXML schema • GML • JSON/GeoJSON
Unstructured data	• Scanned survey plans • Drone imagery • LiDAR point clouds • PDF or images of legal documents (e.g., deeds, abstracts of field records)

the selected DBMS. This step is especially common in relational database design; however, it's not mandatory and some sources may allow direct mapping from the conceptual model to the physical model.

The final phase is the *physical data model*, which converts the conceptual/logical design into an executable database schema. It addresses technical implementation details, including data types, spatial indexing strategies, partitioning, storage optimization, and – when applicable – support for 3D spatial data structures and operations.

Once the database has been implemented, data must be loaded, and the system's performance and efficiency should be evaluated to ensure it meets functional and non-functional requirements.

21.5 Data Migration, ETL/ELT, and Deployment

Following the design and implementation of the 3D cadastral database, the next critical step is migrating existing data from legacy systems, 2D cadastral databases, map-based systems, file-based storage, or other sources into the new system. This process must ensure that data is accurately mapped, transformed, and validated to align with the defined database schema and constraints. Figure 21.2 illustrates the pipeline for data migration and ETL process.

Data entry can be done through various methods, including manual SQL commands, BULK INSERT, DBMS interface, built-in import functions, and programmatic approaches using APIs or ORM tools such as Django ORM (Shahidinejad, Kalantari, and Rajabifard 2025). While these methods are suitable for prototyping and testing phases, production environments require automated and scalable solutions. Typically, data migration is accomplished through an ETL process. Various sources of data are first extracted, including

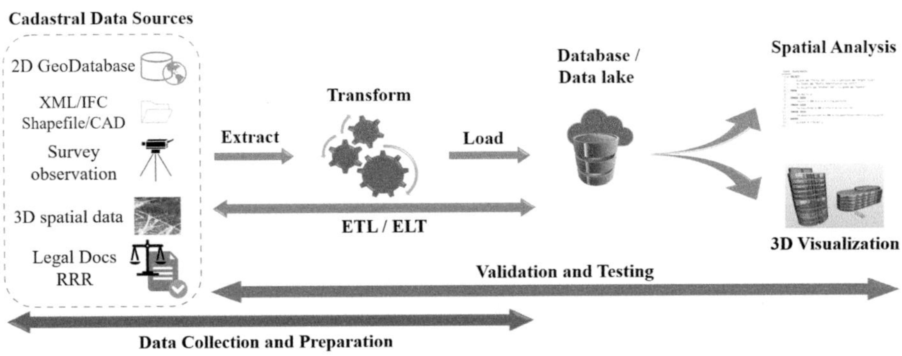

FIGURE 21.2
Cadastral data migration and ETL process.

shapefiles, XML-based formats like ePlan, and IFC files from BIM software like Revit, and unstructured documents like scanned survey plans. These datasets often require a series of preprocessing tasks and transformation such as format conversion, schema mapping, geometry validation, and spatial reference transformation. For instance, a 3D building model exported from Revit in .rvt or .ifc format can be transformed using ETL tools like Feature Manipulation Engine (FME) to align coordinate systems, enrich attribute data, handle null values, and validate data types before being loaded into the target 3D spatial database. ETL platforms are particularly effective for automating the end-to-end processing and integration of complex spatial datasets from diverse sources. In some scenarios, especially when dealing with semi-structured or unstructured data, an ELT (extract, load, transform) approach may be more appropriate. In this model, raw data is first loaded into a data lake and transformation occurs post-loading.

Throughout the process, data validation and quality assurance are critical. Testing should occur at every stage – extraction, transformation, and loading – to ensure accuracy, completeness, and structural integrity and prevent any data loss. Key quality criteria include completeness, conformity, consistency, accuracy, validity, and duplicate detection, particularly in normalized relational schemas where anomalies may occur (Shahidinejad, Kalantari, and Rajabifard 2024a). ETL pipelines, while designed to automate data workflows, can fail with unexpected data inputs and therefore must be thoroughly tested themselves. In addition, database performance must also be evaluated. Plus, optimization must be performed in different levels such as query optimization, spatial indexing, or clustering.

Once the data is verified and correctly inserted, retrieval and query analysis can be performed to ensure the system supports key use cases such as spatial searches, volumetric queries, and legal relationship tracing. Visualization tools (e.g., QGIS, Cesium) can then be used to confirm spatial accuracy and enable stakeholder interaction with 3D cadastral data. Moreover, the system can be integrated with a digital twin to not only visualize results but also support real-time monitoring, simulation, and decision-making based on dynamic cadastral.

The lifecycle concludes with deployment and ongoing maintenance, where the system is moved to production, monitored, and updated to ensure data quality, security, and performance. As new data and use cases emerge, the system must remain flexible and reliable to support long-term operational needs.

21.6 Discussion and Conclusion

The implementation of a 3D cadastral database involves a comprehensive development lifecycle that extends beyond schema design. This chapter outlined a structured approach, starting with requirements analysis and progressing through data modelling, DBMS selection, data migration and ETL, testing, and deployment. A key emphasis was placed on the need to incorporate diverse data structures such as structured, semi-structured, and unstructured, originating from various sources, including current 2D cadastral databases, XML files, scanned documents, and LiDAR point clouds.

Data migration and ETL processes were identified as critical for transforming datasets into a unified, structured format while enabling schema alignment, quality control, and system optimization. The automation of these processes is essential to ensure scalability, reduce manual errors, and support continuous data integration in production environments. ELT approaches and data lakes were also highlighted as effective for managing semi-structured or unstructured data. The importance of validation and testing at each stage in ETL process was also highlighted to ensure the integrity and reliability of 3D cadastral data throughout the lifecycle.

In conclusion, the DDLC offers a structured framework for designing, deploying, and maintaining a 3D cadastral database. By following these steps, developers can ensure that the system remains accurate, adaptable, and capable of supporting evolving cadastral needs over time. Future work should focus on enhancing automation and database optimization across all stages – from schema design and ETL/data migration to query optimization, data retrieval, and visualization – ensuring efficiency, scalability, and usability in managing 3D cadastral data.

References

Atazadeh, Behnam, Leila Halalkhor Mirkalaei, Hamed Olfat, Abbas Rajabifard, and Davood Shojaei. 2021. 'Integration of Cadastral Survey Data into Building Information Models', Geo-Spatial Information Science, 24: 387–402.

Coronel, Carlos, and Steven Morris. 2017. Database Systems: Design, Implementation, & Management. Cengage Learning.

DELWP. 2021. Victorian Cadastral Surveys Practice Directives. Department of Environment, Land, Water and Planning. https://www.land.vic.gov.au/__data/assets/word_doc/0028/762238/victorian-cadastral-surveys-practice-directives-2025.docx

ICSM. 2022. ICSM Conceptual Model for 3D Cadastral Survey Dataset Submissions. Intergovernmental Committee on Surveying and Mapping, Accessed 03 August 2024. https://icsm-au.github.io/3d-csdm-design/2022/spec.html.

ICSM. 2024. Intergovernmental Committee for Surveying and Mapping, Accessed 03 August 2024. https://www.icsm.gov.au/.

ISO. 2024. ISO 19152-1:2024, Geographic Information – Land Administration Domain Model (LADM) – Part 1: Generic Conceptual Model. International Organization for Standardization (ISO), Geneva, Switzerland.

Land Victoria. 2021. Subdivision (Registrar's Requirements) Regulations. Victorian Legislation.

Olfat, Hamed, Behnam Atazadeh, Farshad Badiee, Yiqun Chen, Davood Shojaei, and Abbas Rajabifard. 2021. 'A Proposal for Streamlining 3D Digital Cadastral Data Lifecycle', Land, 10: 642.

Shahidinejad, Javad, Mohsen Kalantari, and Abbas Rajabifard. 2024a. '3D Cadastral Database Systems—A Systematic Literature Review', ISPRS International Journal of Geo-Information, 13: 30.

Shahidinejad, Javad, Mohsen Kalantari, and Abbas Rajabifard. 2024b. 'Practical Approaches to 3D Cadastre Implementation: Database Schemas and Exchange Formats.' In 12th International FIG Land Administration Domain Model & 3D Land Administration Workshop, 369–88. International Federation of Surveyors (FIG).

Shahidinejad, Javad, Mohsen Kalantari, and Abbas Rajabifard. 2025. 'Legal Requirement Analysis for 3D Cadastral Database Design', Transactions in GIS, 29: e70025.

SPEAR. 2024. 'ePlan Overview', Accessed 19 March 2024. https://www.spear.land.vic.gov.au/spear/pages/eplan/about/what-is-eplan.shtml.

22

Graph Databases for Integrating 3D Legal and Physical Dimensions of Underground Utilities

Ensiyeh Javaherian Pour, Behnam Atazadeh, Abbas Rajabifard, Soheil Sabri, and Masoud Kamali

22.1 Introduction

Modern cities depend on the operation of underground utility networks to support essential services such as water, gas, electricity, and sewerage. Urban environments are becoming increasingly complex and densely populated. Managing utility networks is more challenging because most infrastructure is hidden beneath the surface (Lai et al., 2023; Saeidian et al., 2024). Accurate management of underground networks therefore requires not only spatial awareness of physical components but also a coherent understanding of their legal context, such as ownership rights, easements, and property boundaries (Lieberman & Ryan, 2017; Saeidian et al., 2022).

The fragmentation between legal and physical utility data poses a significant challenge. These datasets are often maintained in separate systems and formats, resulting in reduced interoperability, incomplete information, and diminished analytical capability—especially in scenarios involving urban development, emergency response, or infrastructure maintenance (Rathore et al., 2021; Yong-Kang Qiao, 2022). Excavation or construction works carried out without access to integrated data frequently result in service disruptions, legal disputes, and costly delays. The impacts of this poor integration are felt across agencies, affecting utility providers, municipalities, and land administrators alike (Den Duijn et al., 2018; Department for Science, 2022).

This challenge is directly aligned with the broader objectives of land intelligence and geospatial ecosystems—central themes of this volume. By proposing a semantically enriched, digitally transformable data model for underground utility networks, this chapter advances the goals of sustainability, spatial governance, and digital innovation. Through the integration of 3D spatial standards and graph-based data structures, it supports AI-compatible

 DOI: 10.1201/9781003638308-24

reasoning, dynamic infrastructure analysis, and smart city planning. The resulting framework offers a practical contribution to the digital transformation of land and utility systems.

To address the challenge of disconnected legal and physical records, spatial data modelling has undergone significant innovation. Traditional 2D models, while useful for representing horizontal layouts, are limited in capturing vertical geometry, inter-utility topological relationships, or legal encumbrances (Asghari et al., 2020; Sharafat et al., 2021). In response, 3D spatial data models have emerged, offering more detailed, volumetric representations of underground systems. Among these, CityGML, developed by the Open Geospatial Consortium (OGC), stands out as a widely adopted semantic framework for modelling urban infrastructure, including underground utilities (Gröger et al., 2008; Open Geospatial Consortium, 2021a).

CityGML's strength lies in its extensibility through Application Domain Extensions (ADEs), which allow customization for specific domains such as utility networks. Its Utility Network ADE (UNADE) provides a semantically rich structure for representing the geometry, function, and hierarchy of utility assets (Becker et al., 2011; Kutzner & Kolbe, 2016). Compared with other standards like IFC, INSPIRE, SEDRIS, PipelineML, and ArcGIS Utility Network, CityGML supports multi-utility integration, 3D spatial fidelity, and legal information more comprehensively (Hijazi et al., 2017; Javaherian Pour et al., 2025c; Pouliot et al., 2018; Preciados Royano et al., 2020). However, when implemented in relational data structures, CityGML struggles to capture the connectivity and network reasoning needed for advanced query operations, particularly those involving utility–parcel relationships (Javaherian Pour et al., 2025a, 2025b).

Relational data models, though widely used, are fundamentally tabular and ill-suited for representing complex topological relationships. Basic connectivity queries—such as tracing affected parcels or identifying interdependencies—require costly joins and suffer from performance degradation at scale (Hor et al., 2018; Ilonen, 2023). These limitations hinder their applicability in dynamic, high-connectivity systems like underground utilities, particularly in smart city environments demanding real-time, multi-domain integration.

Graph data models offer a promising alternative. By representing entities such as parcels, pipes, and network features as nodes, and the relationships among them, ownership, adjacency, flow as edges, graph databases natively support the kinds of queries and relationships that utility network analysis demands (Ji, 2020; Kankanamge et al., 2017). Labelled property graphs (LPGs), in particular, provide a flexible mechanism for encoding semantics and attributes on both nodes and edges, allowing for multi-dimensional, real-time querying, topological reasoning, and scalable representation of complex systems (Jamkhedkar et al., 2018; Zhu et al., 2023).

This chapter examines the integration of legal and physical aspects of underground utility networks and argues for the use of graph-based data models to overcome the limitations of relational approaches. It begins by discussing the strengths of CityGML as a 3D semantic model, followed by an analysis of the need for legal–physical connectivity in infrastructure and land administration. It then contrasts the limitations of relational databases with the structural advantages of graph data models. Finally, a case study with water utility data illustrates how graph-based integration enables unified representation of water and sewer networks in relation to land parcels, demonstrating the feasibility and benefits of this approach.

22.2 Importance of CityGML for Underground Utility Network Modelling

Modelling underground utility networks requires a data model that supports both the 3D geometric complexity and the semantic richness of subsurface infrastructure. Among available spatial data standards, CityGML, developed by the OGC, has emerged as one of the most comprehensive frameworks for modelling urban infrastructure in 3D environments. It supports detailed representations of city elements, including buildings, terrain, roads, and utility networks, and its extensible ADEs.

The UNADE enables semantic labelling, 3D topological structuring, and functional and legal attribution of utility networks such as water, sewer, gas, and electricity (Gröger et al., 2008; Kutzner & Kolbe, 2016). Unlike rigid models, CityGML's ADE mechanism allows new feature classes and attributes to be added without altering the core schema (Open Geospatial Consortium, 2021b).

Comparative studies show that CityGML offers more robust support for underground utility modelling than alternatives such as IFC, INSPIRE, LandInfra, ArcGIS Utility Network, SEDRIS, and PipelineML. It meets both common requirements, such as 3D coordinate systems, topological relationships, and semantic richness, and utility-specific needs like multi-utility representation, legal connectivity, and vertical depth modelling (Javaherian Pour et al., 2025c). While models like INSPIRE and LandInfra offer basic support for georeferencing and visualization, they lack 3D topological structure and detailed semantic capacity (European Commission Joint Research Centre Luxembourg, 2013; Open Geospatial Consortium, 2016).

CityGML further supports complex spatial relationships, including inter-network connectivity, parcel linkage, and network hierarchies, capabilities not well handled in PipelineML or ArcGIS Utility Network (Esri, 2024; Open Geospatial Consortium, 2019). It accommodates multiple geometry types (e.g., point, line, surface, solid), enabling detailed modelling of assets like pipes and access structures. Importantly, it also supports legal elements such as boundaries, easements, and ownership—typically absent or underdeveloped in IFC and INSPIRE (BuildingSMART, 2024; European Commission Joint Research Centre Luxembourg, 2013).

In summary, CityGML, especially when extended through UNADE, offers a uniquely balanced and extensible framework for modelling the physical and legal dimensions of underground utilities. Its ability to integrate semantic, spatial, and legal elements within a coherent 3D environment makes it particularly suitable for infrastructure management, legal compliance, and smart city applications.

22.3 The Need for Integrated Legal and Physical Connectivity

Integrating legal and physical dimensions of underground utility networks is essential for accurate modelling, reliable analysis, and effective infrastructure governance. However, physical data (e.g., pipe geometry, connectivity) and legal data (e.g., ownership, easements) are often maintained in isolation, resulting in fragmented records that hinder coordinated decision-making. In practice, utilities' geospatial datasets rarely link directly to cadastral legal data, complicating Underground Land Administration and increasing the risk of costly errors during excavation and development (Den Duijn et al., 2018).

CityGML's semantic 3D city model provides a strong foundation for bridging this gap. It already includes cadastral elements such as boundaries and ownership, and its ADE structure allows the seamless integration of additional legal features (e.g., volumetric rights-of-way) without altering the core schema (Javaherian Pour et al., 2025c; Saeidian et al., 2024).

This integration offers several benefits. Ownership tracking becomes immediate, enabling clear responsibility for asset maintenance and incident response (Atazadeh et al., 2017). Easements can be represented as explicit 3D objects, ensuring visibility and enforceability within the urban model (Lieberman & Ryan, 2017). Urban planners gain the ability to perform compliance checks and conflict analysis—for example, identifying potential clashes between development projects and protected infrastructure zones (Aien et al., 2012).

Moreover, shared legal parcels can serve as linking nodes between different networks, enhancing multi-utility coordination. When water, sewer, and other services are managed by separate agencies, aligning them through common parcel references improves planning, reduces redundancy, and lowers operational risks (Youen Pericault & Hedström, 2023). An integrated legal–physical model thus supports ownership clarity, easement enforcement, spatial conflict prevention, and coordinated network management—all of which are critical for safe, efficient, and legally compliant underground operations.

22.4 Challenges of Modelling Legal–Physical Connectivity in UNADE and the Role of Graph Databases

The integration of legal and physical information in underground utility networks demands data models capable of capturing complex, multi-relational, and hierarchical structures. While conventional relational databases have long supported geospatial data management, they are inherently constrained in their ability to model dynamic topologies and semantically rich relationships across domains such as utility assets, parcels, and legal rights. These limitations pose significant challenges for applications that require topological tracing, legal–physical linkage, and 3D network reasoning. In contrast, graph databases, particularly those based on LPG models, offer a structurally and semantically aligned approach to these challenges. By representing infrastructure elements and legal objects as nodes, and their diverse relationships as edges, graph models enable efficient traversal, expressive querying, and adaptive schema evolution. This section critically examines the shortcomings of relational data structures in modelling integrated utility networks and outlines the conceptual and practical advantages of adopting graph-based approaches.

22.4.1 Challenges with Relational Data Models

Relational models are structured as tables and rely on foreign key relationships to express connections between entities. While effective for flat data, they become inefficient and semantically insufficient for representing interconnected networks:

Topological Traversals: Performing upstream or downstream tracing in utility networks (e.g., to detect leak propagation or isolate a service area) requires recursive joins across multiple tables. This approach scales poorly and introduces complexity in both query formulation and execution (Hijazi et al., 2017; Petrasch & Petrasch, 2022).

Multi-Domain Integration: Integrating diverse domains such as water, sewer, and legal parcels into a unified query in a relational model demands complex schema design and nested joins, often resulting in slow performance and fragmented data access (Daulat et al., 2022; Rathore et al., 2021).

Hierarchical Relationships: Relationships like those between networks and subnetworks or shared assets (e.g., manholes or valves serving multiple lines) are cumbersome to model and traverse in relational schemas, often requiring additional associative tables and constraints to simulate relationships.

Semantic Limitations and Join Complexity: The semantics of relationships (e.g., ownership vs. flow) are not inherently captured in relational models. Furthermore, as the network grows, join operations become increasingly expensive, limiting real-time responsiveness (Kolbe et al., 2005; Widl et al., 2021).

22.4.2 Graph Data Models: A Structural and Semantic Fit

Graph databases, particularly LPGs, are purpose-built to model and query highly interconnected data. In LPGs,

1. **Nodes** represent entities such as parcels, pipes, valves, or legal parties.
2. **Edges** represent spatial, topological, functional, network links, or legal relationships, each with their own direction, label, and properties.

This representation allows graph databases to excel in several areas:

Efficient Traversal and Pathfinding: Graphs natively support depth-first and breadth-first searches, making it possible to trace a pipeline through multiple parcels or utilities with minimal overhead (Ji, 2020; Li et al., 2024).

Semantic and Structural Expressiveness: LPGs support labelled and typed relationships, enabling meaningful queries like "Find all parcels supplied by a water pipe owned by entity X and intersecting a sewer line." These semantics are embedded within the graph structure and do not require auxiliary mapping tables (Dunn & Holmes, 2019; Zhu et al., 2023).

Flexible Schema Evolution: New metadata attributes, such as ownership history, maintenance status, or operational hierarchy, can be appended to nodes and edges without schema refactoring. This makes LPGs highly adaptable for evolving urban infrastructure data (Donkers et al., 2020; Gelling et al., 2023).

Multi-Layer Integration: Graphs support layered connectivity. For instance, water and sewer networks, though distinct in function and management, can be linked through parcel nodes, facilitating inter-utility analysis and coordination (Van Nes & Yamu, 2021).

In sum, while relational databases can store utility network data, they struggle to model the dynamic, multi-scale, and legally embedded nature of modern infrastructure systems. Graph databases overcome these challenges by offering a model that is both structurally aligned with real-world connectivity and semantically expressive. This makes them particularly suitable for applications in 3D urban infrastructure management and legal–spatial integration within smart city frameworks.

22.5 Case Study: Parcel-Centred Integration of Water and Sewer Networks

The case study presented here demonstrates the conceptual importance of integrating legal and physical dimensions within underground utility networks. Using real-world datasets provided by a water utility organization (i.e. South East Water, Melbourne), the scenario explores how a CityGML-based modelling perspective can support legal–physical linkage and cross-network integration. The dataset includes three main spatial layers: the water distribution network, the sewerage network, and cadastral parcels (Figure 22.1). Each layer contains 2D geometry and associated semantic information about physical assets and land boundaries. The water and sewer networks are managed separately and stored in different systems, with no direct topological or semantic link between them despite both serving the same parcels.

In the case of water utilities, the water and sewer networks can each be represented as FeatureGraph node within CityGML, complete with 3D geometry, functional roles, and connectivity. Within this structure, parcels can function as common nodes linking both the water and sewer networks. Each parcel can be connected upstream to the water pipes that supply it, and downstream to the sewer mains that collect its discharge. These relationships can be expressed through CityGML's ability

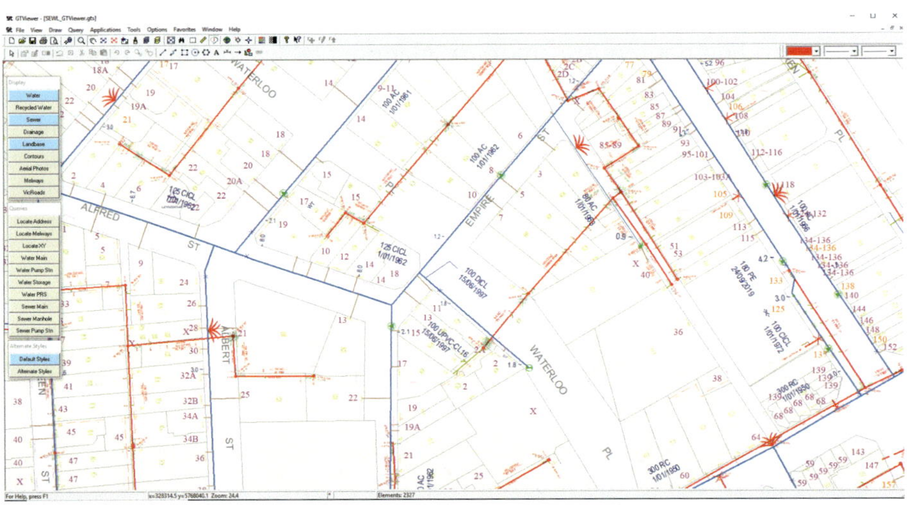

FIGURE 22.1
Screenshot of GTViewer application used by a water utility organization (South East Water, Melbourne), displaying water and sewer networks overlaid with cadastral parcels. The application supports network tracing, maintenance planning, and spatial query operations. (Image by the author.)

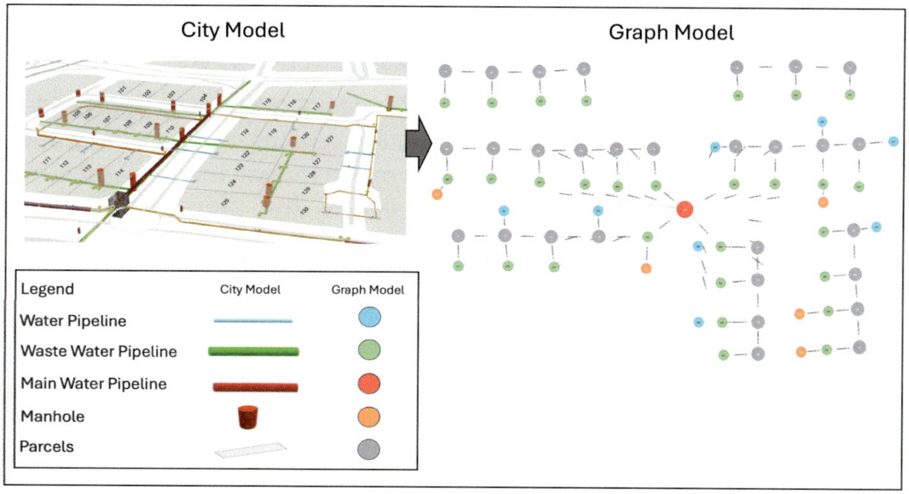

FIGURE 22.2
Illustration of parcel-centred integration of water and sewer networks into a graph data model. The left side shows a 3D CityGML-based visualization of underground infrastructure and land parcels, while the right side depicts the corresponding labelled property graph (LPG) structure, where parcels act as linking nodes across multiple networks. (Image by the author.)

to model associations and topological links between features, allowing for multi-utility integration grounded in legal geography. Figure 22.2 shows the integration of water and sewerage systems within the cadastral information.

For instance, a user could conceptually trace a flow path from a water distribution station, through pressurized mains, into a parcel, and then from that parcel into the sewerage system leading to a pump station. CityGML's semantic labelling, property inheritance, and extensibility make it possible to encode these associations across domains. Furthermore, its ADE structure allows for domain-specific attributes, such as legal ownership, easements, or utility functions, to be attached to each component of the model.

This conceptual integration aligns with the objectives of digital twin frameworks, which require unified models capable of representing not just infrastructure, but the legal, administrative, and operational contexts in which that infrastructure exists. By framing parcels as legally defined interface nodes that bridge two independently managed physical networks, CityGML offers a rich foundation for future models—whether they are implemented relationally or transformed into a graph structure for enhanced query performance. The case study illustrates how the graph data model already supports the semantics and topology needed to conceptually integrate disconnected networks via cadastral elements, even if current data infrastructures have yet to adopt or realize this potential.

22.6 Conclusion and Future Directions

The integration of legal and physical dimensions in underground utility networks is essential for enabling accurate modelling, spatial reasoning, and informed decision-making in modern urban infrastructure. As demonstrated in this chapter, relational data models—while prevalent—struggle to represent the inherent connectivity, semantic complexity, and multi-domain relationships required for managing 3D utility systems. These limitations significantly constrain the ability to perform topological queries, integrate parcel-based legal constraints, and support cross-utility analysis.

Graph databases, particularly those based on the LPG model, offer a more natural and scalable alternative. By representing infrastructure elements and legal constructs as nodes, and their diverse relationships as semantically labelled edges, graph models align structurally with the real-world topology of underground utilities. The case study highlighted how parcels can serve as bridging nodes between water and sewer networks, enabling integrated spatial queries and network tracing that would be infeasible or inefficient in relational systems. This graph-based perspective does not replace CityGML but rather builds upon its semantic richness, offering an implementation pathway to realize its potential in smart city contexts.

Future developments in this domain should focus on extending graph models to incorporate dynamic legal events, such as ownership transfers and easement modifications, enabling temporal queries and compliance tracking. There is also a need to develop support for 3D multi-utility routing, allowing analysts to simulate flows across water, sewer, and other networks through shared parcel nodes. Additionally, graph-based frameworks offer opportunities for integrating zoning rules, building geometries, and sensor data, forming the backbone of a semantically coherent urban digital twin. By advancing graph-based integration of legal and physical infrastructure, we move closer to a future where subsurface systems are not only visible and analysable but also governed with greater precision and accountability.

Reference

Aien, A., Kalantari, M., Rajabifard, A., Williamson, I. P., & Shojaei, D. (2012). Developing and testing A 3D cadastral data model A case study in Australia. *ISPRS Annals of the Photogrammetry, Remote Sensing and Spatial Information Sciences*, I–4, 1–6. https://doi.org/10.5194/isprsannals-I-4-1-2012

Asghari, A., Kalantari, M., & Rajabifard, A. (2020). A structured framework for 3D cadastral data validation – A case study for Victoria, Australia. *Land Use Policy*, 98, 104359.

Atazadeh, B., Kalantari, M., Rajabifard, A., & Ho, S. (2017). Modelling building ownership boundaries within BIM environment: A case study in Victoria, Australia. *Computers, Environment and Urban Systems*, 61, 24–38. https://doi.org/10.1016/j.compenvurbsys.2016.09.001

Becker, T., Nagel, C., & Kolbe, T. H. (2011). Integrated 3D modeling of multi-utility networks and their interdependencies for critical infrastructure analysis. In T. H. Kolbe, G. König, & C. Nagel (Ed.), *Advances in 3D Geo-Information Sciences* (pp. 1–20). Springer Berlin.

BuildingSMART. (2024). *IFC 4.3.x Data Model Documentation*. BuildingSMART.

Daulat, S., Rokstad, M. M., Klein-Paste, A., Langeveld, J., & Tscheikner-Gratl, F. (2022). Challenges of integrated multi-infrastructure asset management: A review of pavement, sewer, and water distribution networks. *Structure and Infrastructure Engineering*, 20(4), 546–565.

Den Duijn, X., Agugiaro, G., & Zlatanova, S. (2018). *Modelling below-and above-ground utility network features with the CityGML Utility Network ADE: Experiences from Rotterdam*. ISPRS Annals of the Photogrammetry, Remote Sensing and Spatial Information Sciences. 3rd International Conference on Smart Data and Smart Cities, Delft, The Netherlands.

Department for Science, Innovation and Technology and Geospatial Commission. (2022). *National Underground Asset Register (NUAR) Economic Benefits Paper*. Department for Science, Innovation and Technology and Geospatial Commission.

Donkers, A., Yang, D., & Baken, N. (2020). *Linked Data for Smart Homes: Comparing RDF and Labeled Property Graphs*. LDAC2020 – 8th Linked Data in Architecture and Construction Workshop, Dublin, Ireland.

Dunn, S., & Holmes, M. (2019). Development of a hierarchical approach to analyse interdependent infrastructure system failures. *Reliability Engineering & System Safety*, 191, 106530. https://doi.org/10.1016/j.ress.2019.106530

Esri. (2024). *ArcGIS Utility Network*. Esri. Retrieved 19 December 2023 from https://www.esri.com/en-us/arcgis/products/arcgis-utility-network/overview

European Commission Joint Research Centre Luxembourg. (2013). INSPIRE Metadata Implementing Rules: Technical Guidelines Based on EN ISO 19115 and EN ISO 19119 (pp. 99). https://knowledge-base.inspire.ec.europa.eu/document/download/6e82ba03-d470-4985-9aa6-355b67dc8940_en?filename=md_ir_and_iso_20131029.pdf

Gelling, E., Fletcher, G., & Schmidt, M. (2023). Bridging graph data models: RDF, RDF-star, and property graphs as directed acyclic graphs. *arXiv preprint arXiv:2304.13097*. https://doi.org/10.48550/arXiv.2304.13097

Gröger, G., Kolbe, T. H., Czerwinski, A., & Nagel, C. (2008). OpenGIS city geography markup language (CityGML) encoding standard, version 1, Open Geospatial Consortium. https://www.ogc.org/standards/citygml/

Hijazi, I., Kutzner, T., & Kolbe, T. H. (2017). *Use cases and their requirements on the semantic modeling of 3D supply and disposal networks. Kulturelles Erbe erfassen und bewahren-Von der Dokumentation zum virtuellen Rundgang*, 37. Wissenschaftlich-Technische Jahrestagung der DGPF.

Hor, A.-H., Sohn, G., Claudio, P., Jadidi, M., & Afnan, A. (2018). A semantic graph database for BIM-GIS integrated information model for an intelligent urban mobility web application. *ISPRS Annals of the Photogrammetry, Remote Sensing and Spatial Information Sciences*, 4, 89–96.

Ilonen, J. (2023). A Case Study on Transitioning from Relational Data models to Graph Data models in an Industrial Context. Master's Thesis in Information Technology, Computer Engineering Åbo Akademi, Faculty of Science and Engineering. https://www.doria.fi/bitstream/handle/10024/187027/ilonen_johanna.pdf?sequence=3

Jamkhedkar, P., Johnson, T., Kanza, Y., Shaikh, A., Shankaranarayanan, N., & Shkapenyuk, V. (2018). *A Graph Database for a Virtualized Network Infrastructure*. Proceedings of the 2018 International Conference on Management of Data.

Javaherian Pour, E., Atazadeh, B., Rajabifard, A., Sabri, S., & Norris, D. (2025a). A Graph Data Model for CityGML Utility Network ADE: A Case Study on Water Utilities. ISPRS International Journal of Geo-Information, 14, 493. https://doi.org/10.3390/ijgi14120493

Javaherian Pour, E., Atazadeh, B., Rajabifard, A., & Sabri, S. (2025b). *Developing a CityGML-Based Graph Data Model for Utility Infrastructure in Smart Cities*. ISPRS Geo-Spatial Week (GSW 2025), Dubai, UAE.

Javaherian Pour, E., Atazadeh, B., Rajabifard, A., & Sabri, S. (2025c). Review and assessment of 3D spatial data models for managing underground utility networks. *Tunnelling and Underground Space Technology*, 157, 106219. https://doi.org/10.1016/j.tust.2024.106219

Ji, Q. (2020). *Geospatial Inference and Management of Utility Infrastructure Networks*. Newcastle University.

Kankanamge, C., Sahu, S., Mhedbhi, A., Chen, J., & Salihoglu, S. (2017). *Graphflow: An Active Graph Database*. Proceedings of the 2017 ACM International Conference on Management of Data.

Kolbe, T. H., Gröger, G., & Plümer, L. (2005). CityGML: Interoperable access to 3D city models. In *Geo-Information for Disaster Management* (pp. 883–899). Springer.

Kutzner, T., & Kolbe, T. H. (2016). *Extending semantic 3D city models by supply and disposal networks for analysing the urban supply situation*. Lösungen für eine Welt im Wandel, Dreiländertagung der SGPF, DGPF und OVG, 36. Wissenschaftlich-Technische Jahrestagung der DGPF, Bern, Switzerland.

Lai, Y., Wang, Y., Cheng, J., Chen, X., & Liu, Q. (2023). Review of constraints and critical success factors of developing urban underground space. *Underground Space*, 12(1), 137–155. https://doi.org/10.1016/j.undsp.2023.03.001

Li, T., Rui, Y., Zhu, H., Lu, L., & Li, X. (2024). Comprehensive digital twin for infrastructure: A novel ontology and graph-based modelling paradigm. *Advanced Engineering Informatics*, 62, 102747. https://doi.org/10.1016/j.aei.2024.102747

Lieberman, J., & Ryan, A. (2017). OGC underground infrastructure concept study engineering report. O. E. Report. https://docs.ogc.org/per/17-048.html

Open Geospatial Consortium. (2016). *OGC® Land and Infrastructure Conceptual Model Standard (LandInfra)Version 1.0* (pp. 315). Wayland, MA: Open Geospatial Consortium.

Open Geospatial Consortium. (2019). *OGC PipelineML Conceptual and Encoding Model Standard*. Open Geospatial Consortium.

Open Geospatial Consortium. (2021a). *CityJSON Community Standard 1.0*. Open Geospatial Consortium.

Open Geospatial Consortium. (2021b). OGC City Geography Markup Language (CityGML) Part 1: Conceptual Model Standard. Open Geospatial Consortium. https://docs.ogc.org/is/20-010/20-010.html.

Petrasch, R. J., & Petrasch, R. R. (2022). Data integration and interoperability: Towards a model-driven and pattern-oriented approach. *Modelling*, 3(1), 105–126. https://doi.org/10.3390/modelling3010008

Pouliot, J., Larrivee, S., Ellul, C., & Boudhaim, A. (2018). *Exploring schema matching to compare geospatial standards: application to underground utility networks* International Archives of the Photogrammetry, Remote Sensing and Spatial Information Sciences-ISPRS Archives. 13th 3D GeoInfo Conference, Delft, The Netherlands.

Preciados Royano, A., Díaz Severiano, J. A., Gómez-Jáuregui, V., Manchado del Val, C., López Iglesias, J., Fernández García, O., & Otero González, C. (2020). *3D Semantic-Rich Modelling of Underground Utility Networks.* Advances in Design Engineering: Proceedings of the XXIX International Congress INGEGRAF, 20–21 June 2019, Logroño, Spain.

Rathore, M. M., Attique Shah, S., Awad, A., Shukla, D., Vimal, S., & Paul, A. (2021). A cyber-physical system and graph-based approach for transportation management in smart cities. *Sustainability*, 13(14), 7606. https://www.mdpi.com/2071-1050/13/14/7606

Saeidian, B., Rajabifard, A., Atazadeh, B., & Kalantari, M. (2022). Development of an LADM-Based Conceptual Data Model for 3D Underground Land Administration in Victoria. 10th International FIG workshop on the Land Administration Domain Model, 31 March - 2 April 2022, Dubrovnik, Croatia. https://doi.org/10.4233/uuid:d745bc28-1031-4f3b-b2a3-d7da69b7aaf1

Saeidian, B., Rajabifard, A., Atazadeh, B., & Kalantari, M. (2024). Managing underground legal boundaries in 3D-extending the CityGML standard. *Underground Space*, 14, 239–262.

Sharafat, A., Khan, M. S., Latif, K., Tanoli, W. A., Park, W., & Seo, J. (2021). BIM-GIS-based integrated framework for underground utility management system for earthwork operations. *Applied Sciences*, 11(12), 5721.

Van Nes, A., & Yamu, C. (2021). Analysing linear spatial relationships: the measures of connectivity, integration, and choice. In A. van Nes & C. Yamu (Eds.), *Introduction to Space Syntax in Urban Studies* (pp. 35–86). https://link.springer.com/chapter/10.1007/978-3-030-59140-3_2

Widl, E., Agugiaro, G., & Peters-Anders, J. (2021). Linking semantic 3D city models with domain-specific simulation tools for the planning and validation of energy applications at district level. *Sustainability*, 13(16), 8782.

Yong-Kang Qiao, F.-L. P., Luan, Y.-P., & Wu, X.-L. (2022). Rethinking underground land value and pricing: A sustainability perspective. *Tunnelling and Underground Space Technology*, 127. https://doi.org/10.1016/j.tust.2022.104573

Youen Pericault, M. V., & Hedström, A. (2023). Modelling the long-term sustainability impacts of coordination policies for urban infrastructure rehabilitation. *Water Research*, 236, 0043–1354. https://doi.org/10.1016/j.watres.2023.119912

Zhu, J., Wu, P., & Lei, X. (2023). IFC-graph for facilitating building information access and query. *Automation in Construction*, 148, 104778.

Part III

Applications and Advanced Services

Empowered by Land and Geospatial Intelligence

23

Energy Entitlement in Multi-Owned Buildings: Resolving Allocation Challenges Through Governance, Policy, and Technology

Aravind Poshnath

23.1 Introduction

The adoption of distributed energy resources (DERs), particularly rooftop solar photovoltaic (PV) systems, is increasingly recognised as essential for enhancing urban energy resilience and supporting the United Nations' Sustainable Development Goals (SDGs). Despite significant progress in deploying rooftop solar PV in detached residential dwellings, the adoption within multi-owned buildings (MOBs), such as apartments, is remarkably low. The unique administration and managerial structures and the multi-ownership arrangements pose considerable challenges to the fair and equitable allocation of renewable energy.

Illustrating this disparity, approximately 25% of detached houses in Australia possess solar panels, yet 0.6% of the apartment buildings benefit from similar installations, leaving an estimated 4GW of solar potential unutilised (Roberts et al., 2018). Similar trends are evident internationally, such as in Germany, where out of 1.9 million suitable MOBs, only 9000 apartments have solar panels installed (McGovan, 2024). These disproportions underscore the need to address institutional and social complexities rather than purely technical hurdles.

A significant share of apartment buildings is concentrated in urban areas, characterised by high-density, multi-cultural dwellings. Though promoting diversity and inclusivity, such arrangements may also lead to complex ownership dilemmas. In apartments, the common areas, particularly the rooftops, are owned and operated collectively through an Owners Corporation (OC). Any structural modifications, such as the installation of rooftop PV, would necessitate the consent of other residents through a formal voting process (Shin et al., 2022). However, disagreements often sprout concerning the perceived equity and fairness in allocating the benefits and liability to the residents, leading to the abandonment of sustainability improvements. These subjective preferences and perceived fairness often create concerns around

DOI: 10.1201/9781003638308-26

distributive justice and highlight the policy and regulatory vacuum concerning equity in shared resources.

Recent advances in land intelligence and geospatial technologies offer transformative opportunities to address complex urban energy challenges, particularly those related to ownership and resource allocation in MOBs. Integrating renewable energy data with property title information in digital cadastral systems makes it possible to accurately map and demarcate energy entitlements at the level of individual apartments or lots. Artificial intelligence (AI)-driven analytics can then be applied to these geospatial datasets to optimise allocation models, predict usage patterns, and ensure equitable distribution of shared renewable resources. Embedding energy entitlement within digital property titles clarifies ownership and rights and paves the way for automated transactions, streamlined governance, and adaptive policy frameworks.

Nevertheless, the underlying institutional and governance challenges remain unsolved without a structured and regulated allocation principle focusing on fairness, transparency, and flexibility. Poshnath et al. (2023a) introduced the concept of 'energy entitlement' as a foundational framework designed to delineate the ownership each apartment unit can manage from shared solar, with policy and regulatory implications (see Figure 23.1). This chapter critically explores the energy entitlement concept through a multi-dimensional lens, examining its applicability and implications across technical, policy, and market contexts, focusing on resolving the energy allocation challenges in MOBs.

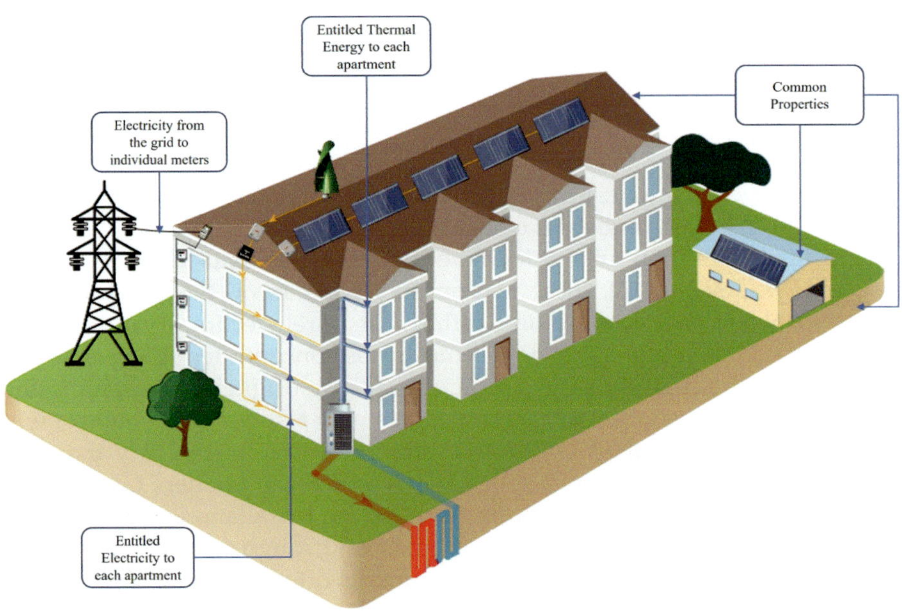

FIGURE 23.1
Schematic representation of energy entitlement concept (Poshnath et al., 2023a).

23.2 Energy Allocation Challenges in MOBs

Equitable energy allocation from shared systems presents multi-faceted legal, regulatory, social, technical, and economic challenges. While the mechanism for delineating costs associated with common property is stipulated in legislation (SCA, 2024), equivalent guidelines for energy allocation are overlooked. For instance, the Subdivision Act (Victoria Government,1988) recommends allocating equally or based on floor area proportion, lot entitlement, or occupancy. However, these principles do not adequately reflect the dynamic nature of energy generation and consumption patterns. Thus, identifying a context-sensitive model catering to the building's physical and social characteristics is critical for effective and just energy distribution.

23.2.1 Legal Challenges

Legal frameworks often impede renewable energy initiatives by mandating strict and higher consent thresholds within OC. For instance, in Victoria, Australia, modifications to the CP require a minimum consent threshold of 75% (Ti, 2022), potentially discouraging proactive residents from pursuing renewable installations (Poshnath et al., 2023b). Additionally, legally permissible allocation models, such as equal or lot-entitlement-based distribution, may inadvertently produce practical inequities. For instance, an apartment with higher lot entitlement due to its favourable location attributes (e.g., sea-facing on a higher floor) might incur disproportionate financial responsibility despite lower energy consumption. These misalignments between legal norms and perceived fairness often disincentivise resident participation, undermining collective decision-making. Furthermore, the recent emergence of solar-sharing technologies outpaces the evolution of existing legal frameworks, highlighting the urgent need for updated regulatory instruments to address complex governance and equity challenges.

23.2.2 Regulatory Challenges

Regulatory challenges, such as the requirement of utility licences to legally engage in energy sharing, significantly limit renewable energy systems (RES) adoption in MOBs. Although certain jurisdictions, such as California, have introduced virtual net energy metering (VNEM) to facilitate solar sharing across multiple meters (Patey, 2018), they are not widespread. Embedded networks were seen as an alternative, where OC assumes the role of energy retailer, purchasing energy wholesale and distributing internally (Roberts et al., 2018). However, this often locks residents into long-term contracts, nullifying their independence in choosing the energy retailer. Additionally, when solar electricity is credited to a single meter, the credits typically benefit the common property account, with residents receiving no direct bill reductions (Roberts et al., 2015). Although novel approaches, such as behind-the-meter distribution technologies, have demonstrated technical feasibility without involving the grid, the fundamental challenge of fair energy allocation remains unanswered.

23.2.3 Social Challenges

Residents often hold diverging perceptions of fairness and equity, influenced by their socioeconomic status, personal values, and backgrounds. Allocation based on investment proportion (Akter et al., 2019) may privilege wealthier owners and exclude low-income earners from accessing solar. Conversely, usage-based allocation rewards high-consumption behaviours (Poshnath et al., 2024). The split incentive dilemma worsens the scenario, particularly in rental units, with landlords lacking the motivation to fund RES installations and renters lacking legal authority to vote on decisions (Castellazzi et al., 2017). Moreover, mandatory participation may provoke dissent among those who perceive an unfair distribution of benefits. These conflicting values necessitate structured and transparent pre-installation agreements concerning energy allocation, cost-sharing, and rights to opt-in or exit.

23.2.4 Technical Challenges

Most apartment buildings are not designed foreseeing the installation of shared RES. Existing building infrastructure typically includes a separate meter for each apartment unit and a common meter for the CP. Furthermore, the available roof space to the number of dwellings ratio is lower in taller buildings. Architectural designs causing shading between clustered apartments further reduce renewable energy potential (Roberts et al., 2015). However, energy entitlement can reduce potential conflicts by specifying the ownership percentages instead of absolute energy availability.

23.2.5 Economic Challenges

Although the cost of solar PV systems has declined over the past decade, the associated capital expenditure concerning storage devices, inverters, and solar-sharing technologies can still pose challenges for the residents. Moreover, the payback period may disincentivise owners planning to sell their units within a few years. Furthermore, the feed-in tariff (FiT) has been declining rapidly globally, with the aim of maximising the self-consumption of energy (Zander, 2020). Facilitating peer-to-peer (P2P) internal energy trade among the residents may improve their economic utility but is regulatorily challenging (Azim et al., 2020). Australian states such as Victoria and New South Wales have introduced subsidies targeting solar installation for apartments. However, the responsibility of deciding the allocation proportion is left to the OC.

23.3 Policy and Market Mechanisms: Support or Hindrance?

Despite global momentum towards renewable energy, the policy and market mechanisms remain fragmented and inconsistent in supporting shared solar adoption in MOBs. Legislative reforms such as the reduction of voting thresholds (NSW Government, 2025), and balcony solar installations (Breit, 2023) are movements towards removing procedural barriers. However, declining FiT in the absence of a P2P trading facility reduces the economic utility of the residents. Nevertheless, Spain's abolishment of the 'sun tax' and the 'Solar for Apartment' subsidy in Australia have lowered the financial entry barriers. While these policies reduce upfront costs, their impact is restrained unless coupled with mechanisms promoting fair and transparent energy allocation.

Community solar models and 'energy communities' allow residents to benefit from off-site renewables. However, administrative issues and the probable involvement of third-party regulators may hinder the progress (Bonfert, 2024). Though individual off-site solar purchase schemes are available, the capital gain is not attractive compared to self-consumption. P2P trading and embedded networks can bypass traditional markets but face regulatory resistance or licensing limits. To enable energy entitlement, reforms must align legal, financial, and market frameworks with the available technological possibilities.

23.4 Role of Emerging Technologies

Technology is central to resolving energy allocation challenges in MOBs. Blockchain enables decentralised, secure energy tracking and automated P2P transactions through smart contracts, facilitating transparency and trust without third-party oversight (Gardner, 2018). IoT (Internet-of-Things) and smart meters provide real-time monitoring and enable granular solar-grid consumption tracking, optimising energy use based on solar availability. Digital twins simulate different allocation models and policy scenarios, aiding building-level planning and city-scale policy targeting. AI further supports multi-objective optimisation of fairness, self-consumption, and behavioural forecasting while automating dynamic pricing in internal energy markets. However, these systems demand robust data and often require a neutral managing entity. Together, these emerging technologies offer a blueprint for scalable, fair, and efficient energy distribution in MOBs if supported by enabling governance.

23.5 Conclusion

The transition to distributed RES in MOBs necessitates technical, institutional, legal, and social support. Despite the growing access to solar-sharing technologies and targeted incentives, equitable energy allocation remains overlooked in policy and practice. Social inequities, legal ambiguities, and lack of policy support further enhance this complexity.

The 'energy entitlement' concept highlights the necessity of a structured pathway to navigate these barriers by transparently delineating the ownership of energy equitability. Technological supports, policy enablers, and market reforms can transform how apartment residents access, manage, and benefit from shared renewable energy. However, realising this underutilised potential requires coordinated action across legal, regulatory, technical, and social domains, ensuring that future energy systems are smart, efficient, fair, and inclusive.

References

Akter, M. N., Mahmud, M. A., Haque, M. E., & Oo, A. M. T. (2019, 4–8 Aug. 2019). A Transactive Energy Trading Framework for Community Microgrids in Residential Multi-Dwelling Apartment Buildings. 2019 IEEE Power & Energy Society General Meeting (PESGM).

Azim, M. I., Tushar, W., & Saha, T. K. (2020, 2–6 Aug. 2020). Regulated P2P Energy Trading: A Typical Australian Distribution Network Case Study. 2020 IEEE Power & Energy Society General Meeting (PESGM).

Bonfert, B. (2024). 'We like sharing energy but currently there's no advantage': Transformative opportunities and challenges of local energy communities in Europe. *Energy Research & Social Science*, *107*, 103351. https://doi.org/10.1016/j.erss.2023.103351

Breit, T. P. (2023). *Balcony PV Power Plants in Austria: Public Support for Adoption and Differences in Implementation Between Austria and Germany*. Technische Universität Wien.

Castellazzi, L., Bertoldi, P., & Economidou, M. (2017). *Overcoming the Split Incentive Barrier in the Building Sector: Unlocking the Energy Efficiency Potential in the Rental & Multifamily Sectors*. Luxembourg: Publications Office of the European Union. https://doi.org/10.2790/912494

Gardner, J. E. (2018). Peer-to-peer energy trading and blockchain: The future of distributed energy resources. *Natural Resources & Environment*, *33*, 8.

McGovan, J. (2024, 24th April). Germany's tenant electricity scheme leaves vast solar PV potential untapped – Report. https://www.cleanenergywire.org/news/germanys-tenant-electricity-scheme-leaves-vast-solar-pv-potential-untapped-report

Patey, A. S. (2018). *Virtual Net Metering Market Development Project Final Technical Report*. https://www.osti.gov/biblio/1462440; https://www.osti.gov/servlets/purl/1462440

Poshnath, A., Rismanchi, B., & Rajabifard, A. (2023a). Adoption of renewable energy systems in common properties of multi-owned buildings: Introduction of 'energy entitlement'. *Energy Policy, 174*. https://doi.org/10.1016/j.enpol.2023.113465

Poshnath, A., Rismanchi, B., & Rajabifard, A. (2023b). Suitability Assessment of Energy Allocation Principles in Multi-Owned Buildings Asia Pacific Solar Research Conference, Melbourne, Australia.

Poshnath, A., Rismanchi, B., & Rajabifard, A. (2024). Renewable energy allocation in multi-owned buildings for sustainable transition: A novel evidence-based decision-making framework. *Energy and Buildings, 320*. https://doi.org/10.1016/j.enbuild.2024.114601.

Roberts, M. B., Bruce, A., & Copper, J. (2018). Analysis of Rooftop Solar Potential on Australian Residential Buildings. Asia Pacific Solar Research Conference, Sydney.

Roberts, M. B., Bruce, A., & MacGill, I. (2015). PV in Australian Apartment Buildings–Opportunities and Barriers. Asia Pacific Solar Research Conference.

Roberts, M. B., Bruce, A., & MacGill, I. (2018). Collective Prosumerism: Accessing the Potential of Embedded Networks to Increase the Deployment of Distributed Generation on Australian Apartment Buildings. 2018 IEEE International Energy Conference (ENERGYCON)

SCA. (2024). *Questions about Strata Schemes*. Strata Community Association. Retrieved 22nd May from https://www.vic.strata.community/vic-faqs

Shin, J., Rajabifard, A., Kalantari, M., & Atazadeh, B. (2022). A BIM-based framework for property dispute minimization – A case study for Victoria, Australia. *Land Use Policy, 119*. https://doi.org/10.1016/j.landusepol.2022.106200

NSW Government (2025), Strata Schemes Management Act 2015 - NSW Legislation. https://legislation.nsw.gov.au/view/whole/html/2025-03-02/act-2015-050

Victorian Government (1988), Subdivision Act 1988 https://content.legislation.vic.gov.au/sites/default/files/2025-05/88-53aa083-authorised.pdf

Ti, E. S. (2022). Strata plan cancellations in Australasia: A comparative analysis of nine jurisdictions. *Monash University Law Review, 48*(1), 209–243.

Zander, K. K. (2020). Unrealised opportunities for residential solar panels in Australia. *Energy Policy, 142*. https://doi.org/10.1016/j.enpol.2020.111508

24

An Intelligent Framework for Developing Nationwide Building Footprint Data Product for Australia

Bipul Neupane, Jagannath Aryal, and Abbas Rajabifard

24.1 Introduction

Building footprint data plays a foundational role in the land intelligence and geospatial ecosystem by enabling accurate, scalable, and timely mapping of urban infrastructure. This data strengthens the sustainability assessments by supporting energy modelling, carbon accounting, urban heat island analysis, and disaster resilience planning. When integrated with digital and AI innovations such as machine learning, digital twins, and predictive analysis, footprint data enhances decision-making in land use planning, smart city development, and infrastructure monitoring (Herfort et al. 2021). Furthermore, building footprints are critical inputs for geospatial applications across property valuation, insurance, telecommunications, and emergency response, reinforcing their value as a core layer in the growing ecosystem of location-based intelligence.

With the advancements in high-resolution remote sensing imagery and deep learning-based extraction methods (such as convolutional neural networks (CNNs)), the accuracy and scalability of building footprint extraction have gradually improved throughout the years. As a result, leading industries like Google, Microsoft, and Esri, along with open data communities like OpenStreetMap (OSM), have generated large-scale building footprint datasets (Haklay and Weber 2008; Microsoft 2022; Sirko et al. 2021). Despite their open and wide availability, these datasets remain incomplete and inconsistent, as shown in Figure 24.1. Moreover, they are unevenly distributed, especially in regions such as Australia. These limitations affect their usability for national-scale planning and decision-making tasks.

These challenges arise in the Australian context. Commercial datasets like Geoscape Buildings provide high spatial and semantic accuracy but are not openly available for model training. Meanwhile, open datasets such as Overture Maps Buildings (https://overturemaps.org/), which is built

 DOI: 10.1201/9781003638308-27

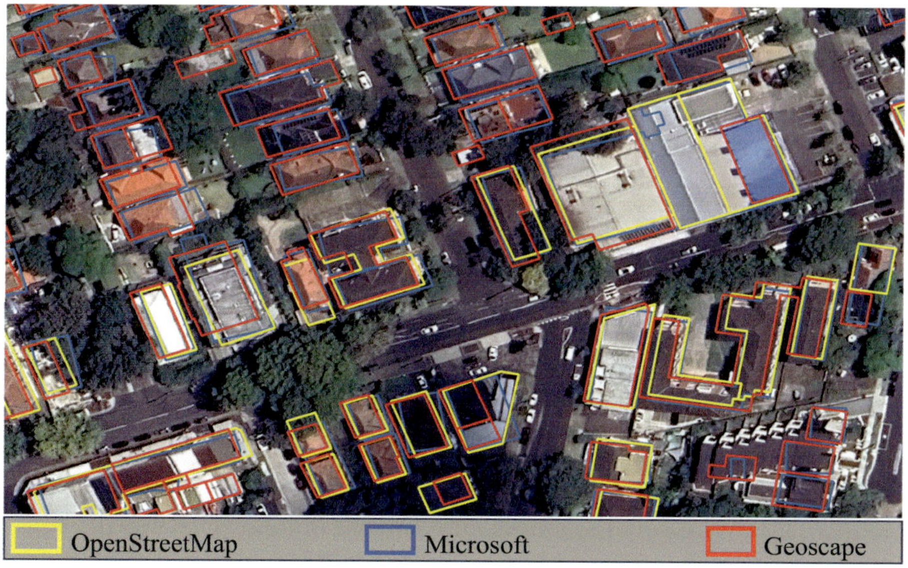

FIGURE 24.1
Building footprint data sample collected from OSM, Microsoft building footprints, and Geoscape (Australian data provider) over Chatswood, NSW, Australia. The datasets reveal the quality of the datasets in terms of completeness and inconsistencies. (The basemap is collected from Google's satellite imagery.)

from a fusion of OSM, Microsoft, and Esri data, lack detailed coverage and verified accuracy across Australian cities and regions. Furthermore, studies suggest that AI-generated features in OSM are disproportionately concentrated in just five countries and infrequently updated by human contributors (Fila et al. 2025). These gaps highlight the need for a locally adaptive, scalable framework that can integrate global datasets and improve their completeness and reliability for regional applications. This chapter addresses this gap by proposing a conceptual framework based on transfer learning (Hussain et al. 2019) for integrating available global and regional building footprint datasets. The goal is to develop a CNN-based model for generating a unified, high-quality data product for Australia.

24.2 Large-Scale Datasets and Transfer Learning Motivations

24.2.1 Global Datasets and AI-Assisted Mapping

Building footprint datasets are available from various global sources. OSM offers freely accessible, community-contributed building footprints, widely used for urban planning, humanitarian response, and AI model training

(Boeing 2017; Haklay and Weber 2008). However, OSM data quality varies regionally due to inconsistent contributor activity (Vargas-Munoz et al. 2020). AI-generated datasets from Microsoft and Google have further expanded coverage using deep learning pipelines trained on high-resolution satellite imagery (Microsoft 2022; Sirko et al. 2021). These efforts evolved from early AI-assisted mapping initiatives led by Facebook and Microsoft, which used tools like Rapid Editor (https://rapideditor.org/) and MapWithAI (https://josm.openstreetmap.de/wiki/Help/Plugin/MapWithAI) to streamline contributions into OSM. These datasets now cover billions of buildings across continents.

Despite the growing availability of large-scale building datasets, major challenges persist. Global data sources are hindered by spatial incompleteness and segmentation inconsistencies (typically 63–68% intersection over union, IoU). The distribution of AI-generated buildings is also highly imbalanced, with over 75% located in just five countries (Fila et al. 2025), highlighting global inequalities in automated mapping. Additionally, AI-derived features in OSM are often modified less frequently than manually mapped buildings, raising questions about their ongoing accuracy and maintenance. These regional quality gaps are difficult to assess due to a lack of verified reference data, as noted in comparative studies focused on Africa and Southeast Asia (Chamberlain et al. 2024; Gevaert et al. 2024).

24.2.2 Large-Scale Datasets in Australia

In the context of Australia, there is a lack of an open-source, comprehensive, and attribute-rich national building footprint dataset that can support intelligent decision-making across planning, environmental, and infrastructure domains. Geoscape Buildings (https://geoscape.com.au/data/buildings/) provides comprehensive data with high spatial accuracy and rich attributes (e.g., height, area, roof type) and integrates well with address and cadastral systems derived from manual annotations and aerial imagery. However, the dataset is commercial and is not available to train deep learning models. The open-source and relatively comprehensive dataset is the Overture Maps Buildings, which merges data from multiple sources, including OSM, Microsoft, and Esri. This product is generated through a structured data conflation process that resolves overlaps, standardized schemas, and ensures internal consistency. OSM contributes community-generated content, Microsoft provides AI-derived footprints, and Esri contributes authoritative geospatial data. Unlike the global product, Google's AI-derived data is not integrated for Australia due to its unavailability for the region. To ensure reliability, Overture prioritizes data quality and enforces validation checks before release. While this unified dataset offers valuable baseline information for global applications, its utility in an Australian context remains limited by regional gaps in coverage and attributes.

Despite advances in integrating multi-source data, two major limitations persist in the current large-scale open building footprint datasets. Firstly, the completeness of AI-generated footprints varies significantly across regions. Regional quality remains uncertain, with known inconsistencies between urban and rural areas and no formal nationwide evaluation. Moreover, the data remains incomplete, especially compared to countries like the United States, due in part to the absence of Google's open building dataset for Australia. Secondly, recent research highlights that AI-generated buildings in OSM are not only unevenly distributed but also modified far less frequently than those mapped manually, which limits their evolution and local relevance over time (Fila et al. 2025). Around 75% buildings are concentrated in just five countries: the United States, Nigeria, Algeria, India, and Kenya. This highlights the global imbalance in AI-generated contributions. Furthermore, the spatial distribution of such data correlates with population size but not with development indicators or community engagement, raising concerns about their completeness and contextual accuracy. These insights highlight the challenges of relying solely on automated mapping outputs and motivate the need for a more intelligent and locally adaptive integration framework.

24.2.3 Transfer Learning-Based Data Integration Framework

As building footprint extraction models trained on these global datasets struggle to generalize across geographies, fine-tuning-based transfer learning (FTL) offers a practical and effective solution. FTL has emerged as a dominant approach for supervised domain adaptation in remote sensing, allowing pre-trained models to be adapted to new regions with limited annotated data (Ma et al. 2024; Oikonomidis et al. 2023). Compared to methods like multi-task or few-shot learning, FTL is more flexible, requires less data, and achieves higher performance with faster convergence. These advantages make FTL well-suited to address regional gaps in building data and motivate the intelligent data integration framework proposed in this chapter, which combines global knowledge with locally adapted model fine-tuning.

24.3 Conceptual Framework for Australian Nationwide Building Footprint Dataset

This chapter proposes a conceptual framework, as illustrated in Figure 24.2, to generate a quasi-complete building footprint dataset. The components of the framework are described below.

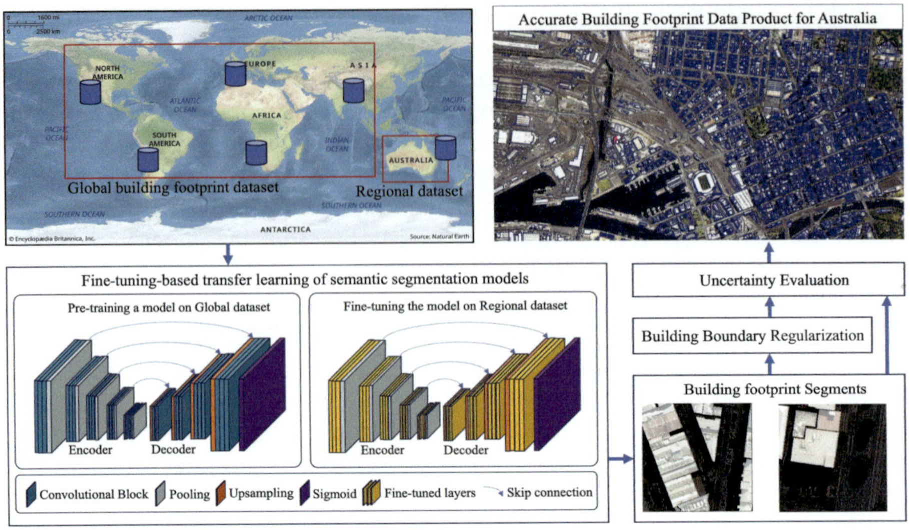

FIGURE 24.2
Conceptual framework for integrating global and regional building footprint data into a semantic segmentation model to produce a quasi-complete dataset for Australia.

24.3.1 Semantic Segmentation Model Selection

Building footprints are extracted from high-resolution aerial or satellite images using deep learning. We adopt an encoder–decoder network, where the encoder can be a CNN or Vision Transformer (ViT). U-Net is used as the base architecture (Ronneberger et al. 2015). The choice of encoder depends on dataset size. EfficientNetV2L and MaxViT are strong encoder candidates for large-scale datasets (Tan and Le 2021; Tu et al. 2022).

24.3.2 Transfer Learning (Global to Regional Knowledge Adaptation)

The framework starts with global training datasets. These include public datasets like WHU, Massachusetts, and RAMP (Neupane et al. 2024). These datasets provide image-label pairs from diverse regions. A pre-trained model is developed using these sources. To adapt the model to Australian conditions, we collect a smaller regional dataset. Images are obtained using the Nearmap Tile API (https://developer.nearmap.com/docs/tile-api) and building labels from local councils such as the City of Melbourne's 2020 Building Footprints (https://data.melbourne.vic.gov.au/explore/dataset/2020-building-footprints/information/). The pre-trained model is then fine-tuned using this local dataset. All layers of the model are made trainable to ensure full adaptation (Neupane et al. 2025).

24.3.3 Boundary Regularization

Predicted segments are converted into polygon features. Post-processing techniques are applied to improve geometric structure. This includes filtering, smoothing, and polygon fitting. The aim is to produce clean, mappable building outlines. These steps follow the methods presented in Neupane et al. (2025). Tools such as GeoAI (https://github.com/opengeos/geoai) can support large-scale processing without image tiling.

24.3.4 Uncertainty Evaluation

Segmentation accuracy is assessed using standard metrics. These include IoU and F1 score for pixel-wise agreement. For boundary accuracy, we use average symmetric surface distance (ASSD) and Hausdorff distance (HD). For shape consistency, we use the structural similarity index (SSIM), completeness, and correctness scores (Neupane et al. 2024, 2025).

24.3.5 Training and Implementation

The models are trained using Neupane's open-source GitHub repository (https://github.com/bipulneupane/EOandAIWorkshop), which allows for training and fine-tuning most segmentation models available through widely popular model databases like HuggingFace. Models are trained with Dice Loss, RMSProp optimizer, 0.0001 learning rate, 16 batch size (depending upon the computing device), and sigmoid classifier, which are all best-performance hyperparameters identified by recent benchmarks (Neupane et al. 2024). The number of training epochs is set as 500 but the training is stopped upon stagnation of loss. The highest segmentation performance is optimized by reducing the learning rate upon the plateau of the loss.

24.4 Conclusion and Way Forward

This chapter proposed a conceptual framework to generate a quasi-complete building footprint dataset for Australia by integrating global and regional data sources using a semantic segmentation model. The motivation arises from the limitations of current public datasets, which are often incomplete, inconsistently distributed, and lacking semantic attributes – especially in the Australian context. By leveraging globally pre-trained models and fine-tuning them with a small, regionally representative dataset, the framework addresses issues of domain shift and reduces the need for costly manual annotation.

Operationalizing this framework involves the use of publicly available remote sensing imagery and locally sourced building labels, which can be accessed through APIs and open government datasets. It enables geospatial industries, local councils, and research institutions in Australia to collaboratively prepare accurate, consistent, and scalable building footprint datasets with reduced computational and resource demands. By applying this approach, stakeholders can build national-scale datasets that support planning, urban monitoring, infrastructure development, and disaster resilience. Future work may focus on extending the framework to include attribute enrichment and integration into broader spatial decision-making systems.

References

Boeing, Geoff. 2017. "OSMnx: New Methods for Acquiring, Constructing, Analyzing, and Visualizing Complex Street Networks." *Computers, Environment and Urban Systems* 65: 126–39.

Chamberlain, Heather R, Edith Darin, Wole Ademola Adewole, Warren C Jochem, Attila N Lazar, and Andrew J Tatem. 2024. "Building Footprint Data for Countries in Africa: To What Extent Are Existing Data Products Comparable?" *Computers, Environment and Urban Systems* 110: 102104.

Fila, Milan, Radim Štampach, and Benjamin Herfort. 2025. "AI-Generated Buildings in OpenStreetMap: Frequency of Use and Differences from Non-AI-Generated Buildings." *International Journal of Digital Earth* 18 (1): 2473637.

Gevaert, Caroline M, Thomas Buunk, and Marc JC Van Den Homberg. 2024. "Auditing Geospatial Datasets for Biases: Using Global Building Datasets for Disaster Risk Management." *IEEE Journal of Selected Topics in Applied Earth Observations and Remote Sensing* 17.

Haklay, Mordechai, and Patrick Weber. 2008. "Openstreetmap: User-Generated Street Maps." *IEEE Pervasive Computing* 7 (4): 12–18.

Herfort, Benjamin, Sven Lautenbach, João Porto de Albuquerque, Jennings Anderson, and Alexander Zipf. 2021. "The Evolution of Humanitarian Mapping within the OpenStreetMap Community." *Scientific Reports* 11 (1): 3037.

Hussain, Mahbub, Jordan J Bird, and Diego R Faria. 2019. "A Study on CNN Transfer Learning for Image Classification." *Advances in Computational Intelligence Systems: Contributions Presented at the 18th UK Workshop on Computational Intelligence, September 5–7, 2018,* Nottingham, UK, 191–202.

Ma, Yuchi, Shuo Chen, Stefano Ermon, and David B Lobell. 2024. "Transfer Learning in Environmental Remote Sensing." *Remote Sensing of Environment* 301: 113924.

Microsoft. 2022. *Microsoft Building Footprints.* Microsoft. https://www.microsoft.com/en-us/maps/bing-maps/building-footprints.

Neupane, Bipul, Jagannath Aryal, and Abbas Rajabifard. 2024. "CNNs for Remote Extraction of Urban Features: A Survey-Driven Benchmarking." *Expert Systems with Applications* 255: 124751.

Neupane, Bipul, Jagannath Aryal, and Abbas Rajabifard. 2025. "Fine-Tuning-Based Transfer Learning for Building Extraction from Off-Nadir Remote Sensing Images." *Remote Sensing* 17 (7).

Oikonomidis, Alexandros, Cagatay Catal, and Ayalew Kassahun. 2023. "Deep Learning for Crop Yield Prediction: A Systematic Literature Review." *New Zealand Journal of Crop and Horticultural Science* 51 (1): 1–26.

Ronneberger, Olaf, Philipp Fischer, and Thomas Brox. 2015. "U-Net: Convolutional Networks for Biomedical Image Segmentation." *International Conference on Medical Image Computing and Computer-Assisted Intervention*, 234–41.

Sirko, Wojciech, Sergii Kashubin, Marvin Ritter, et al. 2021. *Continental-Scale Building Detection from High Resolution Satellite Imagery*, arXiv preprint arXiv:2107.12283. https://doi.org/10.48550/arXiv.2107.12283.

Tan, Mingxing, and Quoc Le. 2021. "Efficientnetv2: Smaller Models and Faster Training." *International Conference on Machine Learning*, 10096–106.

Tu, Zhengzhong, Hossein Talebi, Han Zhang, et al. 2022. "MaxViT: Multi-Axis Vision Transformer." In *Computer Vision – ECCV 2022*, edited by Shai Avidan, Gabriel Brostow, Moustapha Cissé, Giovanni Maria Farinella, and Tal Hassner. Springer Nature Switzerland.

Vargas-Munoz, John E, Shivangi Srivastava, Devis Tuia, and Alexandre X Falcao. 2020. "OpenStreetMap: Challenges and Opportunities in Machine Learning and Remote Sensing." *IEEE Geoscience and Remote Sensing Magazine* 9 (1): 184–99.

25

Adopting the Concept of National Security in the Domain of Land Administration

Kirsikka Riekkinen and Walter Timo de Vries

25.1 Introduction

Land administration systems define and describe the dynamic nature of people-to-land relationships and, thus, evolve over time as the relationship changes (Ting & Williamson 1999). They facilitate the national-level policies, especially by tools for imposing and recording rights, restrictions, and responsibilities related to land, and simultaneously provide infrastructure and tools for supporting land markets, tenure, and land use planning (Enemark et al. 2014). The ultimate goal is to promote social justice and political stability and contribute to sustainable development (Williamson et al. 2010). While land administration literature addresses information security and environmental security, the discussion is usually limited to technical and legal contexts of data sharing and system resilience (see, e.g., Bertino et al. 2008; Assarkhaniki et al. 2020). While national security is considered a cross-cutting theme in public administration, the concept has not yet been widely adopted in the international domain of land administration, but rather in the national context.

The constantly changing geopolitical environment is challenging land administration at multiple levels. Land administration systems are sociotechnical by nature, meaning that adopting a new concept within its domain requires a systems approach, including technical, legal, and institutional aspects (Zevenbergen 2004; Krigsholm et al. 2020). The technical infrastructure provides basis, including geospatial information structure, cadastral information, and their collection and upkeeping. The legal aspect provides requirements for legislation being up to date and able to accommodate the changing Rights, Restrictions, and Responsibilities (RRRs) needed to secure national, or homeland security. Institutional aspect covers the ecosystem of land administration, including various data producers and users, and other actors within the ecosystem.

DOI: 10.1201/9781003638308-28

Conventionally, land administration security is considered an issue which can be handled by mutually agreed national agreements and legislation. The contemporary reality is that security is affected by unknown parties with unknown intentions and actions, which are not necessarily following any consensual activities or procedures. This calls for new ways of conceptualizing land administration security and managing it. The aim of this chapter is to explore a variety of new concepts of national security concerns and discuss about land administration domain's response to these. Therefore, the objective of this chapter is to present and analyze alternative perspectives of perceiving land administration and national security, by critically confronting and challenging existing concepts and assumptions, and by deriving adaptations, extensions, and alternatives.

The rest of this chapter will first introduce the concepts of national security and the theoretical connections between the land administration domain and national security. Then, it introduces the methodology used for collecting and analyzing exploratory narratives for extracting and documenting novel approaches for national security issues in land administration. After presenting the results, we discuss their implications within the domain of land administration and finally conclude by presenting the main findings and avenues for future research.

25.2 Theoretical Background

The concept of national security, or homeland security, has been attempted to be defined by several scholars. The concept originates from protecting nations against external threats and integrates frameworks of political economy combined with military and defense policies. The protection of core national values against external threats is crucial, and for this, one needs to know the external threats. Due to technological development, the external threats come in different forms than what they used to, even though the acquisition of land still plays a key role in several theories of international relations (e.g. conquest, see Fazal 2022; territorial sovereignty and land grabbing, see Jurkevics 2022).

The conventional form of conquest or otherwise infringing the territorial sovereignty of a nation is not the only external threat a country may face. Different hybrid operations and forms of influencing are recognized to create one type of external threat as well (Monaghan 2015). Placing hostile troops or preparing an attack inside a country may be enabled by property ownership and property use providing shelter for this. Hence, it has become extremely important to have the basic data regarding land ownership and use in reliable registers. Reliability in land administration is often compared with digitalization. It can be seen as double-edged sword: on the other hand – digitalization is a prerequisite for developing modern land administration and interoperability between registers and databases, but at the same time it sets the data and system vulnerable to cyberattacks and other data security threats (Blin et al. 2024).

The paradigm of national security relating to only political powers has been challenged by asking an (entitled) question, whether the environment is part of national security (Levy 1995). This ties the previous land administration research regarding climate change and food security together with the national security studies. However, we need to expand the scope of security to reach beyond cybersecurity, climate resilience, and food security. Therefore, accommodating the concept of national security as a broad concept in the domain of land administration is required.

Due to the central role of land administration in supporting the functions and economy of society, evaluating, developing, and standardizing land administration systems accordingly has been an interest of scholars and practitioners alike. For example, Chehrehbargh et al. (2024) have researched and extracted parameters or factors of modern land administration systems that set the scene for the evaluation, both giving attention to a central framework for the evaluation, FELA (Framework for Effective Land Administration (UN-GGIM 2020). It is a UN-endorsed document that provides a comprehensive evaluation and design document for land administration. FELA has been successfully used in several use cases and by scholars alike. For example, Ahsan et al. (2023) used it to discover the factors in land administration needed to support urban land administration and smart cities.

An international standard, Land Administration Domain Model (LADM), was developed to model and understand and efficiently set up the relationship between parties and parcels (Van Oosterom & Lemmen 2015). While designed as generic as possible, its second edition aims at including all the rights, restrictions, and responsibilities that affect the subject–object relationship in the model (Lemmen et al. 2023). This creates an interesting base for the discussion of further development and adoption of the national security concept in the LADM.

25.3 Methodology

Given the exploratory nature of the aim of this book chapter, the methods used were mixed qualitative methods, and the approach to analysis was inductive, following the steps documented by Liu (2016). Data collection was built around multiple case studies, and their content analysis was complemented by document analysis and using an analytical framework conducted from the FELA (UN-GGIM 2020), a UN-endorsed document that provides a comprehensive evaluation and design document for land administration. This choice of analytical framework allows the data analysis to cover multiple aspects of land administration, namely (I) Governance, Institutions and Accountability; (II) Policy and Legal; (III) Financial; (IV) Data (and Processes); (V) Standards; (VI) Innovation; (VII) Partnerships; (VIII) Capacity and Education; (IX) Advocacy and Awareness. The aim of the analysis is to raise awareness, one of the first tasks for which this framework is shown to be useful (Unger et al. 2022).

The research design is built around multiple, documented case study descriptions that were collected from a group of land administration experts representing several nations. The persons were first introduced to the concepts of national and homeland security and were asked to produce a written description of a real-life situation in their countries where they encountered the tools of land administration that were not sufficient to solve a national security concern related to land administration. Altogether, seven cases representing five different countries were analyzed, based on the type of security issue, their main cause mechanism in the national land administration system, and possible suggestions for the solution of the issue. These were then further elaborated with the pathways of FELA, to shed light on the aspects that require special attention in the future, to accommodate national security issues in land administration.

During the analysis, the data was anonymized so that neither the author of the description nor the country of origin can be recognized, to allow the inductive approach to be fully utilized and results to be generalized in an international context.

25.4 Results

The results of the case study analyses are presented in Table 25.1. The case number, identified national security issue, its perceived cause mechanism from the land administration point of view, and suggested possible land administration solutions are compared with FELA pathways. The final column of Table 25.1 represents the pathway/pathways that relate to the solution and/or lack of land administration tools that would prevent the national security issue.

Most of the national security issues relate to the loss of (power over) land rights. This leads to further problems, for example, in terms of loss of control over agricultural land, as a result of which national priorities in food production, food security, and food sovereignty are endangered. Another commonly seen issue is the influence of another, even hostile, nation within the area and its land policy. This can be direct or indirect, by realizing, for example, taking over natural resources and leading to citizens, for example, getting hold of the rights over natural resources or clusters of buildings and constructions. Although often citizens may not be aware of these transfers, and the land administration system often does not explicitly collect or detect data on clusters of transfers by a single party or similar parties, once the

TABLE 25.1

Results of the Case Studies in a Condensed Format

Case	Issue	Cause Mechanism	Possible Solution(s)	FELA Response
1	Loss of citizens' control over land Rising land value Squatting	Policy granting citizenship along with land ownership	Policies not connecting land ownership and citizenship, or restricting ownership right Formalization measures	Policy and Legal
2	Loss of land rights Decreased food production Unrest and agitation among citizens	Land tenure system unable to secure rights Pollution Unequal treatment of right holders	Better recognition of land rights (tenure system) Transparency Participatory approach	Policy and Legal; Governance, Institutions, and Accountability; Partnerships
3	Land grabbing Rise of hostile powers	Complex overlapping land tenure systems Lack of knowledge on local systems	Establishing land tenure system that accommodates multiple tenure types Participatory approach	Policy and Legal; Partnerships
4	Environmental damages Outsider nations interfering in homeland affairs	Weak institutions or governance	Capacity building Participatory approach	Capacity and Education; Partnerships
5	Suspects toward foreign landowners Loss of agricultural land	Lack of transparency in data and processes	Increase transparency	Governance, Institutions, and Accountability
6	Loss of agricultural land	Land use change Overlapping interests on land	Participatory approach, Increased transparency	Partnerships; Governance, Institutions, and Accountability
7	Land used by a hostile nation for hybrid operations	Free land market	Restrict the ownership right of foreigners	Policy and Legal

information comes out, it may lead to leading citizens' unrest. Especially if related to natural resources, or for example, the exploitation of minerals in unauthorized and uncontrolled manners, this issue additionally leads to spatial spillovers and causes environmental hazards.

Although the cause–effect mechanisms between transactions and hazards vary, they can nevertheless often be traced back to policy and political decisions based on inadequate monitoring and information processes. Current information processes tend to focus on the assumptions that land transactions are fairly neutral and that land administration is not aimed at pro-actively monitoring unusual, out-of-the-ordinary, suspicious, wicked, or unintended land transactions. Instead, it neutrally records who and how land is acquired, and how the use of the land is controlled. Such processes are carefully regulated in legislation. But under which circumstances and using which pretenses the actual execution is taking place and what may be the adverse effect or unintended consequence of it is not explicitly defined and thus creates a loophole in the legislative framework. As such, the transparency of governance is missing, and it causes distrust among citizens toward authorities. This is not always the case – for example, in case 7, the authorities noticed this loophole and enacted new laws to restrict the right to purchase land in certain situations.

The levels of measures to improve the situation caused by the mentioned issue of national security are various. They can be of fundamental nature – like accommodating and recognizing all types of land tenures in the land administration system – or more nuanced issues like capacity building among professionals and citizens. Another often recurring solution is the inclusion of the citizens in land administration processes, namely the planning and land development. This participatory approach in land administration also calls for recognition of traditional landowners.

Out of the nine FELA pathways, four were standing out in the results. Out of these, the pathway for policy and legal was the most common pathway that needs to be strengthened when designing land administration systems that also support national security. Other pathways were governance, institutions, and accountability; partnerships; and capacity and education.

25.5 Discussion

The national security issues that relate to land administration, as well as their root causes and possible solutions, relate to the policy and institutional level of land administration, when compared to the FELA pathway. Adoption of the security concepts in the domain of land administration is related to the process of land governance. The collection processes and the distribution possibilities rooted in the technical decisions may influence security. It seems that incorporating the national security issues in models, such as LADM, requires the institutional structures to enable the use of the tools to be developed first. Therefore, the argument based on the results of this chapter is that accommodating the concept of national security in the domain of land administration has prerequisites for governance.

These prerequisites include that the legitimate land tenure institutions (either formal or informal ones) enable transparent and inclusive access to property. They also include a participatory approach for land administration and land development, so that no one is left behind. Third, they call for agile actions for enacting legislation that can be used when foreign interest in land and its resources is overpowering the national interest. Only after that, the integration of the RRRs that are related to securing national security can be integrated in the LADM.

Technical aspects, including technological innovations such as artificial intelligence, challenge digital systems on multiple levels, such as questions related to cybersecurity (Chehrehbargh et al. 2024). Interestingly, the current issues that were concerned as national security issues were connected to other aspects of land administration. These were governance and institutions, policy and legal, and partnership pathways. But obviously, this does not mean that there are no improvement needs regarding the other pathways.

Philosophically, there is a dilemma of the pursuit of transparency and openness of government and government services, rooted in increasing calls from citizens and gradual transitions in how governments and bureaucracies aim to operate, and the pursuit of open societies with limited enforcement and controlling capacities. This results in conflicting and simultaneously emerging claims of both more and less openness on the one hand, as well as simultaneous transformations of both more and less control over sovereignty on the other hand. The concerns over security have thus resulted in the need for new concepts and ideas of what land sovereignty means and how this should be administered and protected.

25.6 Conclusion

The aim of this chapter was to explore a variety of new concepts of national security concerns and discuss the land administration domain's response to these concerns. This was done by analyzing multiple cases describing situations where national security was endangered due to a lack in land administration. These were further elaborated by using FELA, to find areas of land administration requiring special attention when accommodating the concept of national security in the domain of land administration.

While the existence and interconnections of national security issues with land administration are evident, the solutions for accommodating these may vary. Recommendations for adopting the concept of national security call for further research at a more detailed level, namely incorporating this cross-cutting policy issue in the LADM. Technological pathway of FELA, incorporated with current tools, calls for policies and infrastructures for data, innovation, and standards to support national security. Therefore, a discussion of open data, cybersecurity, and possibilities provided by LADM within the field of national security should be opened. This would allow for further study of the necessary RRRs related to land, and the mechanisms of how to integrate these in modern land administration.

References

Ahsan, A., Hussain, E., Ali, Z., Zevenbergen, J., Atif, S., Koeva, M., Waheed, A. (2023). Assessing the Status and Challenges of Urban Land Administration Systems Using Framework for Effective Land Administration (FELA): A Case Study in Pakistan. Land, 12(8), 1560. https://doi.org/10.3390/land12081560.

Assarkhaniki, Z., Rajabifard, A., Sabri, S. (2020). The Conceptualisation of Resilience Dimensions and Comprehensive Quantification of the Associated Indicators: A Systematic Approach. International Journal of Disaster Risk Reduction, 51, 101840. https://doi.org/10.1016/j.ijdrr.2020.101840.

Bertino, E., Thuraisingham, B., Gertz, M. (2008). Security and Privacy for Geospatial Data: Concepts and Research Directions. Cyber Center Publications. https://doi.org/10.1145/1503402.1503406.

Blin, P.-F., Aditya, T., Santosa, P.B., Claramunt, C. (2024). A Methodological Approach Towards Cyber Risk Management in Land Administrations Systems. Land, 13(1), 19. https://doi.org/10.3390/land13010019.

Chehrehbargh, F.J., Rajabifard, A., Atazadeh, B., Steudler, D. (2024). Identifying Global Parameters for Advancing Land Administration Systems. Land Use Policy, 136(2024), 106973. https://doi.org/10.1016/j.landusepol.2023.106973

Enemark, S., Hvingel, L., Galland, D. (2014). Land Administration, Planning and Human Rights. Planning Theory, 13(4), 331–348. https://doi.org/10.1177/1473095213517882

Fazal, T.M. (2022). The Return of Conquest? Why the Future of Global Order Hinges on Ukraine. Foreign Affairs, 101(3), 20–27.

Jurkevics, A. (2022). Land Grabbing and the Perplexities of Territorial Sovereignty. Political Theory, 50(1), 32–58. https://doi.org/10.1177/00905917211008591.

Krigsholm, P., Riekkinen, K., Ståhle, P. (2020). Pathways for a Future Cadastral System: A Socio-Technical Approach. Land Use Policy, 94. https://doi.org/10.1016/j.landusepol.2020.104504.

Lemmen, C., Van Oosterom, P., Kara, A., Kalogianni, E., Alattas, A., Indrajit, A. (2023). Overview of Developments of Edition II of the Land Administration Domain Model. In Proceedings of FIG Working Week 2023, Orlando, USA, 28 May–1 June 2023.

Levy, M.A. (1995). Is the Environment a National Security Issue? International Security, 20(2), 35–62. https://www.muse.jhu.edu/article/447402.

Liu, L. (2016). Using Generic Inductive Approach in Qualitative Educational Research: A Case Study Analysis. Journal of Education and Learning, 5(2). http://dx.doi.org/10.5539/jel.v5n2p129.

Monaghan, A. (2015). The 'War' in Russia's 'Hybrid Warfare'. Parameters 45 (4). https://doi.org/10.55540/0031-1723.2987.

Ting, L., Williamson, I. (1999). Cadastral Trends: A Synthesis. The Australian Surveyor, 4(1), 46–54.

Unger, E.-M., Crompvoets, J., Bennett, R., Cantat, F., Lisec, A. (2022). Advancing FELA – The Framework for Effective Land Administration. In Proceedings of FIG Congress 2022, Warsaw, Poland, 11–15 September 2022.

UN-GGIM (2020). Framework for Effective Land Administration. Framework for Effective Land Administration. United Nations Committee of Experts on Global Geospatial Information Management (UN-GGIM), Department of Economic and Social Affairs – United Nations. https://ggim.un.org/meetings/GGIM-committee/10th-Session/documents/E-C.20-2020-29-Add_2-Framework-for-Effective-Land-Administration.pdf

Van Oosterom, P., Lemmen, C. (2015). The Land Administration Domain Model (LADM): Motivation, Standardisation, Application and Further Development. Land Use Policy, 49(2015), 527–543. https://doi.org/10.1016/j.landusepol.2015.09.032.

Williamson, I., Enemark, S., Wallace, J., Rajabifard, A. (2010). Land Administration for Sustainable Development. ESRI Press Academic Redlands.

Zevenbergen, J. (2004). A Systems Approach to Land Registration and Cadastre. Nordic Journal of Surveying and Real Estate Research, 1(1), 14.

26

BIM Empowering Intelligent Demolition Waste Management to Alleviate Land Pressure

Dongchen Han

26.1 Introduction

The rapid pace of global urbanization intensifies construction and demolition waste (C&DW) generation. Globally, C&DW constitutes approximately 30% of total solid waste streams, mounting environmental challenges, such as land degradation, natural resource depletion, and greenhouse gas emissions (Han et al. 2021). Imprudent demolition waste management (DWM) not only exacerbates the environmental footprint but also imposes tremendous pressure on degrading land, thus undermining sustainable urban development (Carmo et al. 2022; Su et al. 2021; Udawatta et al. 2021).

This chapter directly contributes to the prominent theme of this volume: *Sustainability, Digital and AI Innovations, and Applications on Land Intelligence and Geospatial Ecosystem* – by exploring how digital technologies can transform end-of-life (EoL) demolition planning. Specifically, it positions Building Information Modeling (BIM) as a pivotal digital enabler that supports intelligent land use decisions through sustainability-oriented DWM, in which BIM's attributes, including material quantification, scenario simulation, and 3D visualization, present unique advantages for promoting sustainable management at the building's EoL phase, where selective dismantling and recycling planning are essential for minimizing environmental and land impacts (Han et al. 2024a). By immersing Life Cycle Assessment (LCA) and Multi-Criteria Decision-Aiding (MCDA) attributes and formulas into the BIM-based environment, the chapter illustrates how pairing BIM-based parametric modeling with sustainability analysis can facilitate actionable insights for resource-efficient, low-carbon, and land-friendly demolition practices, which aligns with the book's emphasis on applying digital and Artificial Intelligence (AI) driven solutions to promote smarter, more sustainable land and asset management.

DOI: 10.1201/9781003638308-29

Building upon these insights, this chapter proposes a comprehensive BIM-based sustainability assessment framework for DWM. The framework integrates LCA/Life Cycle Costing (LCC) methodologies, and MCDA techniques to automate the sustainability assessment and ranking of DWM scenarios. A pilot case study demonstrates the framework's applicability, illustrating how intelligent DWM planning can optimize resource efficiency, alleviate the DW impacts on land, and reduce carbon emissions from a life cycle–thinking perspective.

26.2 Background

This section provides the context of the correlations between C&DW generation and the degradation of land resources and introduces BIM technology as a digital innovation for reshaping the workflow of DWM and sustainability assessment by integrating LCA and MCDA functions into the parametric modeling environment.

26.2.1 Construction and Demolition Waste Impacts on Land

The environmental impacts of C&DW are multifaceted. Land degradation stands out as the most conspicuous environmental issue caused by demolition debris or residual construction materials occupying valuable urban and peri-urban land. Additionally, C&DW disposal accounts for soil contamination, leachate generation, and long-term site degradation, often requiring costly remediation (Marrero et al. 2017). Moreover, the embodied carbon arising from both construction material production and landfilled C&DW decomposition contributes to global warming (Xu et al. 2019). These critical issues call for intelligent C&DW management practices to alleviate landfill pressure, promote the circular economy, and reduce the building's life cycle carbon emissions.

26.2.2 BIM Technology for Sustainability Assessment and C&DWM

BIM has demonstrated its capacity for empowering sustainability assessment and enhancement at the design, construction, operation, and EoL phases (Asgari et al. 2023). Specifically, the adoption of BIM is not limited to design and construction coordination but also provides a centralized parametric data repository that documents and updates building assets' environmental and economic profiles throughout the project's life cycle (Han et al. 2021). In the context of C&DWM, BIM enables efficient waste quantification, environmental impact measurement, cost estimation, and visual deconstruction planning (Sanchez et al. 2019). It is proven that BIM applications can be fine-tuned to predict the volume, type, and sequence of DW generated, assess the trade-offs between different waste management strategies, and streamline selective demolition planning via 3D visualization (Han et al. 2024b; Llatas et al. 2020).

26.2.3 Integration of BIM and LCA

LCA as an internationally recognized impact assessment methodology enables the quantification of various environmental indicators such as greenhouse gas emissions, energy consumption, and soil acidification. The integration of LCA with BIM represents a pivotal step toward sustainability-oriented DWM planning (Han et al. 2024a). Three primary approaches to BIM–LCA integration have been identified from prior studies: (1) manual extraction of material quantities from BIM models into external LCA tools; (2) semi-automated BIM-based LCA by linking BIM models with external life cycle inventory databases; and (3) embedding LCA parameters directly within BIM objects to enable BIM-based impact assessments (Santos et al. 2019). Further, automation and user-friendly interfaces are indispensable for BIM-based LCA workflows to be implemented in real-world DWM planning practices. Therefore, developing a user-friendly BIM-based sustainability assessment tool that can streamline the LCA workflows and visualize assessment results is crucial to maximizing the practicability of BIM-based DWM.

The subsequent section illustrates the methodological development of a BIM-based decision-making tool for DWM, integrating LCA, MCDA, and visual decision-support features to advance intelligent C&DW management.

26.3 Methodology

In light of the challenges briefed in the previous section, this section details the design and development of a novel BIM-based sustainability assessment framework for DWM planning. The methodology depicts the integration of LCA and MCDA into the BIM environment to support intelligent and visual demolition planning that aligns with digital transformation goals in the land innovations sector.

26.3.1 BIM-Based DWM Sustainability Assessment Framework Design

To formalize sustainability-oriented DWM, this study develops a BIM-based decision-aiding tool that couples LCA methodologies with MCDA techniques to support the parallel analysis of LCA results for various DWM scenarios and 3D visual planning for the selective dismantling process. The tool is designed to systematically assess and visualize the sustainability performance of different demolition strategies by gauging the embedded sustainability parameters of the BIM elements, thereby supporting the intelligent BIM-based DWM planning process.

The cornerstone is a centralized BIM model with enriched Industry Foundation Classes (IFC) parameters, wherein building components and materials are embedded with additional attributes such as carbon emission factors, energy intensity, recyclability potential, and other impacts associated with DWM processes. Furthermore, the BIM model is equipped with calculation formulas for LCA and MCDA to streamline impact assessment and results interpretation. Indicators like global warming potential, energy consumption, and land use are directly inserted into material properties and linked with external LCA databases through a structured IFC data schema via Dynamo visual programming (Han et al. 2024b).

The workflow starts with material quantity take-off (MQT) directly extracted from the BIM model. These quantities are automatically mapped to calculate environmental impacts, economic costs, and recycling potentials. DWM scenarios – such as baseline landfilling, partial recycling, and adaptive reuse – are formulated by adjusting recycling rates and material recovery strategies within the BIM-based tool. Each scenario's sustainability performance is measured by quantifying and aggregating the cumulative environmental, economic, and social impacts associated with every building component. Ultimately, the integration of BIM, LCA, and MCDA within a unified digital environment ensures seamless data flow, reduces manual errors, and improves the transparency and robustness of DWM planning.

26.3.2 Sustainability Indicators Selection and Weighting

The selection of sustainability indicators for assessing DWM scenarios follows a systematic modified Delphi method, wherein semi-structured interviews and follow-up questionnaire surveys are coupled to elicit opinions from industry experts in relation to the relevance of sustainability indicators for the DWM context (Han et al. 2023).

The results of the indicator prioritization reveal that environmental indicators like global warming potential (32%), energy efficiency (16.1%), and land use (13.5%) are recognized as the prominent ones for DWM sustainability performance (Han et al. 2023). Economic feasibility (i.e. the total cost), albeit important, is assigned lower relative weights (6.54%) compared to environmental factors, reflecting a paradigm shift from financial concerns toward reducing resource exploitation and carbon emission priorities in DWM practices.

26.3.3 Automated Data Interpretation and Visualization

A key innovation of the proposed framework lies in its automated data interpretation and visual decision-support capabilities. By leveraging visual programming environments (i.e. Dynamo for Autodesk Revit), custom Dynamo scripts are developed to sync the automated data extraction, mapping, and calculation of environmental impacts, and recycling values within the BIM environment.

The BIM-based decision-aiding tool computes the sustainability score of each DWM scenario by assessing the weighted sustainability indicators using a hybrid MCDA (i.e. AHP–TOPSIS) model (Han et al. 2024b), which allows for quantitative comparison between DWM scenarios based on a composite sustainability index. To facilitate stakeholder engagement and enhance decision transparency, desired assessment results are visualized directly on the color-coded BIM model. For instance, building components with substantial carbon intensity and recycling potential are highlighted, complying with a gradient color-coding scheme, which enables intuitively visual identification of critical elements for selective dismantling. On the user interface of the Revit Plugin, parallel scenario analysis dashboards display the sustainability scores, carbon emission savings, and material recovery rates associated with each alternative.

26.4 Case Study

This section presents a case study that demonstrates the practical applicability of the proposed framework in real-world demolition projects. The novel BIM-based sustainability assessment approach can be operationalized to generate sustainability insights and inform land-conscious demolition planning.

26.4.1 Pilot Project Description

This study uses a pilot case to validate the practicability of the BIM-based DWM planning tool. The Doug McDonell Building at the University of Melbourne was selected, with updated architectural and structural drawings provided for research use by Kane Group. A BIM model (see Figure 26.1) with Level of Details (LoD) 300 was created to simulate as-is demolition conditions. The case study building (i.e. Doug McDonell) is a mid-rise educational facility with reinforced concrete framing located in The University of Melbourne's Parkville campus, covering 1275 m^2 of land and with a 12,259 m^2 total floor area. To simulate realistic transport distances, the mock demolition site was placed in Baishizhou, Nanshan District, Shenzhen, with logistics data obtained from Baidu Maps.

26.4.2 DWM Scenario Development

DWM schemes were developed within a Revit-based platform (BIM DWM planning plugin) that allows users to choose from different DWM contractors and service providers. The four schemes were differentiated by target

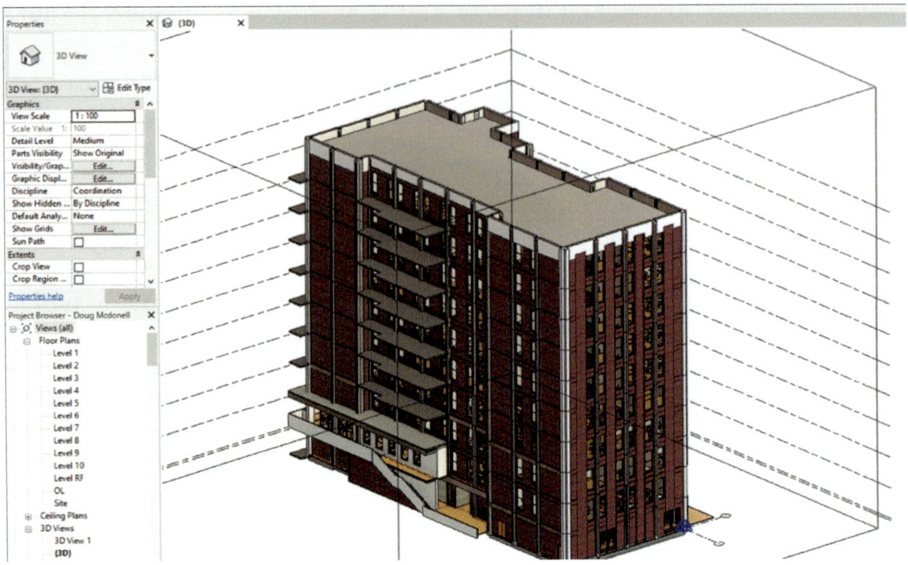

FIGURE 26.1
Rendering 3D view of the Doug McDonnell BIM model.

recycling rates, used for scenario comparison. The key parameters of the case study are summarized below.

Key parameters:

- Demolition contractor: Shenzhen Huatian Construction (80 CNY/m^2)
- Recycling facility: Dinghao Building Materials, 65 km away (9 CNY/t)
- Landfill site: Xiaping Landfill, 22 km away (3 CNY/t)
- Transport: Shenzhen Demolition Hauling Co. (0.4 CNY/t/km)
- Scheme 1: 80% – aligned with Australia's national waste targets
- Scheme 2: 95% – represents best practices from developed countries
- Scheme 3: 50% – benchmark scenario
- Scheme 4: Business-as-usual (10%) – reflects current practice in China

26.4.3 Results

The BIM-based life cycle inventory analysis revealed that the demolition of the Doug McDonell Building would produce approximately 6167 t of demolition waste, where concrete and masonry are the two largest waste streams, contributing to 70% and 27% of total waste generation, respectively. All four DWM schemes produced net negative carbon emissions due to recycling, with Scheme 2 (95% recycling rate) achieving the highest carbon offset of 1,090,988 kg CO_2 eq, primarily from aluminum and concrete, while Scheme

4 (10% recycling) had the least carbon offset contribution owing to the low recycling rate. Scheme 2 consumed the most energy (1,181,669 MJ) due to excessive onsite waste sorting machinery operation.

Cost analysis showed that demolition and transport costs were constant while recycling and landfill fees varied significantly with the recycling rate – Scheme 2 incurred the highest recycling costs and lowest landfill fees. Conversely, the expenditure on the landfill levy is substantially higher for Scheme 4. Landfill cost savings increased with the amount of waste diverted from landfill to recycling facilities, with Scheme 2 saving 69,132 CNY and Scheme 4 only 7277 CNY. After taking into account the priority of DWM (weights of indicators), the overall sustainability rankings placed Scheme 2 as the optimal scenario, offering the best carbon performance despite higher energy use and cost, followed by Scheme 1 with a balanced eco-friendly and low-cost performance. Schemes 3 and 4 ranked significantly lower, with Scheme 4 being the least sustainable. Furthermore, a 3D visual model further helps identify high-value recyclable components – such as metals – using a color gradient scheme to support selective dismantling planning (see Figure 26.2), revealing that most of the concrete and masonry elements offered low recycling returns.

26.5 Discussion

The case study reveals both the strengths and the current limitations of the BIM-based framework. In this section, we critically reflect on the advantages,

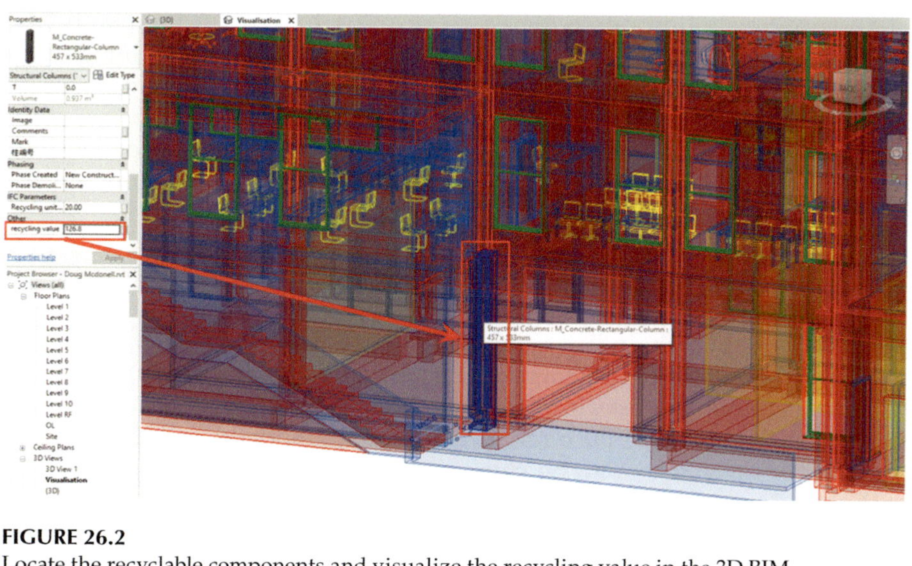

FIGURE 26.2
Locate the recyclable components and visualize the recycling value in the 3D BIM.

practical challenges, and future enhancement of this tool for supporting land intelligence through digital innovation.

26.5.1 Advantages of the BIM-Based Sustainability Framework

The proposed framework enables seamless data integration between BIM and LCA databases by enriching IFC properties with DWM-related and LCA parameters, thus minimizing manual data transfer errors (Han et al. 2024a). Second, the framework facilitates automated sustainability assessment and results interpretation by coupling LCA with hybrid MCDA techniques. Third, the development of visual decision-aiding features within the BIM environment enhances stakeholder engagement and practical usability. Features like color-coded building models, comparative dashboards, and scenario ranking enable dynamic and parallel analysis of DWM scenarios.

Overall, the framework significantly contributes to shifting DWM from reactive waste hauling and disposal toward proactive, data-driven sustainability DWM planning, supporting broader sustainable land development and carbon mitigation agendas.

26.5.2 Barriers and Challenges

The prominent challenge is data availability and quality. In practice, demolition projects often lack comprehensive as-is documentation or rely on generalized material assumptions, compromising the assessment accuracy. A second challenge is data interoperability. Although IFC standards exist, variations in data schema and structure can hinder seamless data exchange. Establishing standardized templates and enriching IFC schemas for demolition-specific applications remains an ongoing process. Additionally, the cost and expertise requirements associated with BIM–LCA integration and visual programming may limit adoption, particularly among small- to medium-sized demolition contractors who may lack digital capabilities. These challenges call for industry collaboration and policy support to enable the wider application of intelligent DWM in real-world practices.

26.5.3 Future Enhancement

Several avenues for enhancing the BIM-based DWM tool are identified based on the research findings and industry trends. First, the automation of sustainability data enrichment remains a key priority. Integrating real-time material condition monitoring technologies (e.g. material tagging, 3D scanning) with BIM could reduce reliance on manual updates and improve the accuracy of inventory analysis. Second, incorporating dynamic LCA databases and location-specific parameters would enhance the contextual relevance of impact assessments. For instance, integrating local transportation emissions factors, regional recycling rates, and market-specific economic information could refine scenario development aligning with local contexts. Third, advancing user-friendly interfaces and decision dashboards is crucial for wider industry adoption. Simplifying the visualization of assessment values and offering modular customization options can make the framework accessible to a broader range of stakeholders with minimum operational knowledge requirements. Finally, scaling the framework from project-level applications to precinct or city-scale planning could support more sustainable urban renewal programs and land conservation initiatives.

26.6 Conclusion

This chapter has proposed and validated an intelligent BIM-based DWM tool by enriching IFC attributes and integrating BIM with LCA methodologies, and hybrid MCDA techniques, to facilitate holistic sustainability assessment, DWM scenario ranking, and intuitive 3D visual decision guidance. Future enhancements should focus on dynamic data synchronization, integration of location-specific project profiles, incorporation of social criteria, and the development of user-friendly interfaces to facilitate broader industry adoption. Overall, this chapter contributes to the evolving discourse on sustainable land management through digital innovation in DWM practices. The proposed framework represents a significant step toward circular economy goals by improving the decision-making at the building's EoL phase and offers a promising foundation for extending sustainability assessments to precinct and city-scale urban renewal initiatives.

References

Asgari, S., S. Haghir, and E. Noorzai. 2023. "Reducing Energy Consumption in Operation and Demolition Phases by Integrating Multi-Objective Optimization with LCA and BIM." *Energy Efficiency* 16 (6): 54.

Carmo, J. L., A. B. Rohden, and M. R. Garcez. 2022. "Recycling Construction and Demolition Waste as Aggregate in Porous Asphalt Pavement for Urban Stormwater Management." *Journal of Materials in Civil Engineering* 34 (10): 04022258.

Han, D., M. Kalantari, and A. Rajabifard. 2021. "Building Information Modeling (BIM) for Construction and Demolition Waste Management in Australia: A Research Agenda." *Sustainability* 13 (23): 12983.

Han, D., M. Kalantari, and A. Rajabifard. 2023. "Identifying and Prioritizing Sustainability Indicators for China's Assessing Demolition Waste Management Using Modified Delphi–Analytic Hierarchy Process Method." *Waste Management & Research* 41 (11): 1649–60.

Han, D., M. Kalantari, and A. Rajabifard. 2024a. "The Development of an Integrated BIM-Based Visual Demolition Waste Management Planning System for Sustainability-Oriented Decision-Making." *Journal of Environmental Management* 351: 119856.

Han, D., M. Kalantari, and A. Rajabifard. 2024b. "Improving the Decision-Making for Sustainable Demolition Waste Management by Combining a BIM-Based Life Cycle Sustainability Assessment Framework and Hybrid MCDA Approach." *Waste Management & Research* 9: 0734242X241291941.

Llatas, C., B. Soust-Verdaguer, and A. Passer. 2020. "Implementing Life Cycle Sustainability Assessment during Design Stages in Building Information Modelling: From Systematic Literature Review to a Methodological Approach." *Building and Environment* 182: 107164.

Marrero, M., M. Puerto, C. Rivero-Camacho, A. Freire-Guerrero, and J. Solís-Guzmán. 2017. "Assessing the Economic Impact and Ecological Footprint of Construction and Demolition Waste during the Urbanization of Rural Land." *Resources, Conservation and Recycling* 117: 160–74.

Sanchez, B., C. Rausch, and C. Haas. 2019. "Deconstruction Programming for Adaptive Reuse of Buildings." *Automation in Construction* 107: 102921.

Santos, R., A. A. Costa, J. D. Silvestre, and L. Pyl. 2019. "Integration of LCA and LCC Analysis within a BIM-Based Environment." *Automation in Construction* 103: 127–49.

Su, S., S. Li, J. Ju, Q. Wang, and Z. Xu. 2021. "A Building Information Modeling-Based Tool for Estimating Building Demolition Waste and Evaluating Its Environmental Impacts." *Waste Management* 134: 159–69.

Udawatta, N., J. Zuo, K. Chiveralls, and G. Zillante. 2021. "From Green Buildings to Living Buildings? Rating Schemes and Waste Management Practices in Australian Educational Buildings." *Engineering, Construction and Architectural Management* 28 (4): 1278–294.

Xu, J., Y. Shi, Y. Xie, and S. Zhao. 2019. "A BIM-Based Construction and Demolition Waste Information Management System for Greenhouse Gas Quantification and Reduction." *Journal of Cleaner Production* 229: 308–24.

27

Smart City Governance Empowering Productivity of Digitalization: Practice in China

Zhixuan Yang

27.1 Introduction

The global governance environment and business landscape have become deeply anchored in digital governance. As cities serve as core drivers of global economic growth, digitalization has emerged as a critical force in reshaping economic growth models. From the perspective of production efficiency enhancement, digital transformation has significantly optimized enterprise workflows. McKinsey research indicates that digital transformation can boost corporate production efficiency by 20–30% while reducing operational costs by 15–25% (McKinsey & Company, 2019). Concurrently, urban governance models have undergone comprehensive transformation through digital technological innovation, which in turn influences the trajectory of digital economic development. Empirical evidence demonstrates that smart city governance offers advantages in fostering high-quality economic growth, enabling effective policy implementation, and improving business environments. These capabilities effectively drive the development of the digital economy, thereby enhancing productivity.

Currently, Chinese cities have integrated smart city governance frameworks, establishing a novel governance paradigm that employs information and communication technologies (ICTs) as primary tools. Unlike traditional governance approaches, smart city governance holistically empowers the digital economy, driving improvements in digital production efficiency. Since initiating smart city pilot programs in 2003, over 200 cities in China have progressively implemented smart city initiatives, making substantial investments in information infrastructure. This has gradually formed an investment-driven model of local digital economic growth—a process spanning nearly two decades. During this period, massive data networks have provided decision-making foundations for smart city governance, enabling more precise data governance processes, enhanced information transparency, and productivity

DOI: 10.1201/9781003638308-30

gains. China's experience stands as an exemplary observational model for global smart city construction and digitally driven development paradigms.

27.2 Effect of Digitalization on Global Economy

Digitalization has played a pivotal role in reshaping global economic growth paradigms, emerging as a critical dimension in assessing competitiveness. Concurrently, the application of digital technologies enhances production efficiency and drives economic growth (MGI, 2016). As industrial digitalization deepens, foundational economic models, competitive market dynamics, and global supply chain systems are undergoing structural transformations. Digitalization significantly boosts productivity by reducing transaction costs, optimizing resource allocation, and fostering technological innovation (Zhu & Yu, 2023). The proliferation of ICTs enables enterprises to achieve economies of scale and scope while enhancing operational efficiency through data analytics and automation (Vasilić et al., 2020). Technological innovation has become a core driver of economic growth, spurring the emergence of novel products, services, and business models (OECD, 2004; Goldfarb & Tucker, 2017). Additionally, digitalization has dissolved traditional industry boundaries, reshaping global supply chains and competitive landscapes. Sectors such as manufacturing, finance, and retail are undergoing modernization through intelligent systems and e-commerce, while traditional industries face mounting pressure to adapt. Furthermore, digitalization lowers market entry barriers, empowering small and medium-sized enterprises (SMEs) and marginalized groups to participate in the global economy. E-commerce platforms and remote work opportunities have unlocked new prospects for citizens, advancing socioeconomic inclusion.

Digitalization reshapes economic growth through efficiency gains, structural shifts, and globalization. However, its efficacy depends on technological adoption capabilities, policy support, and institutional environments. These challenges are supposed to be addressed through the strategic implementation of smart city governance models, which integrate digital infrastructure, data-driven decision-making, and inclusive policy frameworks.

27.3 Smart City Governance

Smart city governance is an ICT-oriented, human-centric governance model applied in precision urban governance scenarios, aiming to efficiently address urban challenges and provide platform-based solutions for public welfare, economic development, urban sustainability, and government administration (Meijer et al., 2016). Rooted in urban governance theories encompassing governance actors, processes, content, and performance, smart governance integrates digital tools to achieve precision management and targeted services through data analytics and platform technologies. This enhances governmental efficacy and offers holistic solutions for urban operations.

Technologically, smart city governance is driven by digital innovations, where information infrastructure and big data processing platforms form its technical foundation. Leveraging these technologies, governments enhance their capacity to manage massive datasets, enabling flexible and citizen-centric governance measures tailored to economic, social, and environmental needs. To achieve this, governance actors must prioritize stakeholder engagement, ICT service delivery, and networked collaborations (Gil-Garcia et al., 2015). Effective outcomes require cross-departmental coordination, data-sharing policies, citizen participation channels, and careful consideration of fiscal autonomy and cultural diversity (Ruhlandt, 2018). Scholarly research highlights three key dimensions of smart governance: technology-driven, citizen-inclusive, and collaboration-oriented. Governance frameworks must balance legitimacy among diverse stakeholders, emphasizing equity and efficiency in decision-making, while ensuring transparency and openness in processes (Bolívar & Meijer, 2016). Notably, governmental roles have shifted from controllers to platform builders.

27.4 Mechanism of Smart City Governance Empowering Productivity of Digitalization

The deep integration of digitalization and intelligence within smart city governance establishes a multidimensional framework for advancing digital productivity. Centered on digital technologies as core drivers, smart city construction not only creates diverse application scenarios but also activates technological innovation and market potential through demand-driven governance models. Functionally, smart city governance encompasses four interrelated domains—economic development, social administration, public services, and livable environments—collectively forming a systemic governance architecture that propels the digital economy (Bolívar & Meijer, 2016; Allam et al., 2022).

In the aspect of economic development, the bidirectional interplay between industrial digitalization and digital industrialization defines the operational logic of smart economies. Concurrently, digital industrialization fosters emerging sectors through innovative business models. Crucially, the massive production, transaction, and user behavioral data generated by smart economies are emerging as novel production factors that sustain industrial upgrading and economic dynamism (Günther et al., 2017). As for the aspect of social administration, digital tools reconstruct the foundational logic of urban governance. Digital platforms integrate cross-departmental data resources, enabling online re-engineering and intelligent optimization of administrative processes and enhancing bureaucratic efficiency, and facilitate market entities' participation in the digital economy. In addition, the digital transformation of public services reflects a value-oriented shift in governance priorities. By digitizing education, healthcare, elderly care, and transportation, smart cities construct an integrated online–offline service ecosystem that transcends spatial–temporal constraints. The technological empowerment not only elevates governance efficiency but also stimulates market demand for green innovations, establishing pathways for converting ecological value into economic value.

27.5 Smart City Governance Practice in China

China's government initiated smart city pilots in 2003. After two decades of smart city construction, urban governance has undergone a digital transformation. Smart city governance is a key area for future urban development in China. The "14th Five-Year Plan for Economic and Social Development of the People's Republic of China and Long-Range Objectives through the Year 2035" emphasizes leveraging digital transformation to drive changes in overall production methods, lifestyles, and governance modes. It promotes deep integration between digital and real economies, strengthens urban infrastructure construction, and aims to create livable, resilient, smart cities. In terms of construction achievements, smart cities and the digital economy have deeply integrated through the effective application of digital governance tools. Nowadays, the digital economy accounts for 41.5% of GDP and serves as the core engine of economic growth (CAICT, 2023).

China's experience demonstrates that new-generation information technologies, represented by the Internet, IoT, big data, and artificial intelligence, are deeply empowering urban transformation and high-quality development. A series of specialized application scenarios and operational models effectively enhance urban governance capabilities and public service levels. This has injected momentum into the development of new technologies, industries, and business models, accelerating the digital economy onto the expressway of development (CSIC, 2023).

27.5.1 Innovations in Governance Models Facilitating Digital Productivity

Chinese cities have developed diverse smart governance models tailored to their unique characteristics. For example, Shanghai pioneered the "One-Net

Integrated Management" system, establishing a "three-tier platform, five-tier application" architecture that enhances administrative efficiency through cross-departmental data integration (Han et al., 2023). Beijing has strengthened grid-based management, creating a novel model of "one network overseeing overall governance and one network coordinating industries". Chengdu achieved breakthroughs in urban safety risk monitoring and early warning, with its innovative practices as exemplary references for mega-city governance (Chen et al., 2023). These practices directly improve urban operational efficiency through refined and intelligent management methods, serving as critical drivers of digital productivity.

27.5.2 Upgrading of Governance Capabilities Empowering Digital Transformation

Smart city construction has significantly optimized foundational governance capacities. For instance, intelligent urban governance interfaces enable real-time monitoring of infrastructure status, facilitating fault prediction and rapid response to mitigate economic losses caused by delayed management (Xu et al., 2022). Shanghai has developed a new smart city paradigm through cross-departmental data governance and analysis, providing technological support for high-quality urban development. Data-driven decision-making models further enhance the scientific allocation of resources, empowering the digital transformation of industries and economic activities (Cao et al., 2019).

27.5.3 Collaborative Mechanisms Unleashing Digital Productivity

Smart city governance emphasizes dismantling institutional barriers to establish collaborative systems. Research indicates that among 35 Chinese cities, technology plus organizational collaborative models effectively improved governance performance (Song et al., 2024). The integration of optimized technological governance systems with organizational mobilization capabilities addressed issues such as insufficient resource allocation flexibility and delayed incident responses in traditional governance. Cities rapidly enhanced cross-domain collaboration by integrating and reusing data resources, with institutional innovations providing systemic safeguards for unleashing digital productivity.

27.5.4 Deep Integration of Digital Economy and Governance Practices

Smart city construction itself has become a key engine for digital economic development. The application of new-generation technologies in governance scenarios has catalyzed new business formats and models. For instance, Hangzhou's urban vital sign prediction and early warning system not only ensures operational safety but also spawns industrial chains in smart security and transportation, demonstrating bidirectional empowerment between governance practices and industrial innovation (Zhao & Zou, 2021).

27.6 Conclusion

The integration of digital governance into urban development has emerged as a transformation force in reshaping global economic growth paradigms and urban governance models. As cities increasingly serve as hubs of economic activity, smart city governance has become a critical mechanism for enhancing productivity, fostering innovation, and driving sustainable development. Through a synthesis of technological innovation, data-driven decision-making, and inclusive policy frameworks, smart cities in China demonstrate how digitalization can optimize resource allocation, streamline administrative efficiency, and catalyze the growth of the digital economy. Chinese cities have evolved governance models from traditional, fragmented systems to technology-enabled, collaborative frameworks. Key achievements include the establishment of digital infrastructure, the development of diversified governance paradigms, and the deep integration of digital tools into economic sectors, elevating China's digital economy and boosting the productivity of digitalization.

Acknowledgment

Humanities and Social Sciences Project of the Chinese Ministry of Education "Chunhui Plan": "Smart City Governance and Sustainable Development Driven by Digital Twins" (Project 202200228, Grant Number: HZKY20220414). Liaoning Province Social Science Planning Fund Project "Smart City Governance: Urban Sustainable Development Facilitated by Digital Twins" (Grant Number: L22BGL022).

References

Albert Meijer, Rodriguez Bolivar, and Manuel Pedro. 2016. "Governing the Smart City: A Review of the Literature on Smart Urban Governance." International Review of Administrative Sciences 82 (2): 392–408.

Avi Goldfarb, and Catherine Tucker. 2017. "Digital Economics." National Bureau of Economic Research, Working Paper 23684, https://doi.org/10.3386/w23684.

China Academy of Information and Communications Technology (CAICT). 2023. "Report on the Development of China's Digital Economy." Beijing: China Academy of Information and Communications Technology.

China State Information Center (CSIC). 2023. "Big Data Industry Development Report 2023." Accessed June 15th, 2024. http://bigdata.sic.gov.cn/sic/595/603/0807/9549_pc.html.

J. Ramon Gil-Garcia, Theresa A. Pardo, and Taewoo Nam. 2015. "What Makes a City Smart? Identifying Core Components and Proposing an Integrative and Comprehensive Conceptualization." Information Polity 20 (1): 61–87.

Manuel Pedro Rodríguez Bolívar, and Albert J. Meijer. 2016. "Smart Governance: Using a Literature Review and Empirical Analysis to Build a Research Model." Social Science Computer Review 34 (6): 673–692.

McKinsey & Company. 2019. "Twenty-Five Years of Digitization: Ten Insights into How to Play It Right." Accessed May 15th, 2024. https://www.mckinsey.com/.

McKinsey Global Institute (MGI). 2016. "Digital Globalization: The New Era of Global Flows." Accessed April 20th, 2025. https://www.mckinsey.com/.

Nana Song, Longshun Xu, Xiansheng Chen, and Huange Xu. 2024. "Research on the Improvement Path of Grassroots Social Governance Innovation Performance in China: Qualitative Comparative Analysis Based on 35 Cases." PLoS ONE 19 (2): e0297587.

Nikola Vasilić, Dušica Semenčenko, and Sanja Popović-Pantić. 2020. "Evaluating ICT Usage in Enterprises in Europe: TOPSIS Approach." Economic Themes 58 (4): 529–544.

OECD. 2004. "The Economic Impact of ICT: Measurement, Evidence and Implications." Paris: OECD Publishing.

Robert Wilhelm Siegfried Ruhlandt. 2018. "The Governance of Smart Cities: A Systematic Literature Review." Cities 81: 1–23.

Wanxia Zhao, and Yonghua Zou. 2021. "Smart Urban Governance in Epidemic Control: Implications of Hangzhou." The Chinese Public Administration Review 12 (1): 51–60.

Wendy Arianne Günther, Mohammad H. Rezazade Mehrizi, Marleen Huysman, and Frans Feldberg. 2017. "Debating Big Data: A Literature Review on Realizing Value from Data." Journal of Strategic Information Systems 26 (3): 191–209.

Xiaohong Chen, Xiangbo Tang, and Xuanhua Xu. 2023. "Digital Technology-Driven Smart Society Governance Mechanism and Practice Exploration." Frontiers of Engineering Management 10 (2): 319–338.

Yan Han, Jianming Cai, Enpu Ma, and Jing Lin. 2023. "Understanding Smart City Practice in Urban China: A Governance Perspective." Sustainability 15 (9): 7034.

Yang Cao, Feng Zhen, and Guangliang Xi. 2019. "The Intelligent Urban Governance Based on Big Data: International Experience and China Strategy." Urban Planning International 34 (3): 71–77.

Yanmin Xu, Wengang Li, Jianjiang Tai, and Chunjiong Zhang. 2022. "A Bibliometric-Based Analytical Framework for the Study of Smart City Lifeforms in China." International Journal of Environmental Research and Public Health 19 (22): 14762.

Yongyi Zhu, and Di Yu. 2023. "Can Government Digitalization Promote Firm Productivity? Evidence from Chinese Listed Firms." Applied Economics 56 (33): 3939–3952.

Zaheer Allam, Ayyoob Sharifi, Simon Elias Bibri, and Didier Chabaud. 2022. "Emerging Trends and Knowledge Structures of Smart Urban Governance." Sustainability 14 (9): 5275.

28

Bridging the Gap: From Planning Permits to Building Permits in Automated Systems

Nikoo Mirhosseini, Davood Shojaei, and Soheil Sabri

28.1 Introduction

The building permit process has traditionally relied on manual, paper-based systems, leading to long processing times, inconsistencies, and bureaucratic inefficiencies across jurisdictions (WithPulley 2023; CityGrows 2023). These systems struggle to handle the increasing complexity of urban projects and multi-stakeholder design submissions. The lack of real-time coordination, opaque workflows, and outdated documentation practices results in delayed approvals and reduced public trust in the permitting system (Nemetschek 2023). As cities grow and construction demands evolve, the limitations of legacy permitting processes have become more pronounced. Governments and regulatory bodies recognize that permitting inefficiencies not only slows down development but also hinders transparency, sustainability, and citizen engagement (Datos.gob.es 2023; CitizenServe 2023). A shift toward modernization is now seen as essential for delivering smarter urban infrastructure.

The automation of building permit compliance, as proposed in this chapter, directly contributes to the broader discourse on land governance, digital transformation, and spatial intelligence. By embedding regulatory reasoning into the early stages of the planning-to-construction continuum, the framework supports informed decision-making that accounts for spatial, environmental, and policy constraints. Such integration is instrumental in improving traceability, reducing procedural ambiguity, and promoting more equitable and efficient land development practices. Moreover, the use of Building Information Modeling (BIM) as a spatial data environment—combined with semantic technologies and artificial intelligence (AI)-driven interpretation of regulatory texts—advances the development of intelligent geospatial infrastructures. This approach not only facilitates proactive compliance and reduces inefficiencies but also aligns sustainability objectives by minimizing material waste and streamlining approval processes. The framework thus exemplifies the convergence of AI innovation and geospatial thinking in the context of urban systems.

 DOI: 10.1201/9781003638308-31

Digital transformation initiatives aim to address these inefficiencies by introducing BIM, cloud-based submission platforms, and AI tools for real-time compliance checks (Archistar 2024; MuniLogic 2024). Municipalities like Austin and Kelowna have adopted integrated platforms that use automation to speed up document review and streamline communication between applicants and reviewers (Plantemoran 2023; Accord Project 2024a). These technologies offer significant advantages: reduced delays, improved accuracy, and better accessibility for stakeholders (Cloudflight 2024). As shown in Figure 28.1, the transition from traditional to intelligent building permit

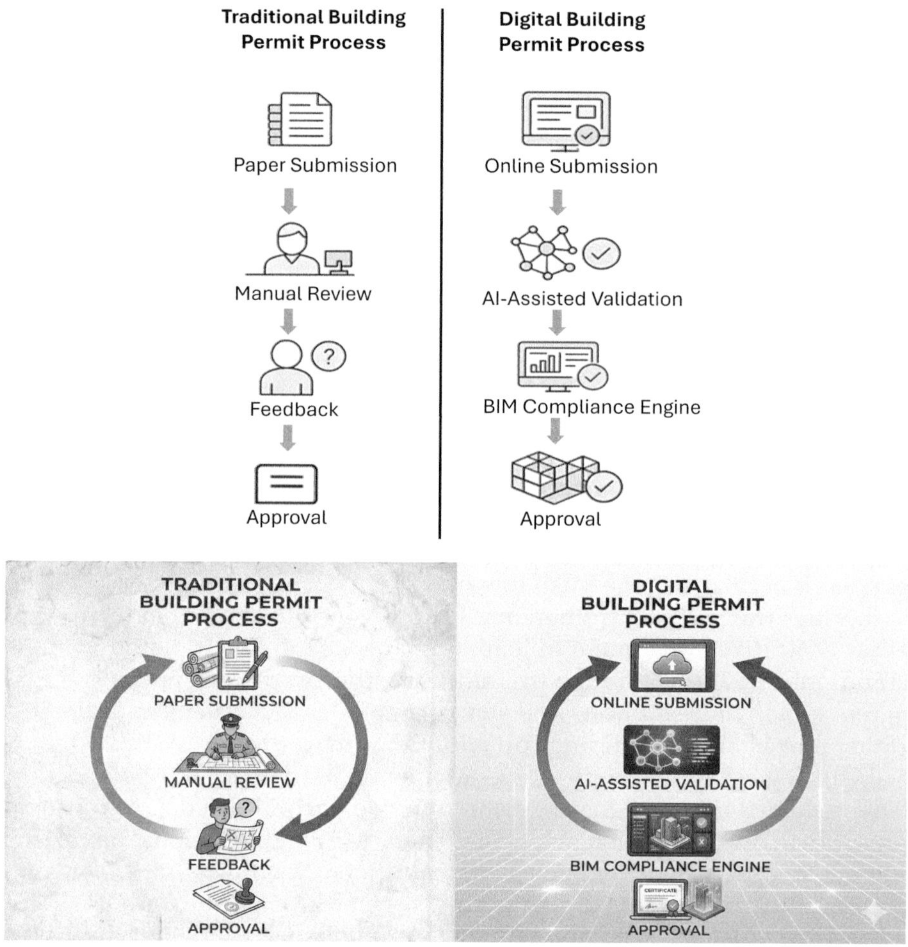

FIGURE 28.1
Comparison of traditional and AI-assisted building permit processes. The manual, paper-based approach is replaced by automated validation and BIM-integrated rule-checking, enabling faster and more accurate compliance verification.

processes introduces automation and semantic validation that eliminate manual bottlenecks and accelerate approval cycles.

This chapter introduces a hybrid, AI-driven framework that bridges the gap between planning and building permits. By combining semantic interpretation of regulations, rule-based compliance engines, and BIM data integration, the proposed model aims to overcome traditional limitations and lay the foundation for intelligent, responsive permitting environments in the digital age.

The history of building permits dates back over 4,000 years, beginning with the Code of Hammurabi in ancient Babylon, which outlined rules to ensure the structural safety of buildings (Lexology 2023). Ancient urban civilizations like Mohenjodaro and Harappa in the Indus Valley exemplified early regulatory systems embedded in architectural planning. As urban societies advanced through the medieval period, formal governance structures increasingly enforced building rules to manage growth, safety, and aesthetics (Investopedia 2023).

By the mid-20th century, the complexity of construction practices led to the emergence of specialized legal structures. Notably, the establishment of the American Bar Association's Forum on Construction Law in the 1970s institutionalized the regulatory oversight of construction projects and shaped the legal foundation of today's permitting systems (Lexology 2023). These frameworks emphasized not just safety compliance but also zoning, liability, and equitable access to development approvals. Building permit procedures vary widely across countries and even between municipalities, driven by institutional, legal, and cultural contexts (Clariti Software 2023). Some regions have streamlined workflows, while others rely on fragmented or outdated systems. This variability causes inconsistencies in documentation, submission requirements, and approval timelines. Applicants often experience delays due to unclear expectations and differing interpretations of building codes (Buena Park 2023; CityGrows 2023). The COVID-19 pandemic accelerated the push toward digital permit systems. Many local authorities transitioned from manual submissions to online platforms that allow centralized document upload, tracking, and automated notifications (Tandfonline 2024). Although this shift has improved efficiency and transparency, many jurisdictions still struggle with legacy systems and organizational resistance, hindering full adoption (Investopedia 2023).

Technological innovations like Computer-Aided Design (CAD) and BIM have reshaped how designs are communicated and validated. CAD ensures precision in documentation, while BIM enables three-dimensional modeling, conflict detection, and compliance visualization (CityGrows 2023). The integration of these tools into digital permit systems has laid the foundation for AI-assisted rule-checking and automated compliance validation discussed in later sections of this chapter.

In response to the increasing complexity and inefficiencies of the current building permit system, this chapter introduces a conceptual framework that integrates AI, semantic technologies, and rule-based reasoning to automate

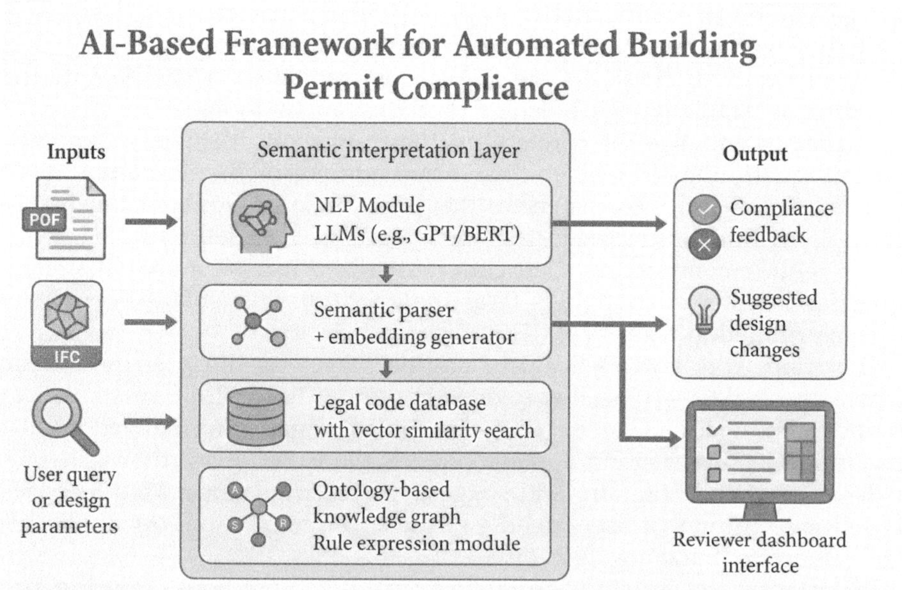

FIGURE 28.2
Conceptual architecture of the proposed AI-semantic framework for automated building permit compliance. The system integrates BIM data, NLP-based regulatory parsing, ontology-driven reasoning, and rule-checking engines to provide scalable, real-time compliance verification.

the transition from planning to building permits (Figure 28.2). The proposed system is designed to address key regulatory challenges—such as interpretive ambiguity, fragmented approval workflows, and delayed compliance verification—by leveraging Natural Language Processing (NLP), ontological modeling, and BIM integration.

At the core of the framework lies an **ontological infrastructure** that captures the semantic structure of building regulations. This enables formal representation of key concepts—such as fire resistance, exit width, or maximum floor area—and their interdependencies. These ontologies are constructed using rule-expression languages such as Shapes Constraint Language (SHACL) or Semantic Web Rule Language (SWRL) to facilitate logic-based reasoning and automated rule-checking (MDPI 2021). While ontology creation is resource-intensive, it provides the necessary scalability for automated compliance across jurisdictions (ResearchGate 2020).

Complementing this foundation is an **NLP module** powered by large language models (LLMs) such as Generative Pre-trained Transformer (GPT) and Bidirectional Encoder Representations from Transformers (BERT). These models are fine-tuned on regulatory corpora to parse legal clauses, extract conditional logic, and detect context-sensitive constraints. Through the generation of vector embeddings, regulatory statements are semantically matched to project specifications derived from planning or BIM data (Warrington 2023; ResearchGate 2020). This facilitates an adaptive understanding of legal language, essential for robust automation.

To further enhance interpretability, the framework includes a **semantic disambiguation layer**. This module is trained to resolve lexical ambiguities and terminological inconsistencies that often arise in regulatory texts. For example, distinctions between "exit" and "egress," or variations in the phrasing of requirements across states, are normalized using contextual embeddings and synonym mappings (Warrington 2023). This ensures consistent rule interpretation regardless of linguistic variability.

The framework's **rule-checking engine** serves as the execution layer, where design data—typically provided in standardized BIM formats such as Industry Foundation Classes (IFC)—is validated against formalized regulatory rules. Upon detecting non-compliance, the system flags the issue, provides a regulatory citation, and suggests corrective actions. This feedback is delivered through a user-friendly dashboard interface for both applicants and reviewers (BlitzPermits 2023).

Importantly, the system is designed to integrate with both **planning systems and downstream BIM platforms**. It allows early-stage validation of planning concepts and seamless progression to building-level compliance checks. By drawing from standardized open data formats and linking to spatial and performance attributes in BIM models, the framework ensures real-time alignment between regulatory requirements and project specifications (Elantis 2023; Accord Project 2024b). For example, consider a case in which a residential building model exceeds 25 meters in height. The system retrieves the corresponding clause—e.g., Clause 3.7.1.3(a), requiring dual egress paths for high-rise buildings—and assesses the BIM model for the presence and spacing of exits. If only one exit is detected, the system flags it as non-compliant, recommends the inclusion of a second stairwell, and documents the reasoning for permit reviewers.

The framework delivers tangible benefits: it reduces approval time by automating rule application, improves compliance by standardizing interpretations, and minimizes rework by detecting violations early in the design cycle. Moreover, it frees human reviewers to focus on complex cases that require qualitative judgment, enabling a more balanced, efficient, and trustworthy regulatory process (CertPro 2023; Frontiers in Built Environment 2025).

28.2 Implementation Scenario

To demonstrate how the proposed AI-semantic framework functions in practice, this section presents a simplified yet representative implementation scenario

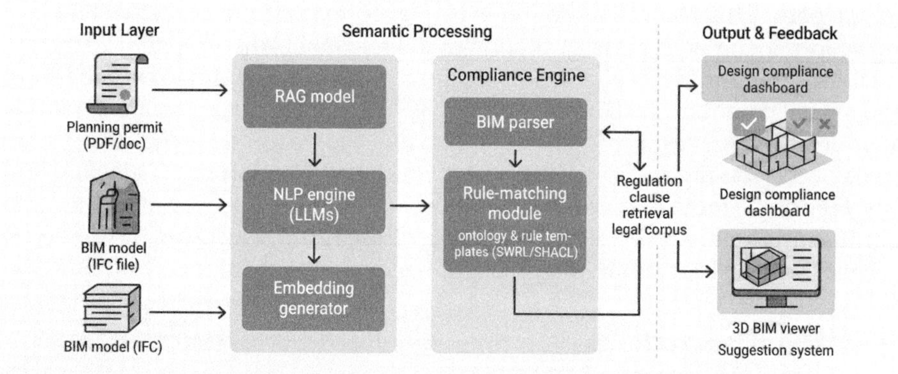

FIGURE 28.3
Workflow of the proposed AI-semantic compliance checking system, showing the data pipeline from planning permit submission to BIM analysis, regulation retrieval, and automated rule-checking with feedback to users.

in the context of a residential building project. The example walks through the transition from **planning permit approval** to **automated building code compliance checking**, highlighting the role of LLMs, Retrieval-Augmented Generation (RAG) pipelines, and rule engines across each step (Figure 28.3).

The process begins with a **planning permit** application that is submitted digitally via a municipal e-planning platform. The AI system verifies that the proposal aligns with zoning regulations, environmental impact assessments, and land-use designations (Archistar 2023b). Once approved, the design advances to the building permit phase, where the system automatically identifies applicable building regulations by extracting semantically relevant clauses using a **RAG** model (Rapid Innovation 2025).

At this stage, the design—typically submitted as an **IFC model**—is parsed to extract geometric, spatial, and functional data (e.g., door types, staircases, room heights). A fine-tuned LLM processes regulatory clauses and generates vector representations that are matched against these design parameters. Semantic disambiguation modules handle lexical variability (e.g., "exit" vs. "egress") using contextual cues, ensuring that code interpretation remains consistent and explainable (Brookings Institution 2024a; CityGrows 2023).

Consider a simplified fire safety regulation:

> Class 1 buildings with a floor height exceeding 12 meters must include a minimum of two independent egress paths.

AI extracts this clause and checks the submitted BIM/IFC model for height metadata and elements labeled IfcStair and IfcDoor. If only one compliant exit is identified, the system flags the issue, highlights the floor in question, and recommends an alternative design such as adding a second stairwell or external escape path. This recommendation is displayed to both the applicant and permit officer, along with a direct citation of the relevant clause (OpalAI 2025).

The system's interface presents flagged issues, regulatory explanations, and visual annotations directly on the 3D model. Applicants can iterate on their designs and resubmit without restarting the process. Permit officers can focus on reviewing complex cases and validating AI outputs. This **continuous feedback loop** fosters collaboration, accelerates decision-making, and significantly reduces revision cycles (CityGrows 2023; Planning.org 2024).

This scenario highlights the practical benefits of the proposed framework:

- **Time Savings**: Processing time is reduced by up to 30% through automation (OpalAI 2025).
- **Accuracy and Transparency**: AI-backed interpretations ensure traceability and justification of each decision (Brookings Institution 2024b).
- **Scalability**: Once regulatory clauses are semantically mapped, the system can be adapted across jurisdictions with minimal retraining.
- **User Experience**: User-friendly dashboards and applicant portals improve transparency and communication (CityGrows 2023).

28.3 Challenges and Limitations

While AI-based rule-checking frameworks offer transformative potential for automating building permit compliance, their deployment presents a complex set of challenges—ranging from technical and semantic barriers to legal, ethical, and operational concerns.

The effectiveness of AI systems hinges on the availability, quality, and interoperability of both regulatory and design data. However, building codes are often fragmented, unstructured, and inconsistent across jurisdictions, making them difficult to translate into machine-readable formats (ASCE 2023). Moreover, design and construction data often reside in disconnected silos across platforms, complicating real-time integration. Middleware tools such as APIs are essential for connecting these data streams but add further complexity and require standardization efforts (Certa 2023).

Interpreting regulatory text remains one of the most persistent challenges. Even the most advanced NLP models struggle with vague or subjective legal language—phrases like "adequate ventilation" or "reasonable egress distance" often require contextual or jurisdictional interpretation that cannot be easily codified. This semantic ambiguity leads to varying outcomes and undermines the transferability of AI systems across legal systems or regions (Archistar 2023a).

AI systems trained on historical or institutional data may unintentionally reinforce systemic biases. For example, approval patterns skewed against certain urban neighborhoods or housing types can be learned and perpetuated by automated systems (HolisticAI 2024). Furthermore, proprietary algorithms often function as "black boxes," offering little transparency into how compliance decisions are reached—raising concerns over procedural fairness and due process.

Current legal frameworks such as the Fair Housing Act or Title VII are not fully equipped to address the complexities introduced by AI in permitting. Legal precedents on disparate impact, where neutral algorithms produce discriminatory effects, are evolving but remain under-enforced (Carnegie Council 2024). Without robust oversight, jurisdictions risk deploying systems that may inadvertently marginalize communities or violate constitutional rights (BearingPoint 2023).

Resistance to technological change remains strong in construction and public sector contexts. Field professionals may distrust AI recommendations or fear automation will reduce their roles (WTW 2024). Additionally, municipalities and planning agencies often lack the resources and training infrastructure needed to implement, monitor, and retrain AI systems over time. Ongoing audits and explainability tools are necessary to ensure model integrity and stakeholder trust (Rapid Innovation 2025).

28.4 Conclusion

This chapter introduced a novel AI-augmented framework designed to close the gap between planning and building permits. By integrating semantic rule interpretation, LLM-assisted NLP, and BIM-based compliance verification, the framework directly addresses the longstanding inefficiencies of manual, fragmented permitting processes. It draws on advanced techniques such as RAG, regulatory ontologies, and IFC-integrated design analysis to automate the evaluation of construction documents in real time (Rapid Innovation 2025).

Through a simulated implementation scenario, this chapter illustrated how the proposed system enhances both procedural efficiency and legal clarity. It reduces approval delays, mitigates ambiguity in regulation interpretation, and improves collaboration between stakeholders. These outcomes align with documented pilot studies in Austin, Kelowna, and Sydney, which observed over 30–40% improvements in turnaround time, stakeholder engagement, and regulatory transparency (Scale AI 2025; BuildBite 2024).

To realize the full potential of AI-based compliance systems, several pathways for further research and system enhancement must be prioritized:

- **Cross-Jurisdictional Adaptability**: Developing modular ontologies and rule engines that can adapt to diverse legal contexts and local codes.
- **Explainability and Human Oversight**: Ensuring transparency in AI decisions by integrating explainable AI (XAI) tools into permit dashboards (Brookings Institution 2024b).
- **Smart City and Digital Twin Integration**: Enabling continuous compliance checks linked to real-time city data and spatial models (UF AI & Cities 2024).
- **Workforce Evolution**: Equipping urban planners and permit officers with skills to supervise, adjust, and co-work with AI systems.
- **Ethical and Legal Governance**: Establishing robust policy frameworks that safeguard against algorithmic bias and ensure accountable decision-making (Carnegie Council 2024).

The shift toward AI-driven, intelligent permitting is not merely a digital upgrade—it marks a profound rethinking of how regulatory systems interact with urban innovation. As cities embrace digital transformation and smart infrastructure, this framework offers a scalable, responsive, and ethically grounded path forward. With proper governance, human–machine collaboration in compliance checking can catalyze a future where development is not only faster but also fairer, more inclusive, and legally transparent.

References

Accord Project. 2024a. Transformation Pathways towards Digital Permitting Processes in Demo Countries. https://accordproject.eu/transformation-pathways

Accord Project. 2024b. Use Cases for AI-Enhanced Rule Checking. https://accordproject.eu/use-cases-2 [accessed May 2025].

Archistar. 2023a. Avoiding Common Pitfalls in Building Permit Compliance. https://www.archistar.ai/resources/errors-to-avoid [accessed May 2025].

Archistar. 2023b. Demystifying the Planning Permit. https://www.archistar.ai/resources/planning [accessed May 2025].

Archistar. 2024. Top 7 Digital Building Permit Tools. Company blog. https://www.archistar.ai/blog/top-7-digital-building-permit-tools [accessed May 2025].

ASCE. 2023. What Does Human-Machine Intelligence Mean for Civil Engineers? https://www.asce.org/articles/human-machine-intelligence [accessed May 2025].

BearingPoint. 2023. The AI Act Requires Human Oversight. https://www.bearingpoint.com/en/insights/ai-act-oversight [accessed May 2025].

BlitzPermits. 2023. AI-Powered Technologies for Faster Building Permit Approvals. https://blitzpermits.ai/features [accessed May 2025].

Brookings Institution. 2024a. Algorithmic Bias Detection and Mitigation. https://www.brookings.edu/research/ai-bias-detection [accessed May 2025].

Brookings Institution. 2024b. The Three Challenges of AI Regulation. https://www.brookings.edu/articles/challenges-in-ai-regulation [accessed May 2025].

BuildBite. 2024. 10 Benefits and Challenges of AI in Construction. https://buildbite.com/insights/ai-construction-benefits [accessed May 2025].

Buena Park. 2023. Building Permit Requirements. https://www.buenapark.com/government/departments/community-development/building-permit [accessed May 2025].

Carnegie Council. 2024. AI and Equality: Algorithmic Bias and Ethical Implications. https://www.carnegiecouncil.org/publications/ai-equality [accessed May 2025].

CertPro. 2023. Continuous vs. Traditional Auditing. https://certpro.com/ai-compliance-building-code [accessed May 2025].

Certa. 2023. Challenges and Solutions in AI-Driven Compliance. https://www.certa.ai/insights/ai-permitting-barriers [accessed May 2025].

CitizenServe. 2023. Benefits of Digital Transformation in Local Government. https://www.citizenserve.com/articles/benefits-of-digital-transformation [accessed May 2025].

CityGrows. 2023. Best Practices in Online Building Permits for Local Governments. https://citygrows.com/background/building-permits [accessed May 2025].

Clariti Software. 2023. The Importance of Building Permitting to a Thriving Economy. https://www.claritisoftware.com/solutions/building-permits [accessed May 2025].

Cloudflight. 2024. Streamlining the Building Permit Process with AI Technology. Company whitepaper. https://www.cloudflight.io/resources/ai-permitting [accessed May 2025].

Construction Physics. 2023. The Ups and Downs of Automated Code Checking. https://www.construction-physics.com/posts/automated-code-checking [accessed May 2025].

Datos.gob.es. 2023. The Digital Transformation of the Construction Sector. Spanish Open Government Portal. https://datos.gob.es/en/blog/digital-construction [accessed May 2025].

Elantis. 2023. Streamlining Building Permit Applications. https://www.elantis.com/articles/building-permit-automation [accessed May 2025].

Fauth, Judith, Peter Nørkjær Gade, Stefanie Kaiser, Kavita Raj, Jonas Goul Pedersen, Per-Ola Olsson, Nicholas Nisbet, et al. 2025. "Investigating Building Permit Processes across Europe: Characteristics and Patterns." *Building Research & Information* 53 (4): 417–34. doi:10.1080/09613218.2024.2400467

Frontiers in Built Environment. 2025. AI-Driven Compliance in the Construction Sector. https://www.frontiersin.org/articles/10.3389/fbuil.2025.00001/full [accessed May 2025].

HolisticAI. 2024. Understanding AI Bias, Risks, and Mitigation Strategies. https://www.holisticai.com/reports/ai-bias-construction [accessed May 2025].

Investopedia. 2023. Building Permits: Types, Effect on Economy, Example. https://www.investopedia.com/terms/b/building-permit.asp [accessed May 2025].

Lexology. 2023. Construction Law – The History is Ancient! https://www.lexology.com/library/detail.aspx?g=12345 [accessed May 2025].

MDPI. 2021. Semantic Rule Checking in Construction. https://www.mdpi.com/xxxx [accessed May 2025].

MuniLogic. 2024. What Digital Transformation Means for Local Governments. https://munilogic.com/blogs/municipal-permitting [accessed May 2025].

Nemetschek. 2023. The Digital Building Permit – A Dream Coming True? Company Article. https://blog.nemetschek.com/digital-building-permit [accessed May 2025].

OpalAI. 2025. Accelerating City Permit Processes by Leveraging AI. https://www.opal-ai.com/solutions/permit-intelligence [accessed May 2025].

Planning.org. 2024. Planning With Artificial Intelligence. https://www.planning.org/publications/planning-ai [accessed May 2025].

PlanteMoran. 2023. Five Benefits of a Permit Agency's Digital Transformation. https://www.plantemoran.com/explore-our-thinking/insight/2023/05/benefits-of-digital-permit-systems [accessed May 2025].

Rapid Innovation. 2025. AI Agents for Permit Applications. https://www.rapidinnovation.io/use-cases/ai-permit [accessed May 2025].

ResearchGate. 2020. Semantic Rule Checking in the AEC Industry. https://www.researchgate.net/publication/123456789 [accessed May 2025].

Scale AI. 2025. Agentic Building Permit Validation. https://scale.com/use-cases/permit-validation [accessed May 2025].

UF AI & Cities. 2024. AI and Cities: International Forum for Innovation. https://dcp.ufl.edu/urp/ai-and-cities [accessed May 2025].

Warrington. 2023. Using AI to Review Construction Plans. University of Florida. https://warrington.ufl.edu/research/ai-construction-review [accessed May 2025].

WithPulley. 2023. Why Is Permitting So Slow? https://www.withpulley.com/resources/why-is-permitting-so-slow [accessed May 2025].

WTW. 2024. AI at Work: Guaranteeing Compliance with Building Codes. https://www.wtwco.com/insights/building-code-compliance [accessed May 2025].

29

Technology-Driven Approaches to Bridge Asset Management Within Geospatial Frameworks

Maryam Nasim

29.1 Introduction

Bridges are foundational components of national and regional economies, providing essential connectivity for communities, freight, and emergency responses (Shukla et al., 2024). Enhancing the resilience and sustainability of bridge infrastructure is a critical priority amid escalating pressures from climate change, such as intensified flooding, thermal extremes, hydrological volatility, and aging assets while rapid urbanization escalates traffic loads and land use conflicts. However, decades of chronic underinvestment have left significant portions of bridge networks vulnerable to deterioration, service disruptions, and environmental hazards (Nasim & Setunge, 2018). For instance, the 2025 Report Card for America's Infrastructure reveals that 49.1% of the nation's 623,000 bridges are currently in "fair" condition, 6.8% remain classified as "poor," and the number of bridges in fair condition continues to rise, indicating a potential future burden of costly rehabilitation or replacement (ASCE, 2025). Although federal programs such as the Infrastructure Investment and Jobs Act (IIJA) have injected critical funding, allocating $27.5 billion for bridge repairs alone, strategic asset management planning remains essential, particularly given an estimated $373 billion funding gap projected over the next decade (ASCE, 2025).

Traditional bridge asset management has largely relied on periodic manual inspections, subjective condition ratings, and reactive maintenance practices. Although foundational, these approaches have proven increasingly inadequate for managing complex, distributed networks subjected to dynamic environmental, load, and aging stresses. Studies have shown that deferred maintenance leads to compounding risks and significantly higher lifecycle costs, with emergency repairs often being several times more expensive than planned interventions, ultimately increasing the total infrastructure costs by over 50% (World Bank, 2021). Furthermore, the lack

DOI: 10.1201/9781003638308-32

of reliable and real-time data, particularly spatially contextualized data, has limited the ability of decision-makers to accurately forecast risks, allocate resources efficiently, and prioritize interventions across expansive bridge networks (Bush et al., 2012).

To address these challenges, a paradigm shift is underway: one that integrates geospatial intelligence (GEOINT), artificial intelligence (AI), and digital technologies into bridge asset management. Tools such as Geographic Information Systems (GIS), remote sensing, Light Detection and Ranging (LiDAR), and geospatial AI (GeoAI) enable a holistic, predictive approach by fusing spatially contextualized data with real-time monitoring and advanced analytics (Otero et al., 2016; Shi et al., 2023; Wei et al., 2021; Zhao et al., 2024).

This chapter explores how these technology-driven approaches can transform bridge asset management from reactive maintenance to proactive, intelligent strategies. Through the integration of digital twins, sensor networks, and spatial analytics, infrastructure managers can gain a comprehensive understanding of asset conditions and environmental interactions, supporting safer, more sustainable, and cost-effective infrastructure decisions.

29.2 Conceptual Foundations of Geospatial Intelligence in Infrastructure Systems

Bridges are inherently spatial assets, whose design, performance, and durability are deeply affected by their geographic and environmental context. Factors such as soil erosion, seismic activity, hydrological volatility, and urban expansion influence their structural integrity and lifecycle costs. Traditional asset management systems, which often lack geographic awareness, rely heavily on scheduled inspections and static records, approaches that fall short in capturing the dynamic, location, dependent risks that modern bridges face.

GEOINT addresses these limitations by embedding geographic context into the decision-making processes of infrastructure management. GEOINT integrates spatial data, advanced analytics, and real-time monitoring to enable adaptive, risk, informed asset stewardship. It enhances situational awareness and predictive capabilities, transforming fragmented infrastructure datasets into actionable insights. At the core of this transformation is the use of GIS, which serve as integrative platforms for managing georeferenced infrastructure data. GIS enables:

1. *Centralized Spatial Data Management*: In order to support inspection and maintenance operations, GIS plays a pivotal role in centralized spatial data management by serving as a digital infrastructure hub.

GIS integrates bridge locations, inspection histories, and condition ratings with environmental overlays (e.g., flood zones and seismic risks), creating a unified spatial database (Tao, 2013). By integrating high-resolution environmental data such as seismic overlays and floodplain maps, GIS provides a comprehensive view of vulnerability, helping infrastructure managers assess and mitigate hazards more effectively. This consolidation supports system-wide monitoring and prioritization of maintenance activities.

2. *Integration of Real-Time Data Streams*: GIS platforms integrate real-time sensor data from Internet of Things (IoT) devices and Supervisory Control and Data Acquisition (SCADA) systems, enabling continuous monitoring and analytics. This fusion of static and dynamic data creates a real-time decision-support platform. Infrastructure managers can detect anomalies through GIS dashboards and initiate proactive interventions. GIS maps asset performance against environmental risks, allowing targeted maintenance of bridges in high-risk zones (Jianbo et al., 2024; Rossi & Bournas, 2023). This feedback loop supports predictive maintenance and improves asset resilience.
3. GIS platforms integrate real-time sensor data from IoT devices and SCADA systems, enabling continuous monitoring and analytics. This fusion of static and dynamic data creates a real-time decision-support platform. Infrastructure managers can detect anomalies through GIS dashboards and initiate proactive interventions. GIS maps asset performance against environmental risks, allowing targeted maintenance of bridges in high-risk zones (Jianbo et al., 2024; Rossi & Bournas, 2023). This feedback loop supports predictive maintenance and improves asset resilience.
4. *Field Efficiency with Mobile GIS*: Mobile GIS applications significantly enhance field efficiency by providing remote data collection capabilities for field crews to collect inspection data offline in remote areas, synchronizing updates with centralized systems once connectivity is restored. These tools also contribute to improved operational efficiency and safety in various industrial applications. Integrating mobile GIS with Computerized Maintenance Management Systems (CMMS) and SCADA platforms further automates maintenance task scheduling and sensor data ingestion (Maldonado-Correa et al., 2024).
5. *Support for Smart City and Urban Planning Integration*: GIS plays a broader role in smart cities by supporting multidisciplinary collaboration and spatial coordination of assets, utilities, and environmental planning. Its ability to model complex urban systems enhances infrastructure sustainability and equity. It supports urban-scale spatial models that improve infrastructure planning, zoning, and

sustainability initiatives. By integrating data across agencies and domains, GIS becomes a critical enabler of resilient, interconnected urban infrastructure (Tao, 2013).

Many public sector agencies still operate with fragmented paper records or legacy systems not interoperable with GIS platforms. This creates data silos and delays, undermining spatial integration benefits. Adoption barriers persist due to workforce skill gaps and institutional inertia, only about 30% of municipalities globally leverage spatial data in infrastructure planning (Jianbo et al., 2024). The integration of geospatial intelligence and GIS in bridge asset management redefines infrastructure monitoring, maintenance, and governance. It enables a shift from reactive workflows to predictive, data-informed strategies that support interventions, extend infrastructure lifespans, and enhance resilience amid environmental and operational uncertainties.

29.3 Remote Sensing and IoT Applications in Structural Health Monitoring (SHM)

The integration of remote sensing technologies and IoT sensor networks has revolutionized Structural Health Monitoring (SHM) in bridge asset management. These technologies provide continuous, high-resolution, and real-time insights into bridge performance, enabling proactive and data-driven maintenance strategies. Remote sensing enables the non-contact monitoring of bridges at various spatial and temporal resolutions, offering insights that are often unattainable through manual inspections.

- LiDAR provides high-resolution 3D models of bridge geometry, enabling the detection of structural misalignments, surface defects, and deformation. When deployed via drones, LiDAR facilitates safe and efficient inspections of hard-to-access areas and reduces inspection time by over 90%, while aligning with Federal Highway Administration (FHWA) and American Association of State Highway and Transportation Officials (AASHTO) standards (Moreu et al., 2020).
- InSAR (Interferometric Synthetic Aperture Radar) leverages satellite imagery to detect millimeter, scale displacements due to thermal expansion, subsidence, or tectonic shifts. InSAR has proven valuable in post-construction monitoring and detecting long-term ground movement that affects bridge alignment (Yang & Xia, 2023).
- Multispectral and high-resolution optical imagery from Unmanned Aerial Vehicles (UAVs) and satellites supports environmental

monitoring, hazard identification, and compliance assessments. When integrated with GIS platforms, these datasets enhance the spatial awareness of asset managers, improving decision-making across entire bridge networks (Shi et al., 2023; Zhao et al., 2024).

Remote sensing tools support condition assessments and enable integration with digital twin models, aiding 3D comparisons between as-designed and as-built states (Otero et al., 2016). Embedded IoT sensors provide continuous data on structural and environmental conditions, converting physical data into actionable intelligence. Key IoT components in bridge SHM include accelerometers for capturing dynamic loads and modal analysis, strain gauges to monitor stress and fatigue, tiltmeters to track angular displacement and foundation movement, temperature, and humidity sensors for assessing environmental effects on materials. These sensors stream data to GIS dashboards and SCADA platforms like ArcGIS Velocity, enabling data alerts and predictive analytics.

The Walkway Bridge project in Connecticut integrated IoT sensors with GIS platforms to monitor structural conditions during construction, enabling real-time tracking of behavior and enhancing safety (Connecticut Department of Transportation, 2024). The Australian Rail Track Corporation implemented IoT-enabled digital twins across heritage rail bridges, reducing unplanned maintenance by 37% through predictive modeling (ARTC, 2024). These systems trigger alerts for structural anomalies, while InSAR and UAV tools confirm damage extent (Saviz et al., 2023). They support efficiency through remote monitoring, minimizing on-site labor and traffic disruptions, while real-time dashboards enable rapid decisions. The combination of LiDAR, InSAR, and sensor feedback allows managers to respond to deterioration signs with precision (Jianbo et al., 2024; Xu & Zhang, 2022).

29.4 Digital Twin Architectures in Bridge Infrastructure Management

Digital twin technology is transforming bridge infrastructure management by creating dynamic, data-rich virtual replicas of physical bridges. These models integrate real-time sensor data, environmental parameters, and engineering simulations to enhance monitoring, risk assessment, and predictive maintenance beyond traditional methods.

A digital twin comprises four key components: a high-fidelity geometric model derived from Building Information Model (BIM), LiDAR, or photogrammetric data; continuous streaming of IoT sensor data (e.g., strain, tilt, vibration, temperature); geospatial intelligence from GIS platforms; and an analytical engine powered by AI and Finite Element Analysis (FEA) (Mousavi et al., 2024). This holistic approach enables proactive infrastructure management, with AI detecting anomalies and machine learning predicting deterioration rates, leading to condition-driven maintenance strategies. Recent research has shown that such models can reach predictive accuracies of up to 89% across various bridge types and materials (Saviz et al., 2023).

Digital twins also gain value through GIS integration, becoming spatially aware and capable of contextualizing maintenance decisions based on environmental factors, such as proximity to corrosive bodies or unstable soils. Scenario simulations allow engineers to test various load conditions, informing design modifications and emergency planning.

However, challenges remain in implementing digital twins, such as data interoperability, lack of sensor coverage, computing power requirements, and the absence of a standardized validation framework. Moreover, there is currently no universally accepted framework for validating digital twin predictions against real-world performance outcomes (Zhao et al., 2024). Figure 29.1 illustrates the GeoAI-driven lifecycle algorithm for bridge infrastructure, showcasing the workflow from data acquisition to AI analytics.

In conclusion, the combination of geospatial intelligence, real-time monitoring, and computational modeling in digital twins represents a significant

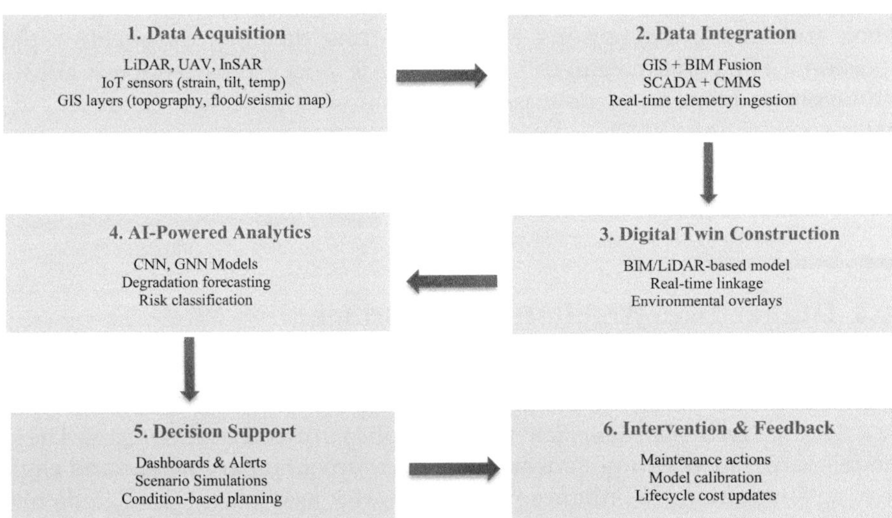

FIGURE 29.1
GeoAI-driven digital twin lifecycle for bridge infrastructure.

advancement in bridge asset management, offering a scalable and adaptive platform for navigating complex operational environments.

29.5 Computational Challenges in GeoAI Deployment

The deployment of GeoAI in bridge infrastructure management promises powerful capabilities, from predictive maintenance and automated inspections to network-level risk modeling. However, realizing these benefits at scale involves a range of computational, technical, and institutional challenges.

One of the most fundamental barriers is data quality and availability. GeoAI models depend heavily on high-resolution, spatially explicit datasets such as LiDAR scans, UAV imagery, IoT sensor feeds, and historical inspection records. Yet, in many cases, these datasets are incomplete, inconsistent, or siloed across agencies. Remote sensing inputs may suffer from low spatial or temporal resolution, obscured imagery due to shadow or weather interference, or insufficient ground truth data for training AI models (Almohsen, 2024).

Compounding this issue is the challenge of interoperability and system integration. Bridge management systems must align BIM models, GIS data layers, sensor telemetry, and legacy asset registries, each potentially using different formats and standards. Without harmonized schemas and open Application Programming Interfaces (APIs), seamless data exchange becomes difficult, limiting the ability of AI engines to operate across diverse information systems. Many public agencies also struggle with connecting their geospatial platforms (e.g., ArcGIS) with computerized CMMS or SCADA platforms in a unified architecture (Wei et al., 2021).

Beyond technical integration, GeoAI models face limitations in computational scalability. Real-time analytics involve large-scale spatial data, such as require significant computing resources. Training deep learning models like convolutional neural networks (CNNs) or graph neural networks (GNNs) on complex infrastructure datasets can overwhelm conventional servers, particularly when models must be retrained frequently or deployed at the network level across hundreds of bridges. Cloud-based solutions offer scalability, but raise concerns over latency, bandwidth, and cost, especially in remote or high, security environments.

Another pressing issue is algorithmic interpretability. Many GeoAI models function as black boxes, providing risk classifications or anomaly alerts without transparent reasoning. Explainable AI (XAI) methods, such as saliency maps or feature importance scores, are emerging solutions, but they are not yet standard practice in infrastructure analytics workflows (Almohsen, 2024). Human and institutional factors further compound these technical hurdles. The absence of regulatory frameworks for validating AI-based infrastructure decisions adds further uncertainty, particularly for risk-sensitive applications such as load rating or failure forecasting (Wu et al., 2024).

29.6 Applications in Alignment and Component Placement

Geospatial intelligence is crucial for the design, alignment, and placement of bridge components, ensuring accurate positioning of structural elements like girders and bearings for long-term performance and seismic resilience. The use of Real-time Kinematic (RTK) – Global Position System (GPS) allows for centimeter-level accuracy, enabling real-time adjustments during construction and improving efficiency by reducing labor costs and project timelines by lowering the risk of human error (De Pace & Kaufmann, 2023).

UAVs enhance these workflows by providing high-resolution imagery to verify site conditions and detect deviations before girder placement. When combined with BIM and GIS, UAV data supports automated validation and real-time feedback for construction processes.

AI-driven validation systems, including CNNs, help identify structural misalignments and defects through site imagery, allowing for immediate alerts on non-conformities. The integration of BIM and GIS further guides alignment decisions based on environmental factors such as terrain profiles and seismic zones. This fusion ensures alignments are geometrically accurate, environmentally and geotechnically sound (Wei et al., 2021). AI models can analyze UAV images to detect incomplete rebar layouts, improper alignment of bearing seats, or surface defects prior to concrete pouring (Xu & Zhang, 2022). These tools can be deployed on mobile devices or drones, giving site engineers instant alerts for non-conformities, such as misplaced formwork, insufficient cover, or uneven load paths (Phan et al., 2024).

Moreover, GIS aids in scenario planning for construction staging, optimizing placements in areas with potential hazards by referencing real-time weather models and historical data. These predictive overlays help contractors avoid delays and mitigate environmental risks during the critical alignment phase (Barazzetti & Banfi, 2017). Overall, the convergence of these technologies transforms infrastructure construction into a more precise, risk-informed, and adaptive process, paving the way for advancements in automation and quality control.

29.7 Conclusion and Future Research Directions

As infrastructure systems face growing challenges from aging assets, climate changes, and increasing demands for service continuity, the need for intelligent, resilient, and proactive asset management has never been greater. This chapter has explored how geospatial intelligence, encompassing GIS, IoT, digital twins, and AI-can reshape the lifecycle management of bridge infrastructure. Through spatially informed decision-making, monitoring, and predictive modeling, these technologies promise to bridge the longstanding gap between inspection data and actionable intervention.

Geospatial systems are no longer passive sources of location data; they now function as integrative platforms that connect design, sensing, analytics, and asset operations. When embedded within broader digital twin architectures, GIS and remote sensing enable not just visualization but spatial context-aware simulations, condition forecasting, and environmentally aligned maintenance strategies. This transition from reactive, schedule-based inspections to proactive, condition-driven asset stewardship represents a profound shift in infrastructure governance.

Nevertheless, several challenges impede full-scale deployment. Issues related to data quality, interoperability, algorithmic transparency, and institutional capacity persist. Many agencies still rely on fragmented systems and paper-based workflows, while advanced analytics tools remain inaccessible to under-resourced asset managers. Moreover, the lack of universally accepted frameworks for validating AI-driven infrastructure predictions limits trust and operational uptake, especially in risk-sensitive scenarios like seismic performance, scour vulnerability, or structural fatigue. To address these gaps and unlock the full potential of geospatial ecosystems in infrastructure, the following research priorities emerge:

1. Standardizing GeoBIM Integration: Develop unified ontologies and data schemas that harmonize BIM, GIS, and SHM systems across platforms and stakeholders.
2. Scalable, Explainable GeoAI Models: Advance lightweight, interpretable machine learning frameworks that can function in low-connectivity environments while maintaining predictive robustness.
3. Real-World Validation of Digital Twins: Establish protocols to benchmark AI-augmented digital twins against long-term field data to ensure trust, safety, and accountability.
4. Edge and Cloud Synergies: Explore architectures that balance edge computing for local analytics with cloud-based platforms for broader scenario modeling and asset coordination.
5. Human-Centric Design and Policy Innovation: Invest in training, governance frameworks, and participatory design approaches that ensure equitable access to spatial intelligence, especially in rural and underserved areas.
6. Ethical and Regulatory Foresight: Investigate the socio-technical implications of AI in infrastructure, including privacy, liability, and transparency in algorithmic decision-making.

Additionally, research on infrastructure monitoring notes that "real-world data often contains errors and discrepancies such as missing data or faulty entries," which significantly impacts the accuracy of deterioration prediction and the ability to optimally schedule preventive interventions (Saviz et al., 2023). This lack of comprehensive, high-quality data directly hampers the effectiveness of risk forecasting and resource allocation strategies for bridge networks. Future success will depend not only on technical integration but on reimagining institutional governance and workforce development strategies aligned with the digital future of infrastructure. The transformation of bridge asset management from static record-keeping to dynamic, data-informed ecosystems requires a collective effort, uniting engineers, data scientists, urban planners, policymakers, and communities. In doing so, we move closer to infrastructure systems that are resilient by design, intelligent by operation, and adaptive by context.

Acknowledgments

The author gratefully acknowledges the support of the Australian Research Council Research Hub for Resilient Intelligent Infrastructure System (ARC-RIIS) and the Centre for Spatial Data Infrastructures and Land Administration (CSDILA) at the University of Melbourne for providing the research environment and resources that contributed to the development of this chapter.

All content and analysis presented are the original work of the author, based on academic research, technical expertise, and practical project experience. Artificial Intelligence tools were used solely to assist with language refinement and formatting; all conceptual development, technical content, and interpretations were generated by the author.

References

Almohsen, A. S. (2024). Challenges Facing the Use of Remote Sensing Technologies in the Construction Industry: A Review. In *Buildings* (Vol. 14, Issue 9). https://doi.org/10.3390/buildings14092861

ARTC. (2024). Structural Health Monitoring of Bridges Using IoT. In *Lecture Notes in Civil Engineering, Track & Civil* (Vol. 457, pp. 487–497). Australian Rail Track Corporation (ARTC). https://doi.org/10.1007/978-981-99-9610-0_39

ASCE. (2025). A Comprehensive Assessment of America's Infrastructure. In *Asce*. https://www.asce.org/topics/report-card-for-americas-infrastructure

Barazzetti, L., & Banfi, F. (2017). BIM and GIS: When parametric modeling meets geospatial data. *ISPRS Annals of the Photogrammetry, Remote Sensing and Spatial Information Sciences*, IV-5/W1. https://doi.org/10.5194/isprs-annals-iv-5-w1-1-2017

Bush, S., Omenzetter, P., Henning, T., & McCarten, P. (2012). Data Collection and Monitoring Strategies for Asset Management of New Zealand Road Bridges. In *NZ Transport Agency Research* (Issue Report 475).

Connecticut Department of Transportation. (2024). *Connecticut Department of Transportation*. https://www.ct.gov/dot

De Pace, F., & Kaufmann, H. (2023). A systematic evaluation of an RTK-GPS device for wearable augmented reality. *Virtual Reality*, 27. https://doi.org/10.1007/s10055-023-00863-3

Lai, J., Zhu, J., Dang, P., Wu, J., Guo, Y., Yang, X., & Li, N. (2024). Knowledge-driven semantic converting method of multimodal models toward a geospatial perspective. *Geo-Spatial Information Science*, 2024, 1–6.

Maldonado-Correa, J., Artigao, E., Gómez-Lázaro, E., Martín-Martínez, S., & Valdiviezo-Condolo, M. (2024). Classification of highly imbalanced supervisory control and data acquisition data for fault detection of wind turbine generators. *Energies*, 17. https://doi.org/10.3390/en17071590

Moreu, F., Lippitt, C. D., Yuan, X. (2020). *Bridge Construction Monitoring Using LIDAR for Quantified, Objective Quality-Control Quality-Assurance (QOQCQA).* Transportation Consortium of South-Central States.

Mousavi, V., Rashidi, M., Mohammadi, M., & Samali, B. (2024). Evolution of digital twin frameworks in bridge management: Review and future directions. *Remote Sensing*, 16(11), 1887.

Nasim, M., & Setunge, S. (2018). Vulnerability Assessment for the Reinforced Concrete Beam Exposed to Monotonic Loading Using Different Damage Indices. *CSTX 2018*, Barcelona, Spain.

Otero, L. D., Peter, A., & Moyou, M. (2016). *Remote Sensing with Mobile LiDAR and Imaging Sensors for Railroad Bridge Inspections. August*, 700.

Phan, T.-N., Nguyen, H.-H., ThAI, H.-T., & Le, K.-H. (2024). Deep Learning Models for UAV-Assisted Bridge Inspection: A YOLO Benchmark Analysis. *2024 International Conference on Advanced Technologies for Communications (ATC)*, 963–968.

Rossi, M., & Bournas, D. (2023). Structural health monitoring and management of cultural heritage structures: A state-of-the-art review. *Applied Sciences*, 13. https://doi.org/10.3390/app13116450

Saviz, M., Yuchuan, J., & Tyrone, A. B. (2023). Estimating the values of missing data related to infrastructure condition states using their spatial correlation. *Journal of Infrastructure Systems*, 29(1), 4022041. https://doi.org/10.1061/(ASCE)IS.1943-555X.0000726

Shi, Z., Zhang, H., Mondal, T. G., Hartnagel, B. A., & Chen, G. (2023). A Streamlining Remote Sensing and Digitalization Process for Bridge Inspection. *Structural Health Monitoring 2023*.

Shukla, R., Pabbisetty, S. K., Jayanthi, S. C., & Janardhanan, K. (2024). Evaluating the damage of collapsed bridges using remote sensing technologies: Case study: Baltimore's Francis Scott key bridge. *Eco Cities*. https://api.semanticscholar.org/CorpusID:272737484

Tao, W. (2013). Interdisciplinary urban GIS for smart cities: advancements and opportunities. *Geo-Spatial Information Science*, 16. https://doi.org/10.1080/10095020.2013.774108

Wei, J., Chen, G., Huang, J., Xu, L., Yang, Y., Wang, J., & Sadick, A.-M. (2021). BIM and GIS Applications in Bridge Projects: A Critical Review. In *Applied Sciences* (Vol. 11, Issue 13). https://doi.org/10.3390/app11136207

World Bank. (2021). *Well Maintained: Economic Benefits from More Reliable and Resilient Infrastructure.* https://ppp.worldbank.org/public-private-partnership/library/well-maintained-economic-benefits-more-reliable-resilient-infrastructure

Wu, Y., Wang, Y., Liu, H., Xie, L., Jiao, L., & Lu, P. (2024). Risk assessment of bridge construction investigated using random forest algorithm. *Scientific Reports*, 14(1), 20964.

Xu, Y., & Zhang, J. (2022). UAV-based bridge geometric shape measurement using automatic bridge component detection and distributed multi-view reconstruction. *Automation in Construction*, 140, 104376.

Yang, H., & Xia, M. (2023). Advancing bridge construction monitoring: AI-based building information modeling for intelligent structural damage recognition. *Applied Artificial Intelligence*, 37(1), 2224995. https://doi.org/10.1080/08839514.2023.2224995

Zhao, T., Wang, S., Ouyang, C., Chen, M., Liu, C., Zhang, J., Yu, L., Wang, F., Xie, Y., & Li, J. (2024). Artificial intelligence for geoscience: Progress, challenges and perspectives. *The Innovation*, 5.

30

Digital Twin Governance for Urban Resilience and Sustainability

Mahsa Moghadas and Abbas Rajabifard

30.1 Introduction

Cities are under increasing pressure from climate change, aging infrastructure, rapid urbanization, and social inequality. Therefore, they need to embrace digital transformation to improve sustainability and resilience, especially by adopting digital twins—virtual replicas of physical urban systems—which can offer dynamic, real-time, data-driven urban modeling, forecasting, and scenario-based planning (Batty 2018; Dembski et al. 2020). However, their real-world impact remains uneven, with notable challenges in equity, scalability, and institutional capacity (Tzachor et al. 2022; Ye et al. 2023). Drawing from polycentric institutionalism and anticipatory governance (Muiderman et al. 2020), this chapter explores the governance dimensions necessary for urban digital twin (UDT) implementation and situates that within evolving urban development agendas beyond 2030, including Sustainable Development Goals (SDGs), the New Urban Agenda, and the Sendai Framework for Disaster Risk Reduction.

30.2 Urban Digital Twin Governance

Effective governance of UDTs requires comprehensive frameworks that address institutional, legal, ethical, and technical dimensions to ensure transparency, accountability, and inclusivity (Batty 2018; Weil et al. 2023; Ye et al. 2023). Key principles include:

- **Data stewardship**: It is fundamental to the governance of digital twins. High-quality, real-time data from multiple sensors, open data platforms, and institutional repositories must be governed through clear protocols for access, security, and interoperability.

DOI: 10.1201/9781003638308-33

For instance, city of Helsinki and Amsterdam have pioneered open urban data ecosystems and illustrate best practices in governance-driven data accessibility (Dembski et al. 2020; Nica et al. 2023).

- **Algorithmic oversight**: Since artificial intelligence (AI) is integral to digital twin analytics, the risks of bias, opacity, and misalignment can grow (Nica et al. 2023; Tambiama 2020). Hence, cities should implement algorithmic impact assessments (Tzachor et al. 2022), create transparent registries of automated systems, and ensure human-in-the-loop decision mechanisms particularly for emergency response, early warnings, and public health. New York's Automated Decision Systems Task Force has pioneered transparent AI governance frameworks for urban data analytics (Ricciardi and Callegari 2023; Ye et al. 2023).
- **Participatory access**: It contributes to enhance the governance of urban UDT by improving transparency, data quality, inclusivity, and institutional responsiveness. Moreover, citizen engagement in decision-making and scenario planning results in having no one left behind. Cities, therefore, should provide accessible visualizations, open simulation platforms, and deliberative tools (Cardullo and Kitchin 2019; Goodspeed 2020). For example, Singapore's Virtual Singapore platform allows for the co-production of urban policy through community-level visualizations (Nica et al. 2023).
- **Institutional intelligence**: UDT should be governed as living systems embedded in municipal institutions to increase the capacity of urban institutions to be able to integrate digital twin technologies via cross-sector coordination, ethical oversight, and adaptive learning leading to harmonize and integrated management of complex urban systems under chronic stresses or acute shocks. Barcelona's Smart Cities model exemplifies structured institutional intelligence in municipal planning (Muiderman et al. 2020; Weil et al. 2023).
- **Regulatory and ethical compliance**: UDT should adhere to evolving legal framework both national and international, such as the General Data Protection Regulation, UNESCO's AI ethics guidelines, and national cybersecurity standards. In addition, beyond legal considerations, ethical governance encompasses inclusivity (e.g., encouraging stakeholder engagement), accountability (e.g., guaranteeing human oversight in urban planning scenarios), transparency (e.g., open access to digital twin models), and privacy protection (e.g., data anonymization) (EU 2024; UNESCO 2021; UN-Habitat 2020).

Table 30.1 provides a summary of the main governance principles discussed above.

TABLE 30.1

Key Governance Dimensions of UDT

Principal	Description	References
Data stewardship	Ensures data quality, accessibility, and privacy in real-time urban systems.	Dembski et al. (2020) and Nica et al. (2023)
Algorithmic oversight	Provides auditability and accountability for AI-driven processes.	Tambiama (2020), Ye et al. (2023), and Nica et al. (2023)
Participatory access	Enables inclusive engagement and public use of simulation tools.	Cardullo and Kitchin (2019) and Goodspeed (2020)
Institutional intelligence	Fosters coordination, ethical oversight, and adaptive learning across institutions.	Muiderman et al. (2020) and Weil et al. (2023)
Regulatory compliance	Aligns UDT operations with legal and ethical frameworks.	UNESCO (2021), UN-Habitat (2020), and EU (2024)

30.3 Operationalizing Digital Twins: Governance Structure and Implementation

This section focuses on applied and operational aspects of digital twin implementation into real-world urban planning and decision-making. In this regard, the integration necessitates a structured framework that enables better coordination, real-time decision-making, and meaningful citizen engagement. A well-designed governance system helps cities to transit to integrated, anticipatory, and inclusive urban planning ensuring resilience and sustainability in increasingly complex urban environments (Goodspeed 2020; Nica et al. 2023; Ricciardi and Callegari 2023).

30.3.1 Multi-Layered Governance Structure

Digital twin governance functions across three interrelated layers—strategic, tactical, and operational—each playing a distinct role in policymaking and urban management.

- **Strategic level**: It supports long-term planning by aligning sustainability and resilience goals with advanced modeling. It enables risk evaluation through simulations and fosters collaboration among urban systems. For example, Melbourne's DT aids resilience planning and supports regeneration and placemaking (Allan et al. 2024; Moghadas et al. 2022; Ricciardi and Callegari 2023).
- **Tactical layer**: It translates strategy into action by coordinating cross-agency collaboration, resource allocation, and policy implementation. It fosters institutional adaptability and integrates DT

insights into governance. Amsterdam's Smart City initiative exemplifies such cross-sectoral coordination (Afif Supianto et al. 2024).

- **Operational layer**: It enables real-time monitoring and adaptive management using live spatio-temporal data from sensors. This supports predictive modeling in areas like transport, environment, and emergency response. Munich's DT, developed through the "Connected Urban Twins" project with Leipzig and Hamburg, illustrates this layer in practice (Moghadas et al. 2023; Nica et al. 2023; Ricciardi and Callegari 2023).

30.3.2 Participatory Governance and Citizen Engagement

A digital twin governance system must prioritize inclusivity by creating mechanisms for public participation and transparency in decision-making. For instance, urban innovation labs can provide a space as collaborative centers where government officials, researchers, and local communities can co-develop solutions through digital twin-based experimentation (Ye et al. 2023). In addition, town halls and public forums facilitate direct dialogue between policymakers and residents, ensuring governance decisions reflect diverse urban perspectives by leveraging digital twin capabilities forecasting models, 3D visualizations, advanced simulations, to name a few (Cardullo and Kitchin 2019). For example, platforms like Singapore's Virtual Singapore enables individuals to engage with digital urban models, hence enhancing participatory decision-making (UN-Habitat 2020).

30.3.3 Scenario-Based Decision-Making and Institutional Adaptability

UDTs help cities prepare for future challenges by simulating multiple scenarios, enabling flexible and informed management of complex urban systems using real-time data and predictive analytics (Goodspeed 2020; Muiderman et al. 2020). These simulations support the assessment of risks related to infrastructure degradation, climate hazards, and socioeconomic inequalities. Moreover, by testing policy interventions in digital environments, planners can anticipate outcomes, refine strategies, and align decisions with resilience, sustainability, and equity goals while reducing unintended consequences (Haraguchi et al. 2024).

To enhance adaptive governance, cities should adopt institutional learning frameworks that promote continuous adaptation, knowledge exchange, and iterative policy refinement. Embedding feedback loops from digital twin simulations into decision-making processes can help maintain responsive and agile governance under evolving urban stresses (Goodspeed 2020; Moghadas et al. 2022).

30.3.4 Implementation Strategy

To ensure institutional maturity and long-term functionality, a phased approach to digital twin implementation is needed:

- **Foundation phase**: It is the phase of establishing governance frameworks through stakeholder mapping, ethical guidelines, digital infrastructure design, and capacity-building programs (Abdeen et al. 2023; Batty 2018; Weil et al. 2023).
- **Integration phase**: It is the phase of deploying pilot projects, activating inter-agency collaboration, optimizing real-time operational capabilities, and standardized interoperability protocols (Afif Supianto et al. 2024; Nica et al. 2023; Temple et al. 2024).
- **Institutionalization phase**: It is the phase of embedding digital twin systems into official urban planning and policy frameworks, expanding their applications, institutionalizing feedback mechanisms for ongoing optimization, and long-term evaluation and adaptation mechanisms (Temple et al. 2024; UN-Habitat 2020).

30.3.5 Ensuring Governance Longevity and Adaptive Capacity

To ensure digital twin systems remain effective in the long term, governance frameworks must embrace continuous adaptation, strategic learning, and cross-sector collaboration. Regular assessments and knowledge-sharing platforms promote policy alignment with evolving urban challenges (UNESCO 2021). Capacity-building should go beyond technical skills to include policy innovation, anticipatory governance, and ethical AI use. Open data systems can further support adaptive governance by enabling transparent monitoring and facilitating shared institutional learning (EU 2024). Figure 30.1 shows the operationalization of UDT schematically.

30.4 Case Study of Digital Twins for Flood Management in Victoria, Australia

The increasing risks posed by climate change, particularly urban flooding, demand innovative tools to inform resilient urban design and planning. UDTs provide a dynamic and data-driven environment for simulating, visualizing, and managing flood-related information across urban landscapes. By integrating hydrological models with spatial data, UDTs enable stakeholders

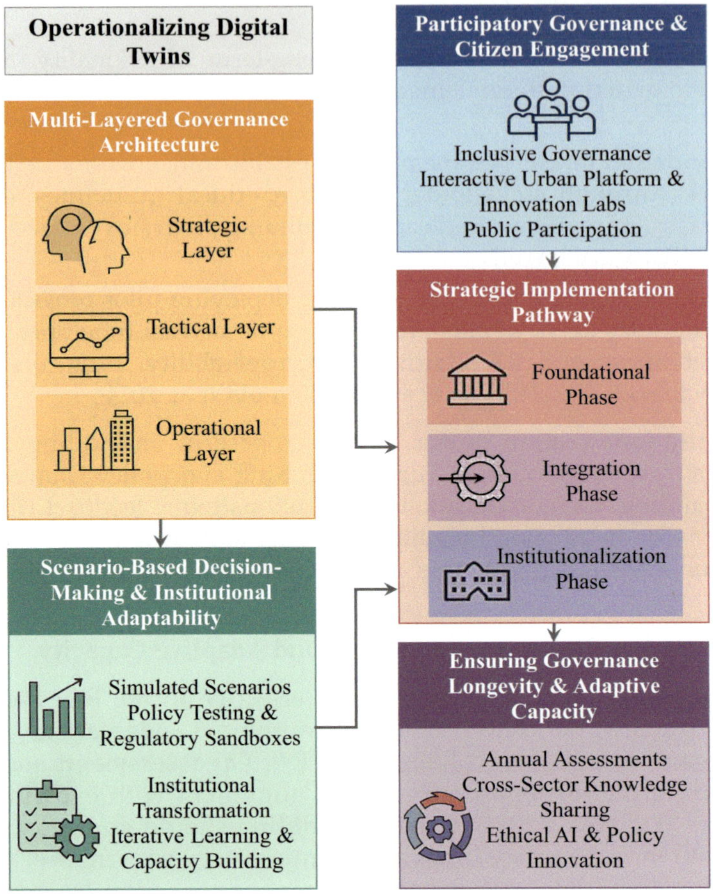

FIGURE 30.1
Operationalizing digital twins: governance system design for future-ready cities.

to assess the implications of planning decisions on flood risks in real time, supporting proactive and adaptive responses to urban water management.

A notable application within this domain is the development of a Decision Support Tool within the Centre for SDIs and Land Administration (CSDILA) Digital Twin platform. This tool specifically addresses flood management by analyzing stormwater capacities in relation to new road networks and urban developments across catchment areas. Through this capability, urban planners can evaluate the impacts of infrastructure proposals on flood risk and water-sensitive urban design objectives, supporting more sustainable and resilient urban futures.

This exemplifies the interdisciplinary collaboration between landscape architects, geospatial scientists, and urban modelers. It delivers an ecological adaptation scenario builder, integrated within UDT environment, to measure and visualize flood risks and the impacts of climate change. This allows for the exploration of incremental flood defense strategies and long-term adaptation measures under varied climate projections.

The Fishermans Bend precinct in Melbourne, which is Australia's largest urban renewal site, was selected as a testbed for this approach. This location, due to its low-lying topography and exposure to future sea-level rise and flooding, exemplifies the critical need for robust UDT capabilities in floodplain redevelopment planning. The platform supports the exploration of near-, mid-, and long-term flood resilience strategies (30, 70, and 100 years horizons), assisting decision-makers in navigating complex trade-offs between urban growth and flood mitigation with the following core capabilities developed:

- A dedicated panel within the CSDILA Digital Twin platform for real-time water-sensitive urban design and flood management scenarios.
- Simulation tools enabling interactive road network design modifications with immediate visualization of stormwater impacts and inundation risks.
- Integrated modeling interfaces linking stormwater systems with buildings, streets, and urban infrastructure for holistic scenario assessment.

Through these capabilities, CSDILA's digital twin platform is not just a visualization tool but an operational instrument for evidence-based flood management, enabling urban governance frameworks to proactively plan for and adapt to climate-induced flood challenges. This reinforces the broader role of UDTs in anticipatory governance, resilience planning, and adaptive urban transformation, directly contributing to post-2030 sustainability and disaster risk reduction agendas.

30.5 Innovation Pathways for Digital Twin

Successful UDTs require institutional and managerial innovation:

- **Institutional intelligence and innovation**: Cities should embed digital transformation within municipal planning through cross-sectoral ecosystems and dedicated digital units. This supports real-time urban insight integration and fosters collaborative governance involving government, academia, industry, and civil society (Haraguchi et al. 2024).
- **Public–private innovation ecosystems**: Co-creation, iterative feedback, and shared governance between public and private actors

should be central. Open innovation platforms can improve stakeholder participation and scalability (Temple et al. 2024).

- **Agile urban management and regulatory innovation**: Adaptive governance using iterative policymaking, real-time data, and regulatory sandboxes enables cities to test and refine technologies and governance models before full deployment (Argota Sánchez-Vaquerizo 2024; Muiderman et al. 2020).
- **Cross-sectoral capacity-building**: Ongoing professional training and interdisciplinary collaboration (via innovation labs, workshops, and leadership development) are necessary to build institutional capacity and sustain UDT systems (Afif Supianto et al. 2024).

Through the mentioned innovation pathways, UDT governance transitions toward holistic, anticipatory urban management from fragmented experimentation, making cities capable to address post-2030 urban complexity.

30.6 Aligning Digital Twin Governance with Post-2030 Agendas

To support the evolving global development agendas such as SDGs, Sendai Framework and New Urban Agenda beyond 2030, digital twin governance should address key limitations identified in recent years, including data inequality, governance gaps, and uneven institutional readiness (Tzachor et al. 2022; Weil et al. 2023). Interlinked priorities are:

- **Climate justice**: Governance systems must foreground the uneven distribution of climate risks and technological capacities (Moghadas et al. 2023). UDT should be used to identify and mitigate environmental injustices across vulnerable urban communities.
- **Circular urbanism**: Integrated models can direct cities toward circular resource use, optimizing energy, water, and material flows within planetary boundaries (Chi et al. 2023).
- **Digital sovereignty**: Cities must govern digital infrastructure, data governance and AI technologies as critical public goods, guaranteeing data autonomy and ethical AI deployment aligned with human rights and international AI ethics standards (Tambiama 2020).
- **Transformative governance**: Digital twin systems should not only manage near-term risks but help envision and assist long-term and systemic urban transformations compatible with post-SDGs

and resilience and regenerative futures (Moghadas et al. 2023; Muiderman et al. 2020).

30.7 Conclusion

Digital twin governance represents a paradigm shift in urban planning. By enabling cities to model complexity, anticipate disruption, and co-create adaptive responses, it transforms UDT from tools into institutional infrastructure that become central to the future of urban resilience and sustainability. To be effective, UDT governance systems must be anticipatory, equitable, and transparent grounded in interdisciplinary theory and designed toward resilient and sustainable urban futures.

References

Abdeen, Fathima Nishara, Sara Shirowzhan, and Samad Sepasgozar. 2023. "Citizen-Centric Digital Twin Development with Machine Learning and Interfaces for Maintaining Urban Infrastructure." *Telematics and Informatics* 84. https://doi.org/10.1016/j.tele.2023.102032.

Afif Supianto, Ahmad, Wajeeha Nasar, Dina Margrethe Aspen, Agus Hasan, Anniken Susanne T. Karlsen, and Ricardo Da Silva Torres. 2024. "An Urban Digital Twin Framework for Reference and Planning." *IEEE Access* 12: 152444–65. https://doi.org/10.1109/ACCESS.2024.3478379.

Allan, Mark, Abbas Rajabifard, and Greg Foliente. 2024. "Urban Regeneration and Placemaking: A Digital Twin Enhanced Performance-Based Framework for Melbourne's Greenline Project?" *Australian Planner* 59 (4): 247–257. https://doi.org/10.1080/07293682.2024.2342793.

Argota Sánchez-Vaquerizo, Javier. 2024. "Urban Digital Twins and Metaverses towards City Multiplicities: Uniting or Dividing Urban Experiences?" *Ethics and Information Technology* 27 (1): 4. https://doi.org/10.1007/s10676-024-09812-3.

Batty, Michael. 2018. "Digital Twins." *Environment and Planning B: Urban Analytics and City Science* 45 (5): 817–20. https://doi.org/10.1177/2399808318796416.

Cardullo, Paolo, and Rob Kitchin. 2019. "Being a 'Citizen' in the Smart City: Up and down the Scaffold of Smart Citizen Participation in Dublin, Ireland." *GeoJournal* 84 (1): 1–13. https://doi.org/10.1007/s10708-018-9845-8.

Chi, Ziyuan, Zhen Liu, Fenghong Wang, and Mohamed Osmani. 2023. "Driving Circular Economy through Digital Technologies: Current Research Status and Future Directions." *Sustainability (Switzerland)* 15 (24). https://doi.org/10.3390/su152416608.

Dembski, Fabian, Uwe Wössner, Mike Letzgus, Michael Ruddat, and Claudia Yamu. 2020. "Urban Digital Twins for Smart Cities and Citizens: The Case Study of Herrenberg, Germany." *Sustainability* 12 (6): 2307. https://doi.org/10.3390/SU12062307.

EU. 2024. *Smart and Sustainable Cities and Communities: Rolling Plan for ICT Standardisation*. European Commission. https://interoperable-europe.ec.europa.eu/collection/rolling-plan-ict-standardisation/smart-and-sustainable-cities-and-communities-rp2024.

Goodspeed, Robert. 2020. *Scenario Planning for Cities and Regions: Managing and Envisioning Uncertain Futures*. Lincoln Institute of Land Policy.

Haraguchi, Masahiko, Tomomi Funahashi, and Filip Biljecki. 2024. "Assessing Governance Implications of City Digital Twin Technology: A Maturity Model Approach." *Technological Forecasting and Social Change* 204 (July): 123409. https://doi.org/10.1016/j.techfore.2024.123409.

Moghadas, Mahsa, Alexander Fekete, Abbas Rajabifard, and Theo Kötter. 2023. "The Wisdom of Crowds for Improved Disaster Resilience: A near-Real-Time Analysis of Crowdsourced Social Media Data on the 2021 Flood in Germany." *GeoJournal*. https://doi.org/10.1007/s10708-023-10858-x.

Moghadas, Mahsa, Abbas Rajabifard, Alexander Fekete, and Theo Kötter. 2022. "A Framework for Scaling Urban Transformative Resilience through Utilizing Volunteered Geographic Information." *ISPRS International Journal of Geo-Information* 22: 114. https://doi.org/10.3390/ijgi11020114.

Muiderman, Karlijn, Aarti Gupta, Joost Vervoort, and Frank Biermann. 2020. "Four Approaches to Anticipatory Climate Governance: Different Conceptions of the Future and Implications for the Present." *WIREs Climate Change* 11 (6): e673. https://doi.org/10.1002/wcc.673.

Nica, Elvira, Gheorghe Popescu, Milos Poliak, Tomas Kliestik, and Oana-Matilda Sabie. 2023. "Digital Twin Simulation Tools, Spatial Cognition Algorithms, and Multi-Sensor Fusion Technology in Sustainable Urban Governance Networks." *Mathematics* 11 (9). https://doi.org/10.3390/math11091981.

Ricciardi, Guglielmo, and Guido Callegari. 2023. "Digital Twins for Climate-Neutral and Resilient Cities. State of the Art and Future Development as Tools to Support Urban Decision-Making." *Urban Book Series*.

Tambiama, Madiega. 2020. "Digital Sovereignty for Europe." *EPRS | European Parliamentary Research Service* PE 651.992.

Temple, Lucy, Gabriela Viale Pereira, and Lukas Daniel Klausner. 2024. "Unravelling the Use of Digital Twins to Assist Decision- and Policy-Making in Smart Cities." *Resilience through Digital Innovation: Enabling the Twin Transition*: 755–64. https://doi.org/10.18690/UM.FOV.4.2024.45.

Tzachor, Asaf, Soheil Sabri, Catherine E. Richards, Abbas Rajabifard, and Michele Acuto. 2022. "Potential and Limitations of Digital Twins to Achieve the Sustainable Development Goals." *Nature Sustainability* 2022. https://doi.org/10.1038/s41893-022-00923-7.

UNESCO. 2021. *Recommendation on the Ethics of Artificial Intelligence*. United Nations Educational, Scientific and Cultural Organization. https://unesdoc.unesco.org/ark:/48223/pf0000381137.

UN-Habitat. 2020. *People-Centered Smart Cities: How to Make Digital Transformation Work for the Benefit of All*. Nairobi: United Nations Human Settlements Programme. https://unhabitat.org/people-centered-smart-cities.

Weil, Charlotte, Simon Elias Bibri, Régis Longchamp, François Golay, and Alexandre Alahi. 2023. "Urban Digital Twin Challenges: A Systematic Review and Perspectives for Sustainable Smart Cities." *Sustainable Cities and Society* 99. https://doi.org/10.1016/j.scs.2023.104862.

Ye, Xinyue, Jiaxin Du, Yu Han, Galen Newman, David Retchless, Lei Zou, Youngjib Ham, and Zhenhang Cai. 2023. "Developing Human-Centered Urban Digital Twins for Community Infrastructure Resilience: A Research Agenda." *Journal of Planning Literature*. https://doi.org/10.1177/08854122221137861.

31

Key Applications of AI and Spatiotemporal Urban Traffic Data in Intelligent Transport Systems: An Overview

Iman Taheri Sarteshnizi

31.1 Introduction

For decades, researchers have sought to understand transport dynamics and to improve its efficiency, safety, and sustainability (Helbing, 1997). With progress in digital technology and infrastructure, Intelligent Transport Systems (ITS) have attracted significant attention as a transformative approach to addressing ongoing challenges. Benefiting from real-time data and automation, ITS enabled transport networks to become more adaptive, allowing infrastructure, vehicles, and users to interact seamlessly. Advancements in Artificial Intelligence (AI), coupled with the rise of big data, have fundamentally reshaped ITS research. The focus in this field has shifted from traditional adaptive modelling to *data-driven* and *learning-based* methodologies. This conversion has not only broadened ITS but also given rise to a new generation of research topics. At the core of these emerging works is "spatiotemporal traffic data" which serves as a critical enabler for many AI-based applications in transport analysis and intelligent system design.

Spatiotemporal traffic data enables an understanding of both the *where* and *when* of traffic dynamics, enabling practitioners to model and analyse traffic behaviour with significantly greater granularity and responsiveness. This type of data can be sourced from a diverse range of technologies, including road-surface loop detectors (LDs), GPS trajectories, Bluetooth/Wi-Fi sensors, traffic cameras, and connected vehicles (CVs) equipped with onboard instruments. These data streams reflect the real-time state of mobility systems and agents, capturing fluctuations in congestion, travel times, and urban traffic events across both spatial and temporal dimensions. In contrast to traditional traffic data, which was often highly aggregated, infrequent, or spatially limited, spatiotemporal real-world data offers the resolution and continuity required for advanced analytics, real-time computations, adaptive and automated decision-making processes.

DOI: 10.1201/9781003638308-34

Building on the latest developments in ITS, this chapter aims to provide an overview of key applications enabled at the intersection of AI and spatiotemporal traffic data, and to highlight the current challenges associated with each application area. It should be noted that many of these applications have already shown great success across diverse contexts and have been applied in real-world settings in multiple countries (Adam Morrissett et al., 2019). Meanwhile, a few others remain in experimental stages, requiring further refinement, validation, and calibration before they can be reliably deployed in real-world scenarios. In this chapter, we first introduce a representative pipeline from data acquisition to decision-making in ITS. We then explain a range of key AI applications that have emerged as essential parts across this workflow. Finally, we also conclude the discussion and provide a snapshot of the information presented.

31.2 Intelligent Transport Systems

Figure 31.1 provides a visual overview of a typical ITS pipeline, illustrating the end-to-end process from the collection of real-world spatiotemporal traffic data to the generation of actionable and automated insights. The pipeline begins with data acquisition across multiple locations and time periods, capturing key traffic attributes such as speed, flow, and events. These data are then passed through a pre-processing stage, where procedures such as

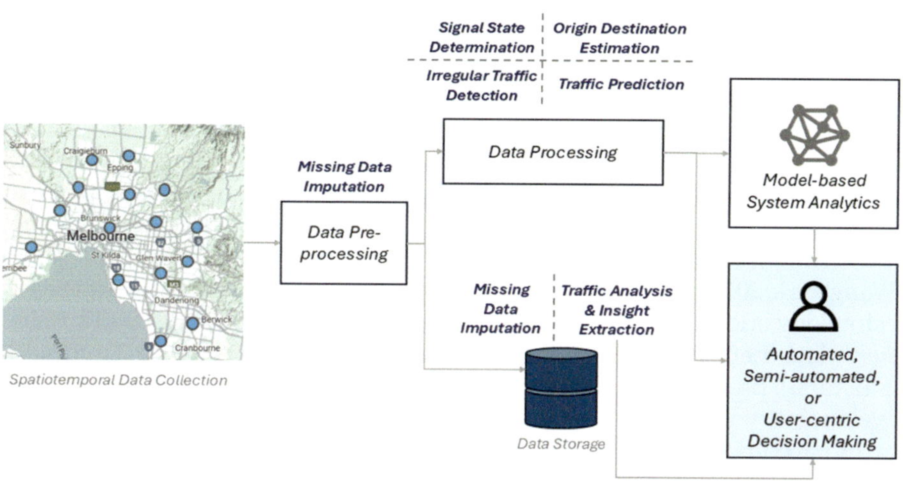

FIGURE 31.1
An illustration of an ITS Pipeline: From real-world data to final decision-making. (Figure by the author.)

cleaning and missing value imputation are applied to prepare the data for further analysis. Then, the data is fed into a set of other processes at the core of the pipeline. The outputs then either interact with model-based system analytics or feed directly into decision-support mechanisms. Depending on the application, decision-making may be fully automated (as in traffic signal control) or involve human oversight (such as in responses to detected anomalies).

31.3 Overview of Spatiotemporal Traffic Data Sources

From the beginning of the pipeline, a variety of traffic data sources are now available to collect spatiotemporal data of traffic systems. *Fixed-point sensors* such as road-surface LDs remain fundamental and are still widely employed in traffic monitoring, providing high-frequency data on vehicle count, speed, and occupancy (Taheri Sarteshnizi et al., 2024). These sensors are particularly effective for flow monitoring and congestion detection. However, their utility is constrained by limited spatial coverage and the inability to capture individuals. However, *GPS trajectory* data (the location of individual vehicles over time) offers a continuous, vehicle-level view of movement with variable, but often high, temporal resolution. This enables detailed analysis of routes, travel times, and behaviours (Kazemeini et al., 2022). Nonetheless, it can present sampling bias and privacy concerns due to the granularity of the information collected.

Additionally, *Bluetooth* and *Wi-Fi* sensors can be used as an intermediate approach by capturing anonymous MAC addresses at specific locations, enabling the estimation of flows and travel times across certain road segments (Pu et al., 2022). However, their detection rates are relatively sparse, and ethical use requires robust anonymisation. *Traffic cameras,* while capable of capturing rich visual data at fixed locations, impose high processing demands and raise privacy concerns. More recently, *CV* data has also emerged as a real-time source of behavioural vehicle-level attributes such as speed, braking, and turning, supporting safety monitoring and driver assistance applications (Wang et al., 2021). Its effectiveness, however, remains yet limited by the level of technology adoption and the lack of standardised data formats. Lastly, *mobile phone* data (network interactions) can also be mentioned (Dypvik Landmark et al., 2021), providing broad coverage and valuable insights into population-level mobility patterns. Its spatial resolution varies, ranging from coarse cell tower-based positioning to more precise app-based location tracking, and its accessibility often depends on agreements with network operators.

The emergence of such data collection systems, despite their strengths and limitations, has enabled a range of critical applications (Figure 31.1) that were previously unattainable. Nevertheless, the literature still highlights several areas where enhancements in data collection could elevate the potential of AI-based applications to the next level (Chevance et al., 2024). One major challenge is the slow and uneven spatial adoption of sensing technologies, such as CV systems, which restricts large-scale implementation. In parallel, the lack of a universally accepted and standardised data format complicates integration across different platforms and limits the scalability of data-driven tools. Data communication security is another growing concern (specifically for vehicles) where breaches could compromise privacy and safety. Moreover, the effectiveness of these systems depends not only on data availability but also on consistent sensor performance, which requires ongoing maintenance and oversight. Addressing these gaps is a key expectation in future and it will create new pathways and unlock further capabilities within ITS.

To provide a structured entry point into the key applications, Table 31.1 presents an overview of the main components. It summarises the application areas based on associated data types, methods, and key challenges. This overview is intended better highlight both shared and distinct characteristics of the applications. A detailed explanation of each application, along with concluding remarks, is provided in the following sections.

TABLE 31.1

Review of the AI-Enabled Applications with Spatiotemporal Traffic Data in ITS

Section	Focus Area	Data Types	Base Methods	Key Challenges
Traffic prediction and imputation	Forecasting and missing value recovery	Loop detector (LD), GPS, CV data	GNNs + RNNs/ TCNs, transformers	Data quality, model generalisability, Context-aware inputs
Bottleneck and abnormality detection	Incident or congestion detection	CCTV, LD, GPS, CV data	CNNs, XGBoost, unsupervised models	Lack of ground truth, validation complexity
Traffic performance and insight gen.	Pattern exploration and strategic planning	LD, GPS, traffic event data	Traditional clustering, AEs/VAEs	Context sensitivity, generalisability across networks
Adaptive traffic signal control	Real-time signal optimisation	LD, CV data, CCTV	Reinforcement learning	Infrastructure limitations, deployment risk, and accountability
OD matrix estimation	Trip-level network demand reconstruction	LD, GPS, CV data	SAE, Bi-level optimisation + DL hybrids	Incomplete data, lack of training data, scalability

31.4 Traffic Prediction and Missing Data Imputation

Among the most explored ITS applications is the modelling of spatiotemporal traffic data to either predict future traffic states or fill in missing observations from sensors. Traffic prediction aims to forecast variables such as traffic volume, speed, or occupancy for upcoming time intervals, enabling proactive traffic management and control strategies. This task is often performed in real time, with forecasting horizons ranging from 5 minutes to several hours. On the other hand, missing data imputation focuses on estimating unrecorded or corrupted sensor data, which may result from failures, transmission errors, or environmental conditions. Imputation can occur both in real time and retrospectively (Figure 31.1). These two topics are closely interrelated as imputation can be framed as a short-term prediction problem, where the goal is to infer the most probable value for missing observations based on the latest spatial and temporal context.

Learning-based methods have significantly outperformed traditional approaches in both traffic prediction and missing data imputation, with numerous studies demonstrating that deep learning delivers substantially better accuracy. The current state-of-the-art techniques in this field often rely on hybrid architectures, which combine multiple model types to better capture the complex spatial and temporal dependencies. Among these, Graph Neural Networks (GNNs) play a central role by modelling the road network as a graph, enabling the understanding of spatial interactions between road segments (Qi et al., 2022). These are frequently integrated with Recurrent Neural Networks (RNNs) or Temporal Convolutional Networks (TCNs) to capture temporal dynamics, from short-term fluctuations or long-term trends, within the data. Recently, attention mechanisms and transformer-based models have also gained traction (Gu et al., 2023), offering enhanced capacity in learning long-range dependencies across both space and time. Latest progress also highlights the critical role of data preprocessing, such as map-matching of GPS trajectories, traffic data cleaning, and information compression, to underline the effectiveness of model inputs. There is a growing emphasis on context-aware modelling, where information such as weather conditions and public events is integrated to improve prediction performance (Nigam & Srivastava, 2023).

31.5 Identifying Traffic Bottlenecks and Irregular Patterns

Alongside traffic prediction, identifying traffic bottlenecks and irregularities has emerged as another key direction. It is well-recognised that the most critical parts of data to detect, predict, and analyse are those with sparse traffic behaviour. These intervals are important, as they represent the most informative segments of data, offering insights that can drive the development of road safety and efficiency interventions. The widely adopted application in this domain is real-time incident detection, often implemented using AI-based image processing techniques and applied to distributed CCTV footages (Pillai et al., 2021). The success of advanced Convolutional Neural Network (CNN) architectures has transformed the detection and analysis of incidents and near-misses in traffic footages, leading to improved policy proposals and interventions. These techniques involve extracting vehicle trajectories from videos and modelling their movement patterns to identify unusual patterns. This progress has been so impactful that it has led to the emergence of startups offering platforms for automated incident and near-miss analysis.

Congestion detection and traffic anomaly detection have also been extensively applied to fixed-point detector data, such as LDs, using a wide range of AI solutions (Taheri Sarteshnizi et al., 2022). These include supervised classifiers like XGBoost, self-supervised predictive models, as well as unsupervised approaches, ranging from deep graph-based architectures to advanced and problem-specific statistical pipelines. The extent to which AI-based solutions outperform traditional methods in this area is still subject to ongoing investigation, as validating the outputs solely through traffic data poses significant challenges. In most cases, there is no definitive ground truth data marking the exact occurrences of incidents. The temporal and spatial boundaries of such events are often ambiguous but can be more reliably verified through visual observation of traffic footage (which is also a time-consuming process).

31.6 Traffic Performance Analysis and Insight Generation

With the support of ITS data collection infrastructure, an ever-growing volume of historical traffic data is being accumulated, intensifying the need for approaches capable of extracting meaningful insights. This includes answering questions such as how traffic congestion propagates through the network, how traffic patterns change in response to specific events or disruptions, or which roads exhibit similar traffic patterns at each time interval. Both traditional clustering algorithms and more recent dimensionality reduction techniques, such as Autoencoders (AEs) and Variational AEs (VAEs), have proven effective in addressing these types of questions, with AI-based solutions showing great potential to better handle large volumes of data and uncover latent hidden patterns (Taheri Sarteshnizi et al., 2023). The insights extracted from such solutions can be instrumental in identifying congestion zones, segmenting the network based on traffic patterns, and enabling proactive strategies for planning and adaptive traffic control.

While insight generation with AI solutions has shown great potential, ensuring the scalability and generalisability of these methods across different spatial regions and temporal conditions remains a key challenge. Traffic patterns are impacted by several dynamic factors, such as roadworks, seasonality, behavioural shifts, and emerging travel modes, which can limit the transferability of models trained on historical data. To address these factors, researchers are increasingly focusing on transfer learning, domain adaptation, and online learning approaches that can adapt to changing environments and extend learned insights (Zhuang et al., 2021). Leveraging universal principles that repeat across different locations is challenging, but doing so could unlock highly transferable insights and significantly improve the scalability and utility of such models.

31.7 Adaptive Traffic Signal Control

Advancements in Reinforcement Learning (RL) and its successful applications across numerous domains have sparked interest in leveraging this technique for improving traffic control and signal timing (Yazdani et al., 2023). Traditionally, two of the prominent approaches to signal control have been fixed time and actuated settings. In fixed time, traffic signals follow a predetermined schedule with specific phase durations; however, actuated signal control dynamically adjusts phase lengths based on data inputs, allowing for a responsive behaviour but relying on rule-based logic. RL approaches, on the other hand, have shown the ability to learn more optimal control policies through interaction with the environment. By continuously observing traffic states and receiving feedback (rewards), such as reductions in waiting time or queue length, RL controllers adapt to changing traffic patterns and enhance overall network efficiency and safety.

Successful traffic control outcomes have been achieved through RL, particularly those leveraging high-quality spatiotemporal traffic data collected from fixed detectors. However, a few critical barriers continue to hinder the deployment of such approaches. Firstly, the data collection infrastructure required to support these solutions often differs from the current setup. High-resolution and real-time data across multiple spatial scales is needed for training these systems, yet most existing traffic control systems lack such capacities. Secondly, RL models are currently validated in simulation environments which cannot fully reflect the uncertainty inherent in real-world traffic systems. Evaluating these solutions in live settings requires extensive real-world experimentation, which not only necessitates stringent protocols for all road users but also raises concerns regarding accountability and the reliability of decisions made by such systems.

31.8 Traffic Origin–Destination Matrix Estimation

Origin–destination (OD) data, which defines the start and end points of all trips within a network, are vital in transport planning and modelling processes, as they enable the evaluation of proposed interventions prior to real-world implementation. Despite their critical importance, current spatio-temporal data collection systems do not offer a complete OD picture under any scenario. As a result, it has also become essential to estimate OD data for the entire network using the available data sources. This problem is usually approached through bi-level optimisation frameworks, which are designed to account for the limitations of observed data. The use of deep learning, however, has been less explored to address this problem. Nonetheless, emerging AI-based perspectives are also showing promising potential for advancing OD estimation as well. Beside the usage of AI to improve data quality, new studies are investigating its direct effectiveness in inferring the relationship between the aggregated or partial traffic data and the target OD matrices. Sparse AEs (SAEs) have shown primary success in learning the aforementioned mapping (fixed-point traffic to OD) by training on a large number of synthetically generated scenarios (Owais, 2024). Such research perspectives are increasingly being explored to shape the future of this field and other similar areas.

31.9 Summary and Conclusions

A diverse range of data sources, from fixed detectors to CVs, now underpin a broad spectrum of AI-based analytical and operational capacities in ITS. These applications include, but are not limited to, traffic state prediction, missing data imputation, traffic anomaly detection, network performance monitoring, adaptive signal control, and OD estimation. Across these domains, machine learning techniques, particularly graph-based deep learning architectures, are showing strong potential in capturing the complex spatial and temporal dependencies of traffic systems. However, some barriers still hinder the widespread deployment of these efforts, including data sparsity, lack of standardisation, infrastructure limitations, and challenges in model generalisability across spatial contexts. Addressing these issues remains critical for scaling the next generation of ITS solutions, where spatiotemporal intelligence empowers both real-time operational decisions and long-term strategic planning. It should be acknowledged that the domains discussed represent a part of main possibilities at the intersection of AI and ITS. Broader use cases continue to emerge beyond the scope of this discussion, underscoring the rapidly expanding frontiers of the field.

References

Adam Morrissett, R. E., Zaman, M., Zohrabi, N., & Sherif, A. (2019). *A Physical Testbed for Intelligent Transportation Systems*. International Conference on Human System Interaction (HSI).

Chevance, G., Nieuwenhuijsen, M., Braga, K., Clifton, K., Hoadley, S., Kaack, L. H., Kaiser, S. K., Lampkowski, M., Lupu, I., Radics, M., Velázquez-Cortés, D., Williams, S., Woodcock, J., & Tonne, C. (2024). Data gaps in transport behavior are bottleneck for tracking progress towards healthy sustainable transport in European cities. *Environmental Research Letters*, 19(5). https://doi.org/10.1088/1748-9326/ad42b3

Dypvik Landmark, A., Arnesen, P., Södersten, C.-J., & Hjelkrem, O. A. (2021). Mobile phone data in transportation research: Methods for benchmarking against other data sources. *Transportation*, 48(5), 2883–2905. https://doi.org/10.1007/s11116-020-10151-7

Gu, B., Zhan, J., Gong, S., Liu, W., Su, Z., & Guizani, M. (2023). A spatial-temporal transformer network for city-level cellular traffic analysis and prediction. *IEEE Transactions on Wireless Communications*, 22(12), 9412–9423. https://doi.org/10.1109/twc.2023.3270441

Helbing, D. (1997). Empirical traffic data and their implications for traffic modeling. *Physical Review E*, 55. https://doi.org/10.1103/PhysRevE.55.R25

Kazemeini, A., Taheri, I., & Samimi, A. (2022). A GPS-based algorithm for brake and turn detection. *International Journal of Intelligent Transportation Systems Research*, 20(2), 433–445. https://doi.org/10.1007/s13177-022-00301-9

Nigam, A., & Srivastava, S. (2023). Hybrid deep learning models for traffic stream variables prediction during rainfall. *Multimodal Transportation*, 2(1). https://doi.org/10.1016/j.multra.2022.100052

Owais, M. (2024). Deep learning for integrated origin–destination estimation and traffic sensor location problems. *IEEE Transactions on Intelligent Transportation Systems*, 25(7), 6501–6513. https://doi.org/10.1109/tits.2023.3344533

Pillai, M. S., Chaudhary, G., Khari, M., & Crespo, R. G. (2021). Real-time image enhancement for an automatic automobile accident detection through CCTV using deep learning. *Soft Computing*, 25(18), 11929–11940. https://doi.org/10.1007/s00500-021-05576-w

Pu, Z., Cui, Z., Tang, J., Wang, S., & Wang, Y. (2022). Multimodal traffic speed monitoring: A real-time system based on passive Wi-Fi and bluetooth sensing technology. *IEEE Internet of Things Journal*, 9(14), 12413–12424. https://doi.org/10.1109/jiot.2021.3136031

Qi, J., Zhao, Z., Tanin, E., Cui, T., Nassir, N., & Sarvi, M. (2022). A graph and attentive multi-path convolutional network for traffic prediction. *IEEE Transactions on Knowledge and Data Engineering* 35(7), 6548–6560. https://doi.org/10.1109/tkde.2022.3179646

Taheri Sarteshnizi, I., Bagloee, S. A., Sarvi, M., & Nassir, N. (2024). Traffic anomaly detection: Exploiting temporal positioning of flow-density samples. *IEEE Transactions on Intelligent Transportation Systems*, 25(5), 4166–4180. https://doi.org/10.1109/tits.2023.3322695

Taheri Sarteshnizi, I., Sarvi, M., Bagloee, S. A., & Nassir, N. (2022). *Abnormality Detection in Urban Traffic Data: A Review*. Australasian Transport Research Forum, Adelaide, Australia. https://australasiantransportresearchforum.org.au/papers/2022

Taheri Sarteshnizi, I., Sarvi, M., Bagloee, S. A., & Nassir, N. (2023). Temporal pattern mining of urban traffic volume data: A pairwise hybrid clustering method. *Transportmetrica B: Transport Dynamics*, 11(1). https://doi.org/10.1080/21680566.2023.2185496

Wang, Q., Yuan, Y., Yang, X., & Huang, Z. (2021). Adaptive and multi-path progression signal control under connected vehicle environment. *Transportation Research Part C: Emerging Technologies*, 124. https://doi.org/10.1016/j.trc.2021.102965

Yazdani, M., Sarvi, M., Asadi Bagloee, S., Nassir, N., Price, J., & Parineh, H. (2023). Intelligent vehicle pedestrian light (IVPL): A deep reinforcement learning approach for traffic signal control. *Transportation Research Part C: Emerging Technologies*, 149. https://doi.org/10.1016/j.trc.2022.103991

Zhuang, F., Qi, Z., Duan, K., Xi, D., Zhu, Y., Zhu, H., Xiong, H., & He, Q. (2021). A comprehensive survey on transfer learning. *Proceedings of the IEEE*, 109(1), 43–76. https://doi.org/10.1109/jproc.2020.3004555

32

Future Directions: Land and Geospatial Information On-Demand

Abbas Rajabifard and Behnam Atazadeh

32.1 Introduction

Land and geospatial systems are rapidly transforming toward an on-demand paradigm at the interface of digital technologies, governance frameworks, and evolving social requirements. Under this new paradigm, land and geospatial information are dynamic, integrated, and constantly updated knowledge infrastructures rather than static and dispersed data repositories. This change will result in a new generation of land administration and geospatial systems that enable decision-making at all levels, from local communities to international policy development, by providing trusted data, predictive analytics, and services in an intelligent and adaptable environment. As stated in Chapter 1, this is consistent with the UN-GGIM's vision of the future geospatial ecosystem as a system-of-systems: an interconnected, interoperable, and adaptive environment that integrates authoritative land and geospatial data with emerging technologies, a wide range of stakeholders, and innovative governance frameworks (UNGGIM 2025). As a result, land and geospatial data will be positioned as an essential part of digital public infrastructure, supporting resilience, evidence-based policymaking, and the accomplishment of sustainable development objectives. In other words, land and geospatial information is a key enabler of digital transformation, supporting future sustainability and resilience.

The future on-demand land and geospatial paradigm is built on a knowledge-based strategy to transform heterogeneous datasets into location-based intelligence through Artificial Intelligence (AI), machine learning (ML), and semantic interoperability. This views land and geospatial information as a public good, which is vital for the sustainability of the economy, society, and environment. It is ecosystem-focused, using feedback loops to connect communities, institutions, and technologies. Automated reasoning, real-time decision-making, and predictive analytics will be made possible by land intelligence and geospatial ecosystems, which can address current and future problems in spatially complex situations.

 DOI: 10.1201/9781003638308-35

On-demand land and geospatial information is considered as the next frontier in research, practice, and innovation in this book's final chapter, which builds on insights derived from chapter contributions ranging from land governance and climate resilience to digital twins, renewable energy, and immersive geospatial technologies. The shift to on-demand systems signifies not only technological advancement but also a paradigm shift in the way we capture, manage, and apply land and geospatial information for sustainability and resilience.

32.2 Enablers of On-Demand Land and Geospatial Information

Enablers of on-demand land and geospatial information include technology, governance and policy, human capacity, market and community. These enablers are critically important for land and geospatial information to be reliable, pertinent, inclusive, publicly available, and future-ready. The following are our thoughts on how these enablers contribute to on-demand land and geospatial information.

32.2.1 Technology

Innovations in data capture, processing, validation, integration, and analytics are technically fundamental to on-demand land and geospatial information. GeoAI and ML advancements allow for automatic integration, anomaly identification, and predictive modeling of land and geospatial data. Digital twins, enabled by cloud and edge computing, provide real-time, dynamic representations of the built and natural environments, enabling scenario-based analysis and adaptive planning while also presenting decision-makers with an integrated perspective of current conditions and potential future scenarios. The expansion of Internet of Things (IoT) and smart sensor technologies assures continuous streams of land and geospatial data, while interoperability standards enable smooth integration across multiple domains. These technological foundations enable the agility and scalability needed for an on-demand land intelligence and geospatial ecosystem.

32.2.2 Governance and Policy

It is critical that institutional and legal frameworks should be future-ready and adapt to keep up with technological advancements related to land intelligence and geospatial ecosystems. More importantly, governance of land and geospatial information should be inclusive, ethical, and rights-protecting. Clear policies should specify who has access to private and sensitive land and geospatial information, and for what purposes. Ethical governance gives the public trust that on-demand land and geospatial systems will be used fairly and responsibly. In addition, public–private partnerships (PPPs) and cross-sector cooperation are essential for driving innovation and investment in land and geospatial activities. Alignment with global initiatives, such as UN-IGIF and FELA, provides both global coherence and local adaptability. The UN-IGIF provides a global overarching strategy, implementation guide, country-level action plan and in particular nine strategic pathways to help countries develop consistent geospatial ecosystems. This framework establishes a set of fundamental principles that each country should follow while developing its own land and geospatial solutions based on jurisdictional requirements. FELA, which is based on UN-IGIF, supports countries to modernize their land administration systems. This framework serves as a guide for establishing and modifying land policy to embrace digital workflows and new practices. Strong governance and policy frameworks, underpinned by updated legislation and privacy protection, active partnerships, and international standards, will assist governments in updating their institutional structures to make on-demand land and geospatial information a reality while maintaining trust and legality. In addition, socio-technical principles such as FAIR data (Findable, Accessible, Interoperable, Reusable) and CARE (Collective Benefit, Authority to Control, Responsibility, Ethics – particularly for indigenous data) are recommended as guidelines for responsible land and geospatial data management (UNGGIM 2025). Robust governance frameworks that prioritize ethics and inclusivity are essential for keeping up with the rapidly developing land intelligence, and geospatial ecosystems. These frameworks will help answer this kind of questions: who benefits from land intelligence? How would land intelligence and geospatial insights benefit all members of the community? Are there safeguards against the misuse of land and geospatial information?

32.2.3 Human Capacity

The transition to on-demand land and geospatial information necessitates a workforce with new skills and multidisciplinary expertise. Technological advancements alone cannot provide value unless individuals understand how to use and interpret land and geospatial data. Traditional skills (such as mapmaking and geographic database management) are still vital, but they are no longer sufficient. Professionals today require new abilities in AI and data science, IoT, digital twins, and domain knowledge in fields such as urban planning and climate science. As a result, land and geospatial specialists should become more adaptable as technology advances. For example, a land surveyor currently requires to not only measure borders but also program a drone, analyze satellite imagery using ML, and explain findings to non-technical stakeholders. Cultivating such a diverse skillset for the workforce remains a challenge that necessitates major investment in education and training. As a result, educational institutions should revise their current land administration and geospatial education curricula to include emerging technologies and interdisciplinary course content. Another essential facet of human capacity is building land and geospatial systems with end-users in mind and actively interacting with stakeholders. User-centric design guarantees that modern land and geospatial systems continue to be accessible and helpful to government, corporate, and community decision-makers. Land and geospatial systems can be designed based on end-user requirements by involving them from the outset through workshops, user testing, and participatory design, increasing end-user adoption. Finally, effective cross-sector collaboration is essential for ensuring that land and geospatial experts collaborate closely with domain experts (e.g., ecologists, economists) to present outputs in a way that is appropriate for the application domain context. Finally, investing in people, including their skills, understanding, and stakeholder engagement, is vital to ensuring that on-demand land and geospatial information is effectively leveraged to drive real-world decisions.

32.2.4 Market

A strong and sustainable market demand underpins the growth of land and geospatial information industry. The demand for land intelligence and geospatial ecosystems is increasing in various sectors. For example, efficient land markets and the property industry benefit greatly from immediate access to land information. Developers, investors, and property owners demand efficient solutions to authenticate ownership titles, determine land and property values, and check zoning restrictions instantaneously. When this information is obtained quickly (in minutes rather than weeks), it lubricates the land market, making transactions faster and more secure, providing a strong economic incentive to use intelligent land administration systems. Likewise, the necessity for real-time infrastructure monitoring drives up the demand for geospatial intelligence. Utilities and transport agencies require real-time mapping of important asset conditions (such as roads, bridges, and pipelines) in order to anticipate maintenance needs and avoid failures. This results in new geospatial services such as live infrastructure dashboards and risk mapping tools. City planners and smart city projects also influence the land and geospatial markets. To develop better cities, urban planners today use up-to-date 3D models, traffic and mobility data, and citizen-generated data. As a result, the market enabler assures that on-demand land and geospatial information is an economically viable and socially desirable service. Strong market pull provides constant incentives for innovation in land and geospatial domains. It promotes public and private investment in land systems and geospatial platforms while also ensuring that the land and geospatial industry stays adaptive and relevant, evolving continuously to meet the economy's on-demand requirements.

32.2.5 Community

The validity and long-term usage of land and geospatial systems is predicated heavily on community trust and engagement. If end-users believe the system is transparent and reliable, they are more likely to utilize and contribute to it. To promote trust, on-demand land and geospatial systems will be transparent in their operations, open in terms of data sources and procedures, and inclusive in their design. For example, integrating local communities into mapping and data collection can significantly enhance data quality as well as community trust. Inclusive design entails making land and geospatial tools available to all members of society. Special outreach efforts would be necessary to reach rural regions, marginalized groups, and those with poor internet access. This requires offering local language interfaces, community geospatial centers, and offline data access. Communities gain trust and a sense of ownership when they recognize that land and geospatial data meet their needs. Crucially, a community-driven strategy guarantees the equitable distribution of the benefits of on-demand land and geospatial data. A "digital divide" could occur in which underrepresented groups are left behind while others benefit from on-demand land and geospatial data. Building capacity and actively engaging with the community will help lessen this difficulty. Specifically, international efforts are working to tackle this issue. The UN-GGIM, for example, places a strong emphasis on international collaboration in order to help all nations and communities advance their geospatial capabilities using the "leave no one behind" principle. The community enabler essentially serves as a reminder that land and geospatial data are ultimately public goods. Greater adoption and more purposeful use of on-demand land and geospatial services will result in promoting trust via openness, guaranteeing inclusive and participatory practices, and enhancing digital literacy in local communities.

32.3 Outlook on Land Intelligence and Geospatial Ecosystems

This book highlighted that there is a significant paradigm change occurring in the way we capture, curate, and communicate land and geospatial data, including the following important elements:

- Moving beyond maps and static datasets to live, dynamic digital twins and predictive intelligence
- Integrated, linked, and interoperable ecosystems replace disjointed systems and siloed data infrastructure
- Immersive and engaging environments for wider communication of land and geospatial information
- From expert-driven assessments to inclusive and participative community involvement and collaboration
- Using land intelligence and geospatial ecosystems to build a sustainable and resilient future.

Here, we will provide an outlook on these crucial elements for the development of geospatial ecosystems and land intelligence (Figure 32.1).

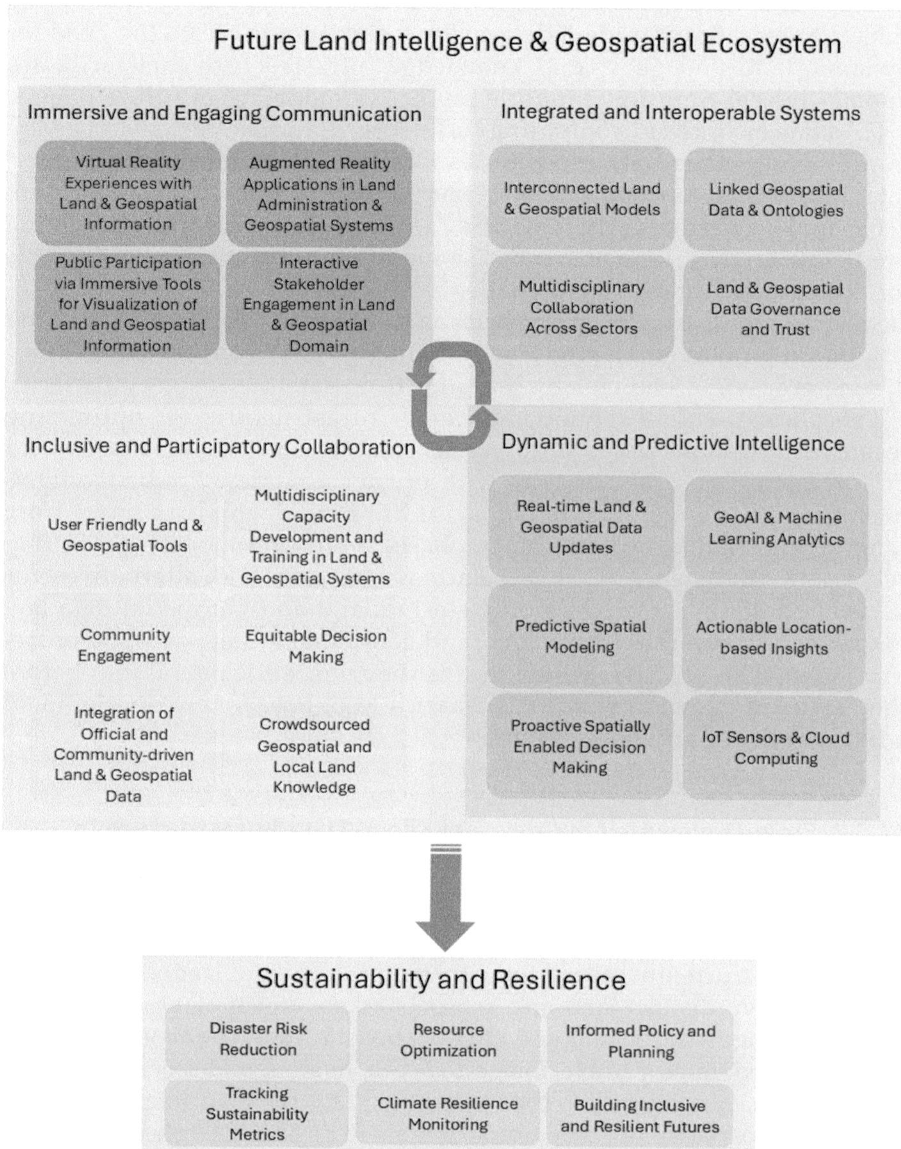

FIGURE 32.1
Conceptual framework illustrating the paradigm shift from current fragmented approaches to future land intelligence and geospatial ecosystem, enabling sustainability and resilience.

32.3.1 From Static Data and Maps to Dynamic, Live Digital Twins and Predictive Intelligence

Traditionally, land and geospatial information have been limited to static datasets and map snapshots captured at discrete points in time, often subject to significant update delays. Emerging trends create the need for dynamic land and geospatial knowledge infrastructure that facilitates updates in real time and supports predictive modeling of future scenarios. Advances in cloud computing, IoT sensors, and AI/ML analytics are transforming static data repositories into live, machine-actionable land intelligence and geospatial ecosystems. United Nations envisions future geospatial ecosystems where data is "acted upon in real-time," enabling predictive analytics, adaptive decision-making, and actionable insights for addressing urgent global challenges (UNGGIM 2025). This will create a wide range of applications such as (1) automatic forecast of flooding risk in a region by analyzing live rainfall and river sensor data; (2) predicting crop yield changes by learning from satellite imagery patterns; (3) enabling predictive maintenance of infrastructure; (4) optimizing resource allocations based on dynamic weather and usage patterns; and (5) forecasting natural disasters with greater accuracy. Predictive intelligence and dynamic, live digital twins are fundamentally a move from "data to knowledge," where the emphasis is on continuously generating actionable intelligence (such as alarms, forecasts, and alternatives for decision-making) rather than only storing land and geospatial data layers. This indicates that future land and geospatial systems will be proactive rather than reactive, which means they will anticipate issues before they arise and assess "what if" scenarios (e.g., how a new development will affect traffic and flood risk).

32.3.2 From Fragmented Systems and Siloed Data Infrastructure to Integrated, Interconnected, and Interoperable Ecosystems

Moving toward a fully integrated land and geospatial ecosystem from the current fragmented, disjointed, and siloed data infrastructure is another significant shift. Currently, several industry sectors and organizations frequently keep their own distinct land and geospatial data, which makes it challenging to exchange information and obtain a holistic view. An evolving web of interconnected, interoperable, integrated data, models, and services that can function as a cohesive unit is the future vision for the land intelligence and geospatial ecosystem. In other words, there will be a single, cohesive system-of-systems rather than numerous disparate datasets. The foundation of this concept is interoperability, which will enable us to combine different software tools and data sources into an integrated framework (AlSayel et al. 2025).

In addition to technology, alignment on semantics and data governance are also necessary to achieve this integration. Consistent interpretation of features across datasets is maintained by linked data techniques and ontologies. Addressing issues related to trust and rights in fully integrated digital environments is another important aspect. Stakeholders require specific information about who owns and manages digital data, who has access to it, and how to guarantee data security and quality. In order to preserve sovereignty and facilitate collaboration, the future ecosystem is anticipated to make use of trusted and safe data hubs where companies exchange data in accordance with established guidelines (AlSayel et al. 2025). Achieving full integration and interoperability requires not just technology but also semantic alignment and data governance. Shared vocabularies (ontologies) and linked data approaches help maintain consistent meaning of features across land and geospatial datasets. This also involves addressing issues associated with rights and trust in fully integrated digital environments. In particular, stakeholders need clarity on who owns and controls digital data, who can access it, and how to ensure data quality and security. The future ecosystem will likely utilize secure data hubs and trusted data spaces where organizations share data under agreed rules, preserving sovereignty and enabling collaboration. Land and geospatial ecosystems that are interconnected, integrated, and interoperable are extremely valuable. Complex and cross-cutting problems that call for interdisciplinary solutions, such as climate resilience, can be addressed when various datasets and models are connected.

32.3.3 Immersive and Engaging Environments for Wider Communication of Land and Geospatial Information

The widespread usage of visualization technologies such as Virtual Reality (VR) and Augmented Reality (AR) will make land and geospatial information more immersive and interesting in the future (Milner 2024). With the help of these technologies, stakeholders may explore and comprehend land and geospatial information intuitively, transforming abstract facts into experiences. As a result, the usage of immersive geospatial tools to engage the public will grow. When viewed through smartphone and smart glasses, AR can superimpose land and geospatial data onto the real world, such as property boundaries and historical imagery of the area. As an additional example, utility workers are increasing productivity and lowering construction hazards by employing mixed-reality headsets in the field to view the precise location of underground cables and pipelines. By allowing community people to virtually experience planned developments (such a new park or building) from a first-person viewpoint, VR improves stakeholders' understanding and input. All stakeholders can interact with land and geospatial data in a meaningful way if it is made visual, interactive, and experiential. Through the appropriate communication of land and geospatial information, immersive experiences essentially establish a common ground where professionals and non-experts will work together and make decisions more effectively.

32.3.4 From Expert-Driven Analyses to Inclusive and Participatory Community Engagement and Collaboration

Land and geospatial information were once the domain of specialists, including cartographers, geospatial analysts, and land surveyors, who generated maps and analytics for the use of other stakeholders. However, the emerging model is now more inclusive and participatory, enabling a wide range of end-users to engage with land and geospatial data and contribute their knowledge. Future intelligent land and geospatial ecosystems are envisioned as human-centric and community-driven, involving many important stakeholders from planners and data scientists to local residents and indigenous communities (UNGGIM 2025). This inclusiveness is facilitated by user-friendly tools that let non-experts visualize data, provide feedback, and contribute to meaningful creation and use of land and geospatial information.

Traditionally, experts in land and geospatial information, such as cartographers, geospatial analysts, and land surveyors, created maps and analytics for other stakeholders to use. However, the paradigm is moving toward more open and participatory, and therefore, a variety of end-users can interact with land and geospatial data and share their expertise. Planners, data scientists, local citizens, and indigenous communities are just a few of the key stakeholders involved in the human-centric and community-driven vision of future intelligent land and geospatial ecosystems (UNGGIM 2025). User-friendly solutions that enable non-experts to display data, offer input, and participate in the meaningful creation and use of land and geospatial information help to promote inclusivity.

Realizing the benefits of land and geospatial information for the broader community, particularly underrepresented and marginalized groups, depends in large part on training programs and capacity building, as we mentioned in Section 32.2. Richer and more accurate insights can be obtained by integrating official land and geospatial datasets with local knowledge and crowdsourced data (such as volunteer real-time updates on wildfires and community reports of land boundary disputes). Communities benefit from this democratization of land and geospatial information, which also increases this information's relevance and reliability. Ultimately, inclusive land and geospatial ecosystems will promote more equitable decision-making, address actual demands, and gain public trust and support.

32.3.5 Building a Sustainable and Resilient Future with Land Intelligence and Geospatial Ecosystems

All of the abovementioned advancements and enablers support resilient and sustainable development. In future geospatial ecosystems, operationalizing "land intelligence" gives societies the ability to foresee threats, maximize resource utilization, and make well-informed decisions for long-term sustainability. Decision-makers will effectively manage resources, prioritize development initiatives, and enhance climate resilience by layering data on risks, infrastructure, and resources using land and geospatial tools. Intelligent land and geospatial ecosystems will contribute to sustainability and resilience significantly as they become more sophisticated and widely available in future. In particular, they are essential for monitoring and accomplishing the Sustainable Development Goals (SDGs) from local to global levels. Countries and communities will be better able to cope with complexity, embrace opportunities, and create a more resilient, inclusive, and sustainable future through targeted development and deployment of land intelligence and geospatial ecosystem. Practically speaking, this will involve being able to accurately target climate adaption strategies and track advancement of sustainability metrics with location-based information.

32.4 Final Remarks

On-demand land and geospatial information represent the next frontier in land administration and geospatial sciences. By leveraging GeoAI, digital twins, immersive technologies, and socio-technically aligned governance frameworks, the future will move beyond fragmented data toward dynamic, intelligent, and community-centered land and geospatial infrastructures. The challenge is predicated on aligning technological innovation with ethical governance and community inclusivity. This new paradigm will ensure that land and geospatial information not only supports sustainability and resilience but also promotes trust, equity, and prosperity for future generations. The pathway forward promises a society with enhanced land and geospatial literacy, in which we will fully leverage the "where" of things to better understand the "why" and "how" of our future.

References

AlSayel, Mohammed Yahya, Ingo Simonis, Zaffar Sadiq Mohamed-Ghouse, and Asim AlGhamdi. 2025. *Moving Forward Together: The Emergence of the Interconnected Geospatial Ecosystem*. nos. 25–017. https://doi.org/10.5281/zenodo.0000000.

Milner, Greg. 2024. *Mapping the Future of Spatial Computing*. Preprint, ESRI, February. https://www.esri.com/about/newsroom/blog/mapping-spatial-computing.

UNGGIM. 2025. *Positioning the Future Geospatial Information Ecosystem*. New York. https://ggim.un.org/meetings/GGIM-committee/15th-session/documents/2025-07_UN-GGIM_FutureGeospatialEcosystem.pdf.

Index

E

M

N

O

P

Q

R

V

W

X

Z